HYPERSPACE

A Scientific Odyssey
Through
Parallel Universes,
Time Warps, and
the Tenth Dimension

Michio Kaku

Illustrations by Robert O'Keefe

D0029581

ANCHOR BOOKS
DOUBLEDAY
New York London Toronto Sydney Auckland

AN ANCHOR BOOK

PUBLISHED BY DOUBLEDAY

a division of Bantam Doubleday Dell Publishing Group, Inc.
1540 Broadway, New York, New York 10036

ANCHOR BOOKS, DOUBLEDAY, and the portrayal of an anchor are
trademarks of Doubleday, a division of Bantam Doubleday Dell
Publishing Group, Inc.

Hyperspace was originally published in hardcover by Oxford University Press in 1994.
The Anchor Books edition is published by arrangement with Oxford University Press.

"Cosmic Gall." From *Telephone Poles and Other Poems* by John Updike. Copyright © 1960
   by John Updike. Reprinted by permission of Alfred A. Knopf, Inc. Originally
   appeared in *The New Yorker*.
Excerpt from "Fire and Ice." From *The Poetry of Robert Frost,* edited by Edward
   Connery Lathem. Copyright 1951 by Robert Frost. Copyright 1923, © 1969 by
   Henry Holt and Company, Inc. Reprinted by permission of Henry Holt and
   Company, Inc.

Library of Congress Cataloging-in-Publication Data

Kaku, Michio.
   Hyperspace: a scientific odyssey through parallel universes, time
warps, and the tenth dimension / Michio Kaku; illustrations by
Robert O'Keefe.
      p.   cm.
   Includes bibliographical references and index.
   1. Physics.   2. Astrophysics.   3. Mathematical physics.
I. Title.
QC21.2.K3   1994
530.1'42—dc20   94-36657
CIP

ISBN 0-385-47705-8
Copyright © 1994 by Oxford University Press
All Rights Reserved
Printed in the United States of America
First Anchor Books Edition: March 1995

10 9 8 7 6 5 4 3 2 1

This book is dedicated
to my parents

Preface

Scientific revolutions, almost by definition, defy common sense.

If all our common-sense notions about the universe were correct, then science would have solved the secrets of the universe thousands of years ago. The purpose of science is to peel back the layer of the appearance of objects to reveal their underlying nature. In fact, if appearance and essence were the same thing, there would be no need for science.

Perhaps the most deeply entrenched common-sense notion about our world is that it is three dimensional. It goes without saying that length, width, and breadth suffice to describe all objects in our visible universe. Experiments with babies and animals have shown that we are born with an innate sense that our world is three dimensional. If we include time as another dimension, then four dimensions are sufficient to record all events in the universe. No matter where our instruments have probed, from deep within the atom to the farthest reaches of the galactic cluster, we have only found evidence of these four dimensions. To claim otherwise publicly, that other dimensions might exist or that our universe may coexist with others, is to invite certain scorn. Yet this deeply ingrained prejudice about our world, first speculated on by ancient Greek philosophers 2 millennia ago, is about to succumb to the progress of science.

This book is about a scientific revolution created by the *theory of hyperspace*,[1] which states that dimensions exist beyond the commonly accepted four of space and time. There is a growing acknowledgment among physicists worldwide, including several Nobel laureates, that the universe may actually exist in higher-dimensional space. If this theory is proved correct, it will create a profound conceptual and philosophical revolution in our understanding of the universe. Scientifically, the hyperspace theory goes by the names of Kaluza–Klein theory and supergravity. But its most advanced formulation is called superstring theory, which even predicts the precise number of dimensions: ten. The usual three dimensions of space (length, width, and breadth) and one of time are now extended by six more spatial dimensions.

We caution that the theory of hyperspace has not yet been experimentally confirmed and would, in fact, be exceedingly difficult to prove in the laboratory. However, the theory has already swept across the major physics research laboratories of the world and has irrevocably altered the scientific landscape of modern physics, generating a staggering number of research papers in the scientific literature (over 5,000 by one count). However, almost nothing has been written for the lay audience to explain the fascinating properties of higher-dimensional space. Therefore, the general public is only dimly aware, if at all, of this revolution. In fact, the glib references to other dimensions and parallel universes in the popular culture are often misleading. This is regrettable because the theory's importance lies in its power to unify all known physical phenomena in an astonishingly simple framework. This book makes available, for the first time, a scientifically authoritative but accessible account of the current fascinating research on hyperspace.

To explain why the hyperspace theory has generated so much excitement within the world of theoretical physics, I have developed four fundamental themes that run through this book like a thread. These four themes divide the book into four parts.

In Part I, I develop the early history of hyperspace, emphasizing the theme that the laws of nature become simpler and more elegant when expressed in higher dimensions.

To understand how adding higher dimensions can simplify physical problems, consider the following example: To the ancient Egyptians, the weather was a complete mystery. What caused the seasons? Why did it get warmer as they traveled south? Why did the winds generally blow in one direction? The weather was impossible to explain from the limited vantage point of the ancient Egyptians, to whom the earth appeared flat, like a two-dimensional plane. But now imagine sending the Egyptians in a rocket into outer space, where they can see the earth as simple and whole in its orbit around the sun. Suddenly, the answers to these questions become obvious.

From outer space, it is clear that the earth's axis is tilted about 23 degrees from the vertical (the 'vertical'' being the perpendicular to the plane of the earth's orbit around the sun). Because of this tilt, the northern hemisphere receives much less sunlight during one part of its orbit than during another part. Hence we have winter and summer. And since the equator receives more sunlight then the northern or southern polar regions, it becomes warmer as we approach the equator. Similarly, since the earth spins counterclockwise to someone sitting on the north pole, the cold, polar air swerves as it moves south toward the equator. The motion of hot and cold masses of air, set in motion by the earth's spin, thus helps to explain why the winds generally blow in one direction, depending on where you are on the earth.

In summary, the rather obscure laws of the weather are easy to understand once we view the earth from space. Thus the solution to the problem is to go *up* into space, into the *third dimension*. Facts that were impossible to understand in a flat world suddenly become obvious when viewing a three-dimensional earth.

Similarly, the laws of gravity and light seem totally dissimilar. They obey different physical assumptions and different mathematics. Attempts to splice these two forces have always failed. However, if we add one more dimension, a *fifth* dimension, to the previous four dimensions of space and time, then the equations governing light and gravity appear to merge together like two pieces of a jigsaw puzzle. Light, in fact, can be explained as vibrations in the fifth dimension. In this way, we see that the laws of light and gravity become simpler in five dimensions.

Consequently, many physicists are now convinced that a conventional four-dimensional theory is "too small" to describe adequately the forces that describe our universe. In a four-dimensional theory, physicists have to squeeze together the forces of nature in a clumsy, unnatural fashion. Furthermore, this hybrid theory is incorrect. When expressed in dimensions beyond four, however, we have "enough room" to explain the fundamental forces in an elegant, self-contained fashion.

In Part II, we further elaborate on this simple idea, emphasizing that the hyperspace theory may be able to unify all known laws of nature into one theory. Thus the hyperspace theory may be the crowning achievement of 2 millennia of scientific investigation: the unification of all known physical forces. It may give us the Holy Grail of physics, the "theory of everything" that eluded Einstein for so many decades.

For the past half-century, scientists have been puzzled as to why the basic forces that hold together the cosmos—gravity, electromagnetism, and the strong and weak nuclear forces—differ so greatly. Attempts by the greatest minds of the twentieth century to provide a unifying picture of all the known forces have failed. However, the hyperspace theory allows the possibility of explaining the four forces of nature as well as the seemingly random collection of subatomic particles in a truly elegant fashion. In the hyperspace theory, "matter" can be also viewed as the vibrations that ripple through the fabric of space and time. Thus follows the fascinating possibility that everything we see around us, from the trees and mountains to the stars themselves, are nothing but *vibrations in hyperspace*. If this is true, then this gives us an elegant, simple, and geometric means of providing a coherent and compelling description of the entire universe.

In Part III, we explore the possibility that, under extreme circumstances, space may be stretched until it rips or tears. In other words, hyperspace may provide a means to tunnel through space and time. Although we stress that this is still highly speculative, physicists are seriously analyzing the properties of "wormholes," of tunnels that link distant parts of space and time. Physicists at the California Institute of Technology, for example, have seriously proposed the possibility of building a time machine, consisting of a wormhole that connects the past with the future. Time machines have now left the realm of speculation and fantasy and have become legitimate fields of scientific research.

Cosmologists have even proposed the startling possibility that our universe is just one among an infinite number of parallel universes. These universes might be compared to a vast collection of soap bubbles suspended in air. Normally, contact between these bubble universes is impossible, but, by analyzing Einstein's equations, cosmologists have shown that there might exist a web of wormholes, or tubes, that connect these parallel universes. On each bubble, we can define our own distinctive space and time, which have meaning only on its surface; outside these bubbles, space and time have no meaning.

Although many consequences of this discussion are purely theoretical, hyperspace travel may eventually provide the most practical application of all: to save intelligent life, including ours, from the death of the universe. Scientists universally believe that the universe must eventually die, and with it all life that has evolved over billions of years. For example, according to the prevailing theory, called the Big Bang, a cosmic explosion 15 to 20 billion years ago set the universe expanding, hurling stars and galaxies away from us at great velocities. However, if the universe one day stops expanding and begins to contract, it will eventually collapse into a fiery cataclysm called the Big Crunch, in which all intelligent life will be vaporized by fantastic heat. Nevertheless, some physicists have speculated that the hyperspace theory may provide the one and only hope of a refuge for intelligent life. In the last seconds of the death of our universe, intelligent life may escape the collapse by fleeing into hyperspace.

In Part IV, we conclude with a final, practical question: If the theory is proved correct, then when will we be able to harness the power of the hyperspace theory? This is not just an academic question, because in the past, the harnessing of just one of the four fundamental forces irrevocably changed the course of human history, lifting us from the ignorance and squalor of ancient, preindustrial societies to modern civilization. In some sense, even the vast sweep of human history can be viewed in a new light, in terms of the progressive mastery of each of the four forces. The history of civilization has undergone a profound change as each of these forces was discovered and mastered.

For example, when Isaac Newton wrote down the classical laws of gravity, he developed the theory of mechanics, which gave us the laws governing machines. This, in turn, greatly accelerated the Industrial Revolution, which unleashed political forces that eventually overthrew the feudal dynasties of Europe. In the mid-1860s, when James Clerk Maxwell wrote down the fundamental laws of the electromagnetic force, he ushered in the Electric Age, which gave us the dynamo, radio, television, radar, household appliances, the telephone, microwaves, consumer electronics, the electronic computer, lasers, and many other electronic marvels. Without the understanding and utilization of the electromagnetic force, civilization would have stagnated, frozen in a time before the discovery of the light bulb and the electric motor. In the mid-1940s, when the nuclear force was harnessed, the world was again turned upside down with the development of the atomic and hydrogen bombs, the most destructive weapons on the planet. Because we are not on the verge of a unified understanding of all the cosmic forces governing the universe, one might expect that any civilization that masters the hyperspace theory will become lord of the universe.

Since the hyperspace theory is a well-defined body of mathematical equations, we can calculate the precise energy necessary to twist space and time into a pretzel or to create wormholes linking distant parts of our universe. Unfortunately, the results are disappointing. The energy required far exceeds anything that our planet can muster. In fact, the energy is a quadrillion times larger than the energy of our largest atom smashers. We must wait centuries or even millennia until our civilization develops the technical capability of manipulating space–time, or hope for contact with an advanced civilization that has already mastered hyperspace. The book therefore ends by exploring the intriguing but speculative scientific question of what level of technology is necessary for us to become masters of hyperspace.

Because the hyperspace theory takes us far beyond normal, common-sense conceptions of space and time, I have scattered throughout the text a few purely hypothetical stories. I was inspired to utilize this pedagogical technique by the story of Nobel Prize winner Isidore I. Rabi addressing an audience of physicists. He lamented the abysmal state of science education in the United States and scolded the physics community for neglecting its duty in popularizing the adventure of science for the general public and especially for the young. In fact, he admonished, science-fiction writers had done more to communicate the romance of science than all physicists combined.

In a previous book, *Beyond Einstein: The Cosmic Quest for the Theory of the Universe* (coauthored with Jennifer Trainer), I investigated superstring theory, described the nature of subatomic particles, and discussed at length the *visible universe* and how all the complexities of matter might be explained by tiny, vibrating strings. In this book, I have expanded on a different theme and explored the *invisible universe*—that is, the world of geometry and space–time. The focus of this book is not the nature of subatomic particles, but the higher-dimensional world in which they probably live. In the process, readers will see that higher-dimensional space, instead of being an empty, passive backdrop against which quarks play out their eternal roles, actually becomes the central actor in the drama of nature.

In discussing the fascinating history of the hyperspace theory, we will see that the search for the ultimate nature of matter, begun by the Greeks 2 millennia ago, has been a long and tortuous one. When the final chapter in this long saga is written by future historians of science, they may well record that the crucial breakthrough was the defeat of common-sense theories of three or four dimensions and the victory of the theory of hyperspace.

New York   M.K.
May 1993

Acknowledgments

In writing this book, I have been fortunate to have Jeffrey Robbins as my editor. He was the editor who skillfully guided the progress of three of my previous textbooks in theoretical physics written for the scientific community, concerning the unified field theory, superstring theory, and quantum field theory. This book, however, marks the first popular science book aimed at a general audience that I have written for him. It has always been a rare privilege to work closely with him.

I would also like to thank Jennifer Trainer, who has been my coauthor on two previous popular books. Once again, she has applied her considerable skills to make the presentation as smooth and coherent as possible.

I am also grateful to numerous other individuals who have helped to strengthen and criticize earlier drafts of this book: Burt Solomon, Leslie Meredith, Eugene Mallove, and my agent, Stuart Krichevsky.

Finally, I would like to thank the Institute for Advanced Study at Princeton, where much of this book was written, for its hospitality. The Institute, where Einstein spent the last decades of his life, was an appropriate place to write about the revolutionary developments that have extended and embellished much of his pioneering work.

Contents

Part III Wormholes: Gateways to Another Universe?

Part IV Masters of Hyperspace

But the creative principle resides in mathematics. In a certain sense, therefore, I hold it true that pure thought can grasp reality, as the ancients dreamed.

Albert Einstein

PART I

Entering
the Fifth Dimension

I
Worlds Beyond Space and Time

I want to know how God created this world. I am not interested in this or that phenomenon. I want to know His thoughts, the rest are details.

Albert Einstein

The Education of a Physicist

TWO incidents from my childhood greatly enriched my understanding of the world and sent me on course to become a theoretical physicist.

I remember that my parents would sometimes take me to visit the famous Japanese Tea Garden in San Francisco. One of my happiest childhood memories is of crouching next to the pond, mesmerized by the brilliantly colored carp swimming slowly beneath the water lilies.

In these quiet moments, I felt free to let my imagination wander; I would ask myself silly questions that a only child might ask, such as how the carp in that pond would view the world around them. I thought, What a strange world theirs must be!

Living their entire lives in the shallow pond, the carp would believe that their "universe" consisted of the murky water and the lilies. Spending most of their time foraging on the bottom of the pond, they would be only dimly aware that an alien world could exist above the surface.

3

The nature of my world was beyond their comprehension. I was intrigued that I could sit only a few inches from the carp, yet be separated from them by an immense chasm. The carp and I spent our lives in two distinct universes, never entering each other's world, yet were separated by only the thinnest barrier, the water's surface.

I once imagined that there may be carp "scientists" living among the fish. They would, I thought, scoff at any fish who proposed that a parallel world could exist just above the lilies. To a carp "scientist," the only things that were real were what the fish could see or touch. The pond was everything. An unseen world beyond the pond made no scientific sense.

Once I was caught in a rainstorm. I noticed that the pond's surface was bombarded by thousands of tiny raindrops. The pond's surface became turbulent, and the water lilies were being pushed in all directions by water waves. Taking shelter from the wind and the rain, I wondered how all this appeared to the carp. To them, the water lilies would appear to be moving around by themselves, without anything pushing them. Since the water they lived in would appear invisible, much like the air and space around us, they would be baffled that the water lilies could move around by themselves.

Their "scientists," I imagined, would concoct a clever invention called a "force" in order to hide their ignorance. Unable to comprehend that there could be waves on the unseen surface, they would conclude that lilies could move without being touched because a mysterious, invisible entity called a force acted between them. They might give this illusion impressive, lofty names (such as action-at-a-distance, or the ability of the lilies to move without anything touching them).

Once I imagined what would happen if I reached down and lifted one of the carp "scientists" out of the pond. Before I threw him back into the water, he might wiggle furiously as I examined him. I wondered how this would appear to the rest of the carp. To them, it would be a truly unsettling event. They would first notice that one of their "scientists" had disappeared from their universe. Simply vanished, without leaving a trace. Wherever they would look, there would be no evidence of the missing carp in their universe. Then, seconds later, when I threw him back into the pond, the "scientist" would abruptly reappear out of nowhere. To the other carp, it would appear that a miracle had happened.

After collecting his wits, the "scientist" would tell a truly amazing story. "Without warning," he would say, "I was somehow lifted out of the universe (the pond) and hurled into a mysterious nether world, with blinding lights and strangely shaped objects that I had never seen before. The strangest of all was the creature who held me prisoner, who did not resemble a fish in the slightest. I was shocked to see that it had no fins whatsoever, but nevertheless could move without them. It struck me that the familiar laws of nature no longer applied in this nether world. Then, just as suddenly, I found myself thrown back into our universe.'' (This story, of course, of a journey beyond the universe would be so fantastic that most of the carp would dismiss it as utter poppycock.)

I often think that we are like the carp swimming contentedly in that pond. We live out our lives in our own ''pond,'' confident that our universe consists of only those things we can see or touch. Like the carp, our universe consists of only the familiar and the visible. We smugly refuse to admit that parallel universes or dimensions can exist next to ours, just beyond our grasp. If our scientists invent concepts like forces, it is only because they cannot visualize the invisible vibrations that fill the empty space around us. Some scientists sneer at the mention of higher dimensions because they cannot be conveniently measured in the laboratory.

Ever since that time, I have been fascinated by the possibility of other dimensions. Like most children, I devoured adventure stories in which time travelers entered other dimensions and explored unseen parallel universes, where the usual laws of physics could be conveniently suspended. I grew up wondering if ships that wandered into the Bermuda Triangle mysteriously vanished into a hole in space; I marveled at Isaac Asimov's Foundation Series, in which the discovery of hyperspace travel led to the rise of a Galactic Empire.

A second incident from my childhood also made a deep, lasting impression on me. When I was 8 years old, I heard a story that would stay with me for the rest of my life. I remember my schoolteachers telling the class about a great scientist who had just died. They talked about him with great reverence, calling him one of the greatest scientists in all history. They said that very few people could understand his ideas, but that his discoveries changed the entire world and everything around us. I didn't understand much of what they were trying to tell us, but what most intrigued me about this man was that he died before he could complete his greatest discovery. They said he spent years on this theory, but he died with his unfinished papers still sitting on his desk.

I was fascinated by the story. To a child, this was a great mystery. What was his unfinished work? What was in those papers on his desk? What problem could possibly be so difficult and so important that such a great scientist would dedicate years of his life to its pursuit? Curious, I decided to learn all I could about Albert Einstein and his unfinished theory. I still have warm memories of spending many quiet hours reading every book I could find about this great man and his theories. When I exhausted the books in our local library, I began to scour libraries and bookstores across the city, eagerly searching for more clues. I soon learned that this story was far more exciting than any murder mystery and more important than anything I could ever imagine. I decided that I would try to get to the root of this mystery, even if I had to become a theoretical physicist to do it.

I soon learned that the unfinished papers on Einstein's desk were an attempt to construct what he called the unified field theory, a theory that could explain all the laws of nature, from the tiniest atom to the largest galaxy. However, being a child, I didn't understand that perhaps there was a link between the carp swimming in the Tea Garden and the unfinished papers lying on Einstein's desk. I didn't understand that higher dimensions might be the key to solving the unified field theory.

Later, in high school, I exhausted most of the local libraries and often visited the Stanford University physics library. There, I came across the fact that Einstein's work made possible a new substance called antimatter, which would act like ordinary matter but would annihilate upon contact with matter in a burst of energy. I also read that scientists had built large machines, or "atom smashers," that could produce microscopic quantities of this exotic substance in the laboratory.

One advantage of youth is that it is undaunted by worldly constraints that would ordinarily seem insurmountable to most adults. Not appreciating the obstacles involved, I set out to build my own atom smasher. I studied the scientific literature until I was convinced that I could build what was called a betatron, which could boost electrons to millions of electron volts. (A million electron volts is the energy attained by electrons accelerated by a field of a million volts.)

First, I purchased a small quantity of sodium-22, which is radioactive and naturally emits positrons (the antimatter counterpart of electrons). Then I built what is called a cloud chamber, which makes visible the tracks left by subatomic particles. I was able to take hundreds of beautiful photographs of the tracks left behind by antimatter. Next, I scavenged around large electronic warehouses in the area, assembled the necessary hardware, including hundreds of pounds of scrap transformer steel, and built a 2.3-million-electron-volt betatron in my garage that would be powerful enough to produce a beam of antielectrons. To construct the monstrous magnets necessary for the betatron, I convinced my parents to help me wind 22 miles of cooper wire on the high-school football field.

We spent Christmas vacation on the 50-yard line, winding and assembling the massive coils that would bend the paths of the high-energy electrons.

When finally constructed, the 300-pound, 6-kilowatt betatron consumed every ounce of energy my house produced. When I turned it on, I would usually blow every fuse, and the house would suddenly became dark. With the house plunged periodically into darkness, my mother would often shake her head. (I imagined that she probably wondered why she couldn't have a child who played baseball or basketball, instead of building these huge electrical machines in the garage.) I was gratified that the machine successfully produced a magnetic field 20,000 times more powerful than the earth's magnetic field, which is necessary to accelerate a beam of electrons.

Confronting the Fifth Dimension

Because my family was poor, my parents were concerned that I wouldn't be able to continue my experiments and my education. Fortunately, the awards that I won for my various science projects caught the attention of the atomic scientist Edward Teller. His wife generously arranged for me to receive a 4-year scholarship to Harvard, allowing me to fulfill my dream.

Ironically, although at Harvard I began my formal training in theoretical physics, it was also where my interest in higher dimensions gradually died out. Like other physicists, I began a rigorous and thorough program of studying the higher mathematics of each of the forces of nature separately, in complete isolation from one another. I still remember solving a problem in electrodynamics for my instructor, and then asking him what the solution might look like if space were curved in a higher dimension. He looked at me in a strange way, as if I were a bit cracked. Like others before me, I soon learned to put aside my earlier, childish notions about higher-dimensional space. Hyperspace, I was told, was not a suitable subject of serious study.

I was never satisfied with this disjointed approach to physics, and my thoughts would often drift back to the the carp living in the Tea Garden. Although the equations we used for electricity and magnetism, discovered by Maxwell in the nineteenth century, worked surprisingly well, the equations seemed rather arbitrary. I felt that physicists (like the carp) invented these "forces" to hide our ignorance of how objects can move each other without touching.

In my studies, I learned that one of the great debates of the nine-teenth century had been about how light travels through a vacuum. (Light from the stars, in fact, can effortlessly travel trillions upon trillions of miles through the vacuum of outer space.) Experiments also showed beyond question that light is a wave. But if light were a wave, then it would require something to be "waving." Sound waves require air, water waves require water, but since there is nothing to wave in a vacuum, we have a paradox. How can light be a wave if there is nothing to wave? So physicists conjured up a substance called the aether, which filled the vacuum and acted as the medium for light. However, experiments con-clusively showed that the "aether" does not exist.*

Finally, when I became a graduate student in physics at the University of California at Berkeley, I learned quite by accident that there was an alternative, albeit controversial, explanation of how light can travel through a vacuum. This alternative theory was so outlandish that I received quite a jolt when I stumbled across it. That shock was similar to the one experienced by many Americans when they first heard that President John Kennedy had been shot. They can invariably remember the precise moment when they heard the shocking news, what they were doing, and to whom they were talking at that instant. We physicists, too, receive quite a shock when we first stumble across Kaluza–Klein theory for the first time. Since the theory was considered to be a wild specula-tion, it was never taught in graduate school; so young physicists are left to discover it quite by accident in their casual readings.

This alternative theory gave the simplest explanation of light: that it was really a vibration of the fifth dimension, or what used to called the fourth dimension by the mystics. If light could travel through a vacuum, it was because the vacuum itself was vibrating, because the "vacuum" really existed in four dimensions of space and one of time. By adding the fifth dimension, the force of gravity and light could be unified in a startlingly simple way. Looking back at my childhood experiences at the Tea Garden, I suddenly realized that this was the mathematical theory for which I had been looking.

The old Kaluza–Klein theory, however, had many difficult, technical problems that rendered it useless for over half a century. All this, how-ever, has changed in the past decade. More advanced versions of the theory, like supergravity theory and especially superstring theory, have

---

*Surprisingly, even today physicists still do not have a real answer to this puzzle, but over the decades we have simply gotten used to the idea that light can travel through a vacuum even if there is nothing to wave.

finally eliminated the inconsistencies of the theory. Rather abruptly, the theory of higher dimensions is now being championed in research laboratories around the globe. Many of the world's leading physicists now believe that dimensions beyond the usual four of space and time might exist. This idea, in fact, has become the focal point of intense scientific investigation. Indeed, many theoretical physicists now believe that higher dimensions may be the decisive step in creating a comprehensive theory that unites the laws of nature—a theory of hyperspace.

If it proves to be correct, then future historians of science may well record that one of the great conceptual revolutions in twentieth-century science was the realization that hyperspace may be the key to unlock the deepest secrets of nature and Creation itself.

This seminal concept has sparked an avalanche of scientific research: Several thousand papers written by theoretical physicists in the major research laboratories around the world have been devoted to exploring the properties of hyperspace. The pages of *Nuclear Physics* and *Physics Letters,* two leading scientific journals, have been flooded with articles analyzing the theory. More than 200 international physics conferences have been sponsored to explore the consequences of higher dimensions.

Unfortunately, we are still far from experimentally verifying that our universe exists in higher dimensions. (Precisely what it would take to prove the correctness of the theory and possibly harness the power of hyperspace will be discussed later in this book.) However, this theory has now become firmly established as a legitimate branch of modern theoretical physics. The Institute for Advanced Study at Princeton, for example, where Einstein spent the last decades of his life (and where this book was written), is now one of the active centers of research on higher-dimensional space–time.

Steven Weinberg, who won the Nobel Prize in physics in 1979, summarized this conceptual revolution when he commented recently that theoretical physics seems to be becoming more and more like science fiction.

Why Can't We See Higher Dimensions?

These revolutionary ideas seem strange at first because we take for granted that our everyday world has three dimensions. As the late physicist Heinz Pagels noted, "One feature of our physical world is so obvious that most people are not even puzzled by it—the fact that space is three-dimensional."[1] Almost by instinct alone, we know that any object can be described by giving its height, width, and depth. By giving three numbers, we can locate any position in space. If we want to meet someone for lunch in New York, we say, "Meet me on the twenty-fourth floor of the building at the corner of Forty-second Street and First Avenue." Two numbers provide us the street corner; and the third, the height off the ground.

Airplane pilots, too, know exactly where they are with three numbers—their altitude and two coordinates that locate their position on a grid or map. In fact, specifying these three numbers can pinpoint any location in our world, from the tip of our nose to the ends of the visible universe. Even babies understand this: Tests with infants have shown that they will crawl to the edge of a cliff, peer over the edge, and crawl back. In addition to understanding "left" and "right" and "forward" and "backward" instinctively, babies instinctively understand "up" and "down." Thus the intuitive concept of three dimensions is firmly embedded in our brains from an early age.

Einstein extended this concept to include time as the fourth dimension. For example, to meet that someone for lunch, we must specify that we should meet at, say, 12:30 P.M. in Manhattan; that is, to specify an event, we also need to describe its fourth dimension, the *time* at which the event takes place.

Scientists today are interested in going beyond Einstein's conception of the fourth dimension. Current scientific interest centers on the fifth dimension (the spatial dimension beyond time and the three dimensions of space) and beyond. (To avoid confusion, throughout this book I have bowed to custom and called the fourth dimension the *spatial* dimension beyond length, breadth, and width. Physicists actually refer to this as the fifth dimension, but I will follow historical precedent. We will call time the fourth *temporal* dimension.)

How do we see the fourth spatial dimension?

The problem is, we can't. Higher-dimensional spaces are impossible to visualize; so it is futile even to try. The prominent German physicist Hermann von Helmholtz compared the inability to "see" the fourth dimension with the inability of a blind man to conceive of the concept of color. No matter how eloquently we describe "red" to a blind person, words fail to impart the meaning of anything as rich in meaning as color. Even experienced mathematicians and theoretical physicists who have worked with higher-dimensional spaces for years admit that they cannot visualize them. Instead, they retreat into the world of mathematical equations. But while mathematicians, physicists, and computers have no problem solving equations in multidimensional space, humans find it impossible to visualize universes beyond their own.

At best, we can use a variety of mathematical tricks, devised by mathematician and mystic Charles Hinton at the turn of the century, to visualize shadows of higher-dimensional objects. Other mathematicians, like Thomas Banchoff, chairman of the mathematics department at Brown University, have written computer programs that allow us to manipulate higher-dimensional objects by projecting their shadows onto flat, two-dimensional computer screens. Like the Greek philosopher Plato, who said that we are like cave dwellers condemned to see only the dim, gray shadows of the rich life outside our caves, Banchoff's computers allow only a glimpse of the shadows of higher-dimensional objects. (Actually, we cannot visualize higher dimensions because of an accident of evolution. Our brains have evolved to handle myriad emergencies in three dimensions. Instantly, without stopping to think, we can recognize and react to a leaping lion or a charging elephant. In fact, those humans who could better visualize how objects move, turn, and twist in three dimensions had a distinct survival advantage over those who could not. Unfortunately, there was no selection pressure placed on humans to master motion in four spatial dimensions. Being able to see the fourth spatial dimension certainly did not help someone fend off a charging saber-toothed tiger. Lions and tigers do not lunge at us through the fourth dimension.)

The Laws of Nature Are Simpler in Higher Dimensions

One physicist who delights in teasing audiences about the properties of higher-dimensional universes is Peter Freund, a professor of theoretical physics at the University of Chicago's renowned Enrico Fermi Institute. Freund was one of the early pioneers working on hyperspace theories when it was considered too outlandish for mainstream physics. For years, Freund and a small group of scientists dabbled in the science of higher dimensions in isolation; now, however, it has finally become fashionable and a legitimate branch of scientific research. To his delight, he is finding that his early interest is at last paying off.

Freund does not fit the traditional image of a narrow, crusty, disheveled scientist. Instead, he is urbane, articulate, and cultured, and has a sly, impish grin that captivates nonscientists with fascinating stories of fast-breaking scientific discoveries. He is equally at ease scribbling on a blackboard littered with dense equations or exchanging light banter at a cocktail party. Speaking with a thick, distinguished Romanian accent, Freund has a rare knack for explaining the most arcane, convoluted concepts of physics in a lively, engaging style.

Traditionally, Freund reminds us, scientists have viewed higher dimensions with skepticism because they could not be measured and did not have any particular use. However, the growing realization among scientists today is that any three-dimensional theory is "too small" to describe the forces that govern our universe.

As Freund emphasizes, one fundamental theme running through the past decade of physics has been that *the laws of nature become simpler and elegant when expressed in higher dimensions,* which is their natural home. The laws of light and gravity find a natural expression when expressed in higher-dimensional space–time. The key step in unifying the laws of nature is to increase the number of dimensions of space–time until more and more forces can be accommodated. In higher dimensions, we have enough "room" to unify all known physical forces.

Freund, in explaining why higher dimensions are exciting the imagination of the scientific world, uses the following analogy: "Think, for a moment, of a cheetah, a sleek, beautiful animal, one of the fastest on earth, which roams freely on the savannas of Africa. In its natural habitat, it is a magnificent animal, almost a work of art, unsurpassed in speed or grace by any other animal. Now," he continues,

> think of a cheetah that has been captured and thrown into a miserable cage in a zoo. It has lost its original grace and beauty, and is put on display for our amusement. We see only the broken spirit of the cheetah in the cage, not its original power and elegance. The cheetah can be compared to the laws of physics, which are beautiful in their natural setting. The natural habitat of the laws of physics is higher-dimensional space–time. However, we can only measure the laws of physics when they have been broken and placed on display in a cage, which is our three-dimensional laboratory. We only see the cheetah when its grace and beauty have been stripped away.[2]

For decades, physicists have wondered why the four forces of nature appear to be so fragmented—why the "cheetah" looks so pitiful and broken in his cage. The fundamental reason why these four forces seem so dissimilar, notes Freund, is that we have been observing the "caged cheetah." Our three-dimensional laboratories are sterile zoo cages for the laws of physics. But when we formulate the laws in higher-dimensional space–time, their natural habitat, we see their true brilliance and power; the laws become simple and powerful. The revolution now sweeping over physics is the realization that the natural home for the cheetah may be hyperspace.

To illustrate how adding a higher dimension can make things simpler, imagine how major wars were fought by ancient Rome. The great Roman wars, often involving many smaller battlefields, were invariably fought with great confusion, with rumors and misinformation pouring in on both sides from many different directions. With battles raging on several fronts, Roman generals were often operating blind. Rome won its battles more from brute strength than from the elegance of its strategies. That is why one of the first principles of warfare is to seize the high ground—that is, to go *up* into the third dimension, above the two-dimensional battlefield. From the vantage point of a large hill with a panoramic view of the battlefield, the chaos of war suddenly becomes vastly reduced. In other words, viewed from the third dimension (that is, from the top of the hill), the confusion of the smaller battlefields becomes integrated into a coherent single picture.

Another application of this principle—that nature becomes simpler when expressed in higher dimensions—is the central idea behind Einstein's special theory of relativity. Einstein revealed time to be the fourth dimension, and he showed that space and time could conveniently be unified in a four-dimensional theory. This, in turn, inevitably led to the unification of all physical quantities measured by space and time, such as matter and energy. He then found the precise mathematical expression for this unity between matter and energy: $E = mc^2$, perhaps the most celebrated of all scientific equations.*

To appreciate the enormous power of this unification, let us now describe the four fundamental forces, emphasizing how different they are, and how higher dimensions may give us a unifying formalism. Over the past 2,000 years, scientists have discovered that all phenomena in our universe can be reduced to four forces, which at first bear no resemblance to one another.

The Electromagnetic Force

The electromagnetic force takes a variety of forms, including electricity, magnetism, and light itself. The electromagnetic force lights our cities, fills the air with music from radios and stereos, entertains us with television, reduces housework with electrical appliances, heats our food with

---

*The theory of higher dimensions is certainly not merely an academic one, because the simplest consequence of Einstein's theory is the atomic bomb, which has changed the destiny of humanity. In this sense, the introduction of higher dimensions has been one of the pivotal scientific discoveries in all human history.

microwaves, tracks our planes and space probes with radar, and electrifies our power plants. More recently, the power of the electromagnetic force has been used in electronic computers (which have revolutionized the office, home, school, and military) and in lasers (which have introduced new vistas in communications, surgery, compact disks, advanced Pentagon weaponry, and even the check-out stands in groceries). More than half the gross national product of the earth, representing the accumulated wealth of our planet, depends in some way on the electromagnetic force.

The Strong Nuclear Force

The strong nuclear force provides the energy that fuels the stars; it makes the stars shine and creates the brilliant, life-giving rays of the sun. If the strong force suddenly vanished, the sun would darken, ending all life on earth. In fact, some scientists believe that the dinosaurs were driven to extinction 65 million years ago when debris from a comet impact was blown high into the atmosphere, darkening the earth and causing the temperature around the planet to plummet. Ironically, it is also the strong nuclear force that may one day take back the gift of life. Unleashed in the hydrogen bomb, the strong nuclear force could one day end all life on earth.

The Weak Nuclear Force

The weak nuclear force governs certain forms of radioactive decay. Because radioactive materials emit heat when they decay or break apart, the weak nuclear force contributes to heating the radioactive rock deep within the earth's interior. This heat, in turn, contributes to the heat that drives the volcanoes, the rare but powerful eruptions of molten rock that reach the earth's surface. The weak and electromagnetic forces are also exploited to treat serious diseases: Radioactive iodine is used to kill tumors of the thyroid gland and fight certain forms of cancer. The force of radioactive decay can also be deadly: It wreaked havoc at Three Mile Island and Chernobyl; it also creates radioactive waste, the inevitable by-product of nuclear weapons production and commercial nuclear power plants, which may remain harmful for millions of years.

The Gravitational Force

The gravitational force keeps the earth and the planets in their orbits and binds the galaxy. Without the gravitational force of the earth, we would be flung into space like rag dolls by the spin of the earth. The air we breathe would be quickly diffused into space, causing us to asphyxiate and making life on earth impossible. Without the gravitational force of the sun, all the planets, including the earth, would be flung from the solar system into the cold reaches of deep space, where sunlight is too dim to support life. In fact, without the gravitational force, the sun itself would explode. The sun is the result of a delicate balancing act between the force of gravity, which tends to crush the star, and the nuclear force, which tends to blast the sun apart. Without gravity, the sun would detonate like trillions upon trillions of hydrogen bombs.

The central challenge of theoretical physics today is to unify these four forces into a single force. Beginning with Einstein, the giants of twentieth-century physics have tried and failed to find such a unifying scheme. However, the answer that eluded Einstein for the last 30 years of his life may lie in hyperspace.

The Quest for Unification

Einstein once said, "Nature shows us only the tail of the lion. But I do not doubt that the lion belongs to it even though he cannot at once reveal himself because of his enormous size."[3] If Einstein is correct, then perhaps these four forces are the "tail of the lion," and the "lion" itself is higher-dimensional space–time. This idea has fueled the hope that the physical laws of the universe, whose consequences fill entire library walls with books densely packed with tables and graphs, may one day be explained by a single equation.

Central to this revolutionary perspective on the universe is the realization that higher-dimensional *geometry* may be the ultimate source of unity in the universe. Simply put, the matter in the universe and the forces that hold it together, which appear in a bewildering, infinite variety of complex forms, may be nothing but different vibrations of hyperspace. This concept, however, goes against the traditional thinking among scientists, who have viewed space and time as a passive stage on which the stars and the atoms play the leading role. To scientists, the visible universe of matter seemed infinitely richer and more diverse than the empty, unmoving arena of the invisible universe of space–time. Almost all the intense scientific effort and massive government funding in particle physics has historically gone to cataloging the properties of subatomic particles, such as "quarks" and "gluons," rather than fathoming the nature of geometry. Now, scientists are realizing that the "use-less" concepts of space and time may be the ultimate source of beauty and simplicity in nature.

The first theory of higher dimensions was called *Kaluza–Klein theory*, after two scientists who proposed a new theory of gravity in which light could be explained as vibrations in the fifth dimension. When extended to *N*-dimensional space (where *N* can stand for any whole number), the clumsy-looking theories of subatomic particles dramatically take on a startling symmetry. The old Kaluza–Klein theory, however, could not determine the correct value of *N*, and there were technical problems in describing all the subatomic particles. A more advanced version of this theory, called *supergravity theory*, also had problems. The recent interest in the theory was sparked in 1984 by physicists Michael Green and John Schwarz, who proved the consistency of the most advanced version of Kaluza–Klein theory, called *superstring theory*, which postulates that all matter consists of tiny vibrating strings. Surprisingly, the superstring theory predicts a precise number of dimensions for space and time: ten.*

The advantage of ten-dimensional space is that we have "enough room" in which to accommodate all four fundamental forces. Further-more, we have a simple physical picture in which to explain the confusing jumble of subatomic particles produced by our powerful atom smash-ers. Over the past 30 years, hundreds of subatomic particles have been carefully cataloged and studied by physicists among the debris created by smashing together protons and electrons with atoms. Like bug collectors patiently giving names to a vast collection of insects, physicists have at times been overwhelmed by the diversity and complexity of these subatomic particles. Today, this bewildering collection of subatomic particles can be explained as mere vibrations of the hyperspace theory.

Traveling Through Space and Time

The hyperspace theory has also reopened the question of whether hyperspace can be used to travel through space and time. To understand this

*Freund chuckles when asked when we will be able to see these higher dimensions. We cannot see these higher dimensions because they have "curled up" into a tiny ball so small that they can no longer be detected. According to Kaluza–Klein theory, the size of these curled up dimensions is called the *Planck length*,[4] which is 100 billion billion times smaller than the proton, too small to be probed by even by our largest atom smasher. High-energy physicists had hoped that the $11 billion superconducting supercollider (SSC) (which was canceled by Congress in October 1993) might have been able to reveal some indirect glimmers of hyperspace.

concept, imagine a race of tiny flatworms living on the surface of a large apple. It's obvious to these worms that their world, which they call Apple-world, is flat and two dimensional, like themselves. One worm, however, named Columbus, is obsessed by the notion that Appleworld is somehow finite and curved in something he calls the third dimension. He even invents two new words, *up* and *down,* to describe motion in this invisible third dimension. His friends, however, call him a fool for believing that Appleworld could be bent in some unseen dimension that no one can see or feel. One day, Columbus sets out on a long and arduous journey and disappears over the horizon. Eventually he returns to his starting point, proving that the world is actually curved in the unseen third dimension. His journey proves that Appleworld is curved in a higher unseen dimension, the third dimension. Although weary from his trav-els, Columbus discovers that there is yet another way to travel between distant points on the apple: By burrowing into the apple, he can carve a tunnel, creating a convenient shortcut to distant lands. These tunnels, which considerably reduce the time and discomfort of a long journey, he calls *wormholes.* They demonstrate that the shortest path between two points is not necessarily a straight line, as he's been taught, but a worm-hole.

One strange effect discovered by Columbus is that when he enters one of these tunnels and exits at the other end, he finds himself back in the past. Apparently, these wormholes connect parts of the apple where time beats at different rates. Some of the worms even claim that these wormholes can be molded into a workable time machine.

Later, Columbus makes an even more momentous discovery—his Appleworld is actually not the only one in the universe. It is but one apple in a large apple orchard. His apple, he finds out, coexists with hundreds of others, some with worms like themselves, and some without worms. Under certain rare circumstances, he conjectures, it may even be possible to journey between the different apples in the orchard.

We human beings are like the flatworms. Common sense tells us that our world, like their apple, is flat and three dimensional. No matter where we go with our rocket ships, the universe seems flat. However, the fact that our universe, like Appleworld, is curved in an unseen dimension beyond our spatial comprehension has been experimentally verified by a number of rigorous experiments. These experiments, performed on the path of light beams, show that starlight is bent as it moves across the universe.

Multiply Connected Universes

When we wake up in the morning and open the window to let in some fresh air, we expect to see the front yard. We do *not* expect to face the towering pyramids of Egypt. Similarly, when we open the front door, we expect to see the cars on the street, not the craters and dead volcanoes of a bleak, lunar landscape. Without even thinking about it, we assume that we can safely open windows or doors without being scared out of our wits. Our world, fortunately, is not a Steven Spielberg movie. We act on a deeply ingrained prejudice (which is invariably correct) that our world is *simply connected,* that our windows and doorways are not entrances to wormholes connecting our home to a far-away universe. (In ordinary space, a lasso of rope can always be shrunk to a point. If this is possible, then the space is called simply connected. However, if the lasso is placed around the entrance of the wormhole, then it cannot be shrunk to a point. The lasso, in fact, enters the wormhole. Such spaces, where lassos are not contractible, are called *multiply connected.* Although the bending of our universe in an unseen dimension has been experimentally measured, the existence of wormholes and whether our universe is multiply connected or not is still a topic of scientific controversy.)

Mathematicians dating back to Georg Bernhard Riemann have studied the properties of multiply connected spaces in which different regions of space and time are spliced together. And physicists, who once thought this was merely an intellectual exercise, are now seriously studying multiply connected worlds as a practical model of our universe. These models are the scientific analogue of Alice's looking glass. When Lewis Carroll's White Rabbit falls down the rabbit hole to enter Wonderland, he actually falls down a wormhole.

Wormholes can be visualized with a sheet of paper and a pair of scissors: Take a piece of paper, cut two holes in it, and then reconnect the two holes with a long tube (Figure 1.1). As long as you avoid walking into the wormhole, our world seems perfectly normal. The usual laws of geometry taught in school are obeyed. However, if you fall into the wormhole, you are instantly transported to a different region of space and time. Only by retracing your steps and falling back into the wormhole can you return to your familiar world.

Time Travel and Baby Universes

Although wormholes provide a fascinating area of research, perhaps the most intriguing concept to emerge from this discussion of hyperspace

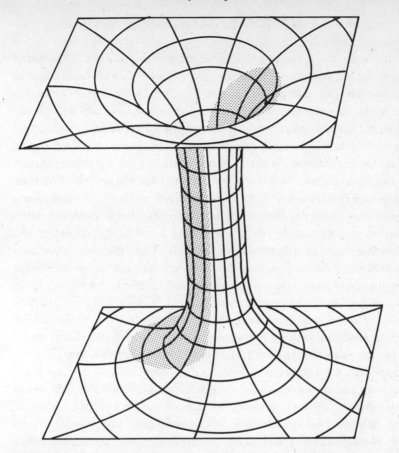

Figure 1.1. Parallel universes may be graphically represented by two parallel planes. Normally, they never interact with each other. However, at times wormholes or tubes may open up between them, perhaps making communication and travel possible between them. This is now the subject of intense interest among theoretical physicists.

is the question of time travel. In the film *Back to the Future*, Michael J. Fox journeys back in time and meets his parents as teenagers before they were married. Unfortunately, his mother falls in love with *him* and spurns his father, raising the ticklish question of how he will be born if his parents never marry and have children.

Traditionally, scientists have held a dim opinion of anyone who raised the question of time travel. Causality (the notion that every effect is preceded, not followed, by a cause) is firmly enshrined in the foundations of modern science. However, in the physics of wormholes, "acausal" effects show up repeatedly. In fact, we have to make strong assumptions in order to prevent time travel from taking place. The main problem is that wormholes may connect not only two distant points in space, but also the future with the past.

In 1988, physicist Kip Thorne of the California Institute of Technology and his collaborators made the astonishing (and risky) claim that time travel is indeed not only possible, but probable under certain conditions. They published their claim not in an obscure "fringe" journal, but in the prestigious *Physical Review Letters*. This marked the first time that reputable physicists, and not crackpots, were scientifically advancing a claim about changing the course of time itself. Their announcement was based on the simple observation that a wormhole connects two regions that exist in different time periods. Thus the wormhole may connect the present to the past. Since travel through the wormhole is nearly instantaneous, one could use the wormhole to go backward in time. Unlike the machine portrayed in H. G. Wells's *The Time Machine*, however, which could hurl the protagonist hundreds of thousands of years into England's distant future with the simple twist of a dial, a wormhole may require vast amounts of energy for its creation, beyond what will be technically possible for centuries to come.

Another bizarre consequence of wormhole physics is the creation of "baby universes" in the laboratory. We are, of course, unable to re-create the Big Bang and witness the birth of our universe. However, Alan Guth of the Massachusetts Institute of Technology, who has made many important contributions in cosmology, shocked many physicists a few years ago when he claimed that the physics of wormholes may make it possible to create a baby universe of our own in the laboratory. By concentrating intense heat and energy in a chamber, a wormhole may eventually open up, serving as an umbilical cord connecting our universe to another, much smaller universe. If possible, it would give a scientist an unprecedented view of a universe as it is created in the laboratory.

Mystics and Hyperspace

Some of these concepts are not new. For the past several centuries, mystics and philosophers have speculated about the existence of other universes and tunnels between them. They have long been fascinated by the possible existence of other worlds, undetectable by sight or sound, yet coexisting with our universe. They have been intrigued by the possibility that these unexplored, nether worlds may even be tantalizingly close, in fact surrounding us and permeating us everywhere we move, yet just beyond our physical grasp and eluding our senses. Such idle talk, however, was ultimately useless because there was no practical way in which to mathematically express and eventually test these ideas.

Gateways between our universe and other dimensions are also a favorite literary device. Science-fiction writers find higher dimensions to be an indispensable tool, using them as a medium for interstellar travel. Because of the astronomical distances separating the stars in the heavens, science-fiction writers use higher dimensions as a clever shortcut between the stars. Instead of taking the long, direct route to other galaxies, rockets merely zip along in hyperspace by warping the space around them. For instance, in the film *Star Wars,* hyperspace is a refuge where Luke Skywalker can safely evade the Imperial Starships of the Empire. In the television series "Star Trek: Deep Space Nine," a wormhole opens up near a remote space station, making it possible to span enormous distances across the galaxy within seconds. The space station suddenly becomes the center of intense intergalactic rivalry over who should control such a vital link to other parts of the galaxy.

Ever since Flight 19, a group of U.S. military torpedo bombers, vanished in the Caribbean 30 years ago, mystery writers too have used higher dimensions as a convenient solution to the puzzle of the Bermuda Triangle, or Devil's Triangle. Some have conjectured that airplanes and ships disappearing in the Bermuda Triangle actually entered some sort of passageway to another world.

The existence of these elusive parallel worlds has also produced endless religious speculation over the centuries. Spiritualists have wondered whether the souls of departed loved ones drifted into another dimension. The seventeenth-century British philosopher Henry More argued that ghosts and spirits did indeed exist and claimed that they inhabited the fourth dimension. In *Enchiridion Metaphysicum* (1671), he argued for the existence of a nether realm beyond our tangible senses that served as a home for ghosts and spirits.

Nineteenth-century theologians, at a loss to locate heaven and hell, pondered whether they might be found in a higher dimension. Some wrote about a universe consisting of three parallel planes: the earth, heaven, and hell. God himself, according to the theologian Arthur Willink, found his home in a world far removed from these three planes; he lived in infinite-dimensional space.

Interest in higher dimensions reached its peak between 1870 and 1920, when the "fourth dimension" (a spatial dimension, different from what we know as the fourth dimension of time) seized the public imagination and gradually cross-fertilized every branch of the arts and sciences, becoming a metaphor for the strange and mysterious. The fourth dimension appeared in the literary works of Oscar Wilde, Fyodor Dostoyevsky, Marcel Proust, H. G. Wells, and Joseph Conrad; it inspired some of the musical works of Alexander Scriabin, Edgard Varèse, and George Antheil. It fascinated such diverse personalities as psychologist William James, literary figure Gertrude Stein, and revolutionary socialist Vladimir Lenin.

The fourth dimension also inspired the works of Pablo Picasso and Marcel Duchamp and heavily influenced the development of Cubism and Expressionism, two of the most influential art movements in this century. Art historian Linda Dalrymple Henderson writes, "Like a Black Hole, 'the fourth dimension' possessed mysterious qualities that could not be completely understood, even by the scientists themselves. Yet, the impact of 'the fourth dimension' was far more comprehensive than that of Black Holes or any other more recent scientific hypothesis except Relativity Theory after 1919."[5]

Similarly, mathematicians have long been intrigued by alternative forms of logic and bizarre geometries that defy every convention of common sense. For example, the mathematician Charles L. Dodgson, who taught at Oxford University, delighted generations of schoolchildren by writing books—as Lewis Carroll—that incorporate these strange mathematical ideas. When Alice falls down a rabbit hole or steps through the looking glass, she enters Wonderland, a strange place where Cheshire cats disappear (leaving only their smile), magic mushrooms turn children into giants, and Mad Hatters celebrate "unbirthdays." The looking glass somehow connects Alice's world with a strange land where everyone speaks in riddles and common sense isn't so common.

Some of the inspiration for Lewis Carroll's ideas most likely came from the great nineteenth-century German mathematician Georg Bernhard Riemann, who was the first to lay the mathematical foundation of geometries in higher-dimensional space. Riemann changed the course of mathematics for the next century by demonstrating that these universes, as strange as they may appear to the layperson, are completely self-consistent and obey their own inner logic. To illustrate some of these ideas, think of stacking many sheets of paper, one on top of another. Now imagine that each sheet represents an entire world and that each world obeys its own physical laws, different from those of all the other worlds. Our universe, then, would not be alone, but would be one of many possible parallel worlds. Intelligent beings might inhabit some of these planes, completely unaware of the existence of the others. On one sheet of paper, we might have Alice's bucolic English countryside. On another sheet might be a strange world populated by mythical creatures in the world of Wonderland.

Normally, life proceeds on each of these parallel planes independent of the others. On rare occasions, however, the planes may intersect and, for a brief moment, tear the fabric of space itself, which opens up a hole—or gateway—between these two universes. Like the wormhole appearing in "Star Trek: Deep Space Nine," these gateways make travel possible between these worlds, like a cosmic bridge linking two different universes or two points in the same universe (Figure 1.2). Not surprisingly, Carroll found children much more open to these possibilities than adults, whose prejudices about space and logic become more rigid over time. In fact, Riemann's theory of higher dimensions, as interpreted by Lewis Carroll, has become a permanent part of children's literature and folklore, giving birth to other children's classics over the decades, such as Dorothy's Land of Oz and Peter Pan's Never Never Land.

Without any experimental confirmation or compelling physical motivation, however, these theories of parallel worlds languished as a branch of science. Over 2 millennia, scientists have occasionally picked up the notion of higher dimensions, only to discard it as an untestable and therefore silly idea. Although Riemann's theory of higher geometries was mathematically intriguing, it was dismissed as clever but useless. Scientists willing to risk their reputations on higher dimensions soon found themselves ridiculed by the scientific community. Higher-dimensional space became the last refuge for mystics, cranks, and charlatans.

In this book, we will study the work of these pioneering mystics, mainly because they devised ingenious ways in which a nonspecialist could "visualize" what higher-dimensional objects might look like. These tricks will prove useful to understand how these higher-dimensional theories may be grasped by the general public.

By studying the work of these early mystics, we also see more clearly what was missing from their research. We see that their speculations lacked two important concepts: a physical and a mathematical principle. From the perspective of modern physics, we now realize that the missing *physical* principle is that hyperspace simplifies the laws of nature, providing the possibility of unifying all the forces of nature by purely geometric arguments. The missing *mathematical* principle is called *field theory*, which is the universal mathematical language of theoretical physics.

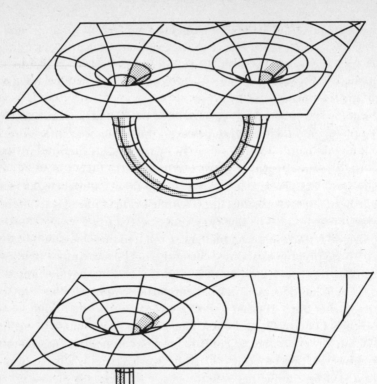

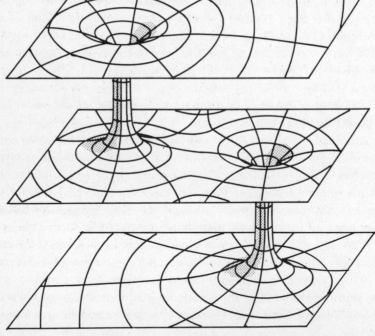

Figure 1.2. Wormholes may connect a universe with itself, perhaps providing a means of interstellar travel. Since wormholes may connect two different time eras, they may also provide a means for time travel. Wormholes may also connect an infinite series of parallel universes. The hope is that the hyperspace theory will be able to determine whether wormholes are physically possible or merely a mathematical curiosity.

Field Theory: The Language of Physics

Fields were first introduced by the great nineteenth-century British scientist Michael Faraday. The son of a poor blacksmith, Faraday was a self-taught genius who conducted elaborate experiments on electricity and magnetism. He visualized "lines of force" that, like long vines spreading from a plant, emanated from magnets and electric charges in all directions and filled up all of space. With his instruments, Faraday could measure the strength of these lines of force from a magnetic or an electric charge at any point in his laboratory. Thus he could assign a series of numbers (the strength and direction of the force) to that point (and any point in space). He christened the totality of these numbers at any point in space, treated as a single entity, a field. (There is a famous story concerning Michael Faraday. Because his fame had spread far and wide, he was often visited by curious bystanders. When one asked what his work was good for, he answered, "What is the use of a child? It grows to be a man." One day, William Gladstone, then Chancellor of the Exchequer, visited Faraday in his laboratory. Knowing nothing about science, Gladstone sarcastically asked Faraday what use the huge electrical contraptions in his laboratory could possibly have for England. Faraday replied, "Sir, I know not what these machines will be used for, but I am sure that one day you will tax them." Today, a large portion of the total wealth of England is invested in the fruit of Faraday's labors.)

Simply put, a *field* is a collection of numbers defined at every point in space that completely describes a force at that point. For example, three numbers at each point in space can describe the intensity and direction of the magnetic lines of force. Another three numbers everywhere in space can describe the electric field. Faraday got this concept when he thought of a "field" plowed by a farmer. A farmer's field occupies a two-dimensional region of space. At each point in the farmer's field, one can assign a series of numbers (which describe, for example, how many seeds there are at that point). Faraday's field, however, occupies a three-dimensional region of space. At each point, there is a series of six numbers that describes both the magnetic and electric lines of force.

What makes Faraday's field concept so powerful is that all forces of nature can be expressed as a field. However, we need one more ingredient before we can understand the nature of any force: We must be able to write down the equations that these fields obey. The progress of the past hundred years in theoretical physics can be succinctly summarized as the search for the *field equations* of the forces of nature.

For example, in the 1860s, Scottish physicist James Clerk Maxwell wrote down the field equations for electricity and magnetism. In 1915, Einstein discovered the field equations for gravity. After innumerable false starts, the field equations for the subatomic forces were finally written down in the 1970s, utilizing the earlier work of C. N. Yang and his student R. L. Mills. These fields, which govern the interaction of all subatomic particles, are now called *Yang–Mills fields*. However, the puzzle that has stumped physicists within this century is why the subatomic field equations look so vastly different from the field equations of Einstein—that is, why the nuclear force seems so different from gravity. Some of the greatest minds in physics have tackled this problem, only to fail.

Perhaps the reason for their failure is that they were trapped by common sense. Confined to three or four dimensions, the field equations of the subatomic world and gravitation are difficult to unify. The advantage of the hyperspace theory is that the Yang–Mills field, Maxwell's field, and Einstein's field can all be placed comfortably within the hyperspace field. We see that these fields fit together precisely within the hyperspace field like pieces in a jigsaw puzzle. The other advantage of field theory is that it allows us to calculate the precise energies at which we can expect space and time to form wormholes. Unlike the ancients, therefore, we have the mathematical tools to guide us in building the machines that may one day bend space and time to our whims.

The Secret of Creation

Does this mean that big-game hunters can now start organizing safaris to the Mesozoic era to bag large dinosaurs? No. Thorne, Guth, and Freund will all tell you that the energy scale necessary to investigate these anomalies in space is far beyond anything available on earth. Freund reminds us that the energy necessary to probe the tenth dimension is a quadrillion times larger than the energy that can be produced by our largest atom smasher.

Twisting space–time into knots requires energy on a scale that will not be available within the next several centuries or even millennia—if ever. Even if all the nations of the world were to band together to build a machine that could probe hyperspace, they would ultimately fail. And, as Guth points out, the temperatures necessary to create a baby universe in the laboratory is 1,000 trillion trillion degrees, far in excess of anything available to us. In fact, that temperature is much greater than anything found in the interior of a star. So, although it is possible that

Einstein's laws and the laws of quantum theory might allow for time travel, this is not within the capabilities of earthlings like us, who can barely escape the feeble gravitational field of our own planet. While we can marvel at the implications of wormhole research, realizing its potential is strictly reserved for advanced extraterrestrial civilizations.

There was only one period of time when energy on this enormous scale was readily available, and that was at the instant of Creation. In fact, the hyperspace theory cannot be tested by our largest atom smashers because the theory is really a theory of Creation. Only at the instant of the Big Bang do we see the full power of the hyperspace theory coming into play. This raises the exciting possibility that the hyperspace theory may unlock the secret of the origin of the universe.

Introducing higher dimensions may be essential for prying loose the secrets of Creation. According to this theory, before the Big Bang, our cosmos was actually a perfect ten-dimensional universe, a world where interdimensional travel was possible. However, this ten-dimensional world was unstable, and eventually it "cracked" in two, creating two separate universes: a four- and a six-dimensional universe. The universe in which we live was born in that cosmic cataclysm. Our four-dimensional universe expanded explosively, while our twin six-dimensional universe contracted violently, until it shrank to almost infinitesimal size. This would explain the origin of the Big Bang. If correct, this theory demonstrates that the rapid expansion of the universe was just a rather minor aftershock of a much greater cataclysmic event, the cracking of space and time itself. The energy that drives the observed expansion of the universe is then found in the collapse of ten-dimensional space and time. According to the theory, the distant stars and galaxies are receding from us at astronomical speeds because of the original collapse of ten-dimensional space and time.

This theory predicts that our universe still has a dwarf twin, a companion universe that has curled up into a small six-dimensional ball that is too small to be observed. This six-dimensional universe, far from being a useless appendage to our world, may ultimately be our salvation.

Evading the Death of the Universe

It is often said that the only constants of human society are death and taxes. For the cosmologist, the only certainty is that the universe will one day die. Some believe that the ultimate death of the universe will come in the form of the Big Crunch. Gravitation will reverse the cosmic expansion generated by the Big Bang and pull the stars and galaxies back, once again, into a primordial mass. As the stars contract, temperatures will rise dramatically until all matter and energy in the universe are concentrated into a colossal fireball that will destroy the universe as we know it. All life forms will be crushed beyond recognition. There will be no escape. Scientists and philosophers, like Charles Darwin and Bertrand Russell, have written mournfully about the futility of our pitiful existence, knowing that our civilization will inexorably die when our world ends. The laws of physics, apparently, have issued the final, irrevocable death warrant for all intelligent life in the universe.

According to the late Columbia University physicist Gerald Feinberg, there is one, and perhaps only one, hope of avoiding the final calamity. He speculated that intelligent life, eventually mastering the mysteries of higher-dimensional space over billions of years, will use the other dimensions as an escape hatch from the Big Crunch. In the final moments of the collapse of our universe, our sister universe will open up once again, and interdimensional travel will become possible. As all matter is crushed in the final moments before doomsday, intelligent life forms may be able to tunnel into higher-dimensional space or an alternative universe, avoiding the seemingly inevitable death of our universe. Then, from their sanctuary in higher-dimensional space, these intelligent life forms may be able to witness the death of the collapsing universe in a fiery cataclysm. As our home universe is crushed beyond recognition, temperatures will rise violently, creating yet another Big Bang. From their vantage point in hyperspace, these intelligent life forms will have front-row seats to the rarest of all scientific phenomena, the creation of another universe and of their new home.

Masters of Hyperspace

Although field theory shows that the energy necessary to create these marvelous distortions of space and time is far beyond anything that modern civilization can muster, this raises two important questions: How long will it take for our civilization, which is growing exponentially in knowledge and power, to reach the point of harnessing the hyperspace theory? And what about other intelligent life forms in the universe, who may already have reached that point?

What makes this discussion interesting is that serious scientists have tried to quantify the progress of civilizations far into the future, when space travel will have become commonplace and neighboring star systems or even galaxies will have been colonized. Although the energy scale necessary to manipulate hyperspace is astronomically large, these scientists point out that scientific growth will probably continue to rise exponentially over the next centuries, exceeding the capabilities of human minds to grasp it. Since World War II, the sum total of scientific knowledge has doubled every 10 to 20 or so years, so the progress of science and technology into the twenty-first century may surpass our wildest expectations. Technologies that can only be dreamed of today may become commonplace in the next century. Perhaps then one can discuss the question of when we might become masters of hyperspace.

Time travel. Parallel universes. Dimensional windows.

By themselves, these concepts stand at the edge of our understanding of the physical universe. However, because the hyperspace theory is a genuine field theory, we eventually expect it to produce numerical answers determining whether these intriguing concepts are possible. If the theory produces nonsensical answers that disagree with physical data, then it must be discarded, no matter how elegant its mathematics. In the final analysis, we are physicists, not philosophers. But if it proves to be correct and explains the symmetries of modern physics, then it will usher in a revolution perhaps equal to the Copernican or Newtonian revolutions.

To have an intuitive understanding of these concepts, however, it is important to start at the beginning. Before we can feel comfortable with ten dimensions, we must learn how to manipulate four spatial dimensions. Using historical examples, we will explore the ingenious attempts made by scientists over the decades to give a tangible, visual representation of higher-dimensional space. The first part of the book, therefore, will stress the history behind the discovery of higher-dimensional space, beginning with the mathematician who started it all, Georg Bernhard Riemann. Anticipating the next century of scientific progress, Riemann was the first to state that nature finds its natural home in the geometry of higher-dimensional space.

2
Mathematicians
and Mystics

Magic is any sufficiently advanced technology.

Arthur C. Clarke

O N June 10, 1854, a new geometry was born.

The theory of higher dimensions was introduced when Georg Bernhard Riemann gave his celebrated lecture before the faculty of the University of Göttingen in Germany. In one masterful stroke, like opening up a musty, darkened room to the brilliance of a warm summer's sun, Riemann's lecture exposed the world to the dazzling properties of higher-dimensional space.

His profoundly important and exceptionally elegant essay, "On the Hypotheses Which Lie at the Foundation of Geometry," toppled the pillars of classical Greek geometry, which had successfully weathered all assaults by skeptics for 2 millennia. The old geometry of Euclid, in which all geometric figures are two or three dimensional, came tumbling down as a new Riemannian geometry emerged from its ruins. The Riemannian revolution would have vast implications for the future of the arts and sciences. Within 3 decades of his talk, the "mysterious fourth dimension" would influence the evolution of art, philosophy, and literature in Europe. Within 6 decades of Riemann's lecture, Einstein would use four-dimensional Riemannian geometry to explain the creation of the universe and its evolution. And 130 years after his lecture, physicists would use ten-dimensional geometry to attempt to unite all the laws of the physical universe. The core of Riemann's work was the realization that physical laws simplify in higher-dimensional space, the very theme of this book.

Brilliance Amid Poverty

Ironically, Riemann was the least likely person to usher in such a deep and thorough-going revolution in mathematical and physical thought. He was excruciatingly, almost pathologically, shy and suffered repeated nervous breakdowns. He also suffered from the twin ailments that have ruined the lives of so many of the world's great scientists throughout history: abject poverty and consumption (tuberculosis). His personality and temperament showed nothing of the breath-taking boldness, sweep, and supreme confidence typical of his work.

Riemann was born in 1826 in Hanover, Germany, the son of a poor Lutheran pastor, the second of six children. His father, who fought in the Napoleonic Wars, struggled as a country pastor to feed and clothe his large family. As biographer E. T. Bell notes, "the frail health and early deaths of most of the Riemann children were the result of under-nourishment in their youth and were not due to poor stamina. The mother also died before her children were grown."[1]

At a very early age, Riemann exhibited his famous traits: fantastic calculational ability, coupled with timidity, and a life-long horror of any public speaking. Painfully shy, he was the butt of cruel jokes by other boys, causing him to retreat further into the intensely private world of mathematics.

He also was fiercely loyal to his family, straining his poor health and constitution to buy presents for his parents and especially for his beloved sisters. To please his father, Riemann set out to become a student of theology. His goal was to get a paying position as a pastor as quickly as possible to help with his family's abysmal finances. (It is difficult to imagine a more improbable scenario than that of a tongue-tied, timid young boy imagining that he could deliver fiery, passionate sermons railing against sin and driving out the devil.)

In high school, he studied the Bible intensely, but his thoughts always drifted back to mathematics; he even tried to provide a mathematical proof of the correctness of Genesis. He also learned so quickly that he kept outstripping the knowledge of his instructors, who found it impossible to keep up with the boy. Finally, the principal of his school gave

Riemann a ponderous book to keep him occupied. The book was Adrien-Marie Legendre's *Theory of Numbers,* a huge 859-page master-piece, the world's most advanced treatise on the difficult subject of number theory. Riemann devoured the book in 6 days.

When his principal asked, "How far did you read?" the young Riemann replied, "That is certainly a wonderful book. I have mastered it." Not really believing the bravado of this youngster, the principal several months later asked obscure questions from the book, which Riemann answered perfectly.[2]

Beset by the daily struggle to put food on the table, Riemann's father might have sent the boy to do menial labor. Instead, he scraped together enough funds to send his 19-year-old son to the renowned University of Göttingen, where he first met Carl Friedrich Gauss, the acclaimed "Prince of Mathematicians," one of the greatest mathematicians of all time. Even today, if you ask any mathematician to rank the three most famous mathematicians in history, the names of Archimedes, Isaac Newton, and Carl Gauss will invariably appear.

Life for Riemann, however, was an endless series of setbacks and hardships, overcome only with the greatest difficulty and by straining his frail health. Each triumph was followed by tragedy and defeat. For example, just as his fortunes began to improve and he undertook his formal studies under Gauss, a full-scale revolution swept Germany. The working class, long suffering under inhuman living conditions, rose up against the government, with workers in scores of cities throughout Germany taking up arms. The demonstrations and uprisings in early 1848 inspired the writings of another German, Karl Marx, and deeply affected the course of revolutionary movements throughout Europe for the next 50 years.

With all of Germany swept up in turmoil, Riemann's studies were interrupted. He was inducted into the student corps, where he had the dubious honor of spending 16 weary hours protecting someone even more terrified than he: the king, who was quivering with fear in his royal palace in Berlin, trying to hide from the wrath of the working class.

Beyond Euclidean Geometry

Not only in Germany, but in mathematics, too, fierce revolutionary winds were blowing. The problem that riveted Riemann's interest was the impending collapse of yet another bastion of authority, Euclidean geometry, which holds that space is three dimensional. Furthermore, this three-dimensional space is "flat" (in flat space, the shortest distance between two points is a straight line; this omits the possibility that space can be curved, as on a sphere).

In fact, after the Bible, Euclid's *Elements* was probably the most influential book of all time. For 2 millennia, the keenest minds of Western civilization have marveled at its elegance and the beauty of its geometry. Thousands of the finest cathedrals in Europe were erected according to its principles. In retrospect, perhaps it was too successful. Over the centuries, it became something of a religion; anyone who dared to propose curved space or higher dimensions was relegated to the ranks of crackpots or heretics. For untold generations, schoolchildren have wrestled with the theorems of Euclid's geometry: that the circumference of a circle is π times the diameter, and that the angles within a triangle add up to 180 degrees. However, try as they might, the finest mathematical minds for several centuries could not prove these deceptively simple propositions. In fact, the mathematicians of Europe began to realize that even Euclid's *Elements,* which had been revered for 2,300 years, was incomplete. Euclid's geometry was still viable if one stayed within the confines of flat surfaces, but if one strayed into the world of curved surfaces, it was actually incorrect.

To Riemann, Euclid's geometry was particularly sterile when compared with the rich diversity of the world. Nowhere in the natural world do we see the flat, idealized geometric figures of Euclid. Mountain ranges, ocean waves, clouds, and whirlpools are not perfect circles, triangles, and squares, but are curved objects that bend and twist in infinite diversity.

The time was ripe for a revolution, but who would lead it and what would replace the old geometry?

The Rise of Riemannian Geometry

Riemann rebelled against the apparent mathematical precision of Greek geometry, whose foundation, he discovered, ultimately was based on the shifting sand of common sense and intuition, not the firm ground of logic.

It is obvious, said Euclid, that a point has no dimension at all. A line has one dimension: length. A plane has two dimensions: length and breadth. A solid has three dimensions: length, breadth, and height. And there it stops. Nothing has four dimensions. These sentiments were echoed by the philosopher Aristotle, who apparently was the first person to state categorically that the fourth spatial dimension is impossible. In *On Heaven,* he wrote, "The line has magnitude in one way, the plane in two ways, and the solid in three ways, and beyond these there is no other magnitude because the three are all." Furthermore, in A.D. 150, the astronomer Ptolemy from Alexandria went beyond Aristotle and offered, in his book *On Distance,* the first ingenious "proof" that the fourth dimension is impossible.

First, he said, draw three mutually perpendicular lines. For example, the corner of a cube consists of three mutually perpendicular lines. Then, he argued, try to draw a fourth line that is perpendicular to the other three lines. No matter how one tries, he reasoned, four mutually perpendicular lines are impossible to draw. Ptolemy claimed that a fourth perpendicular line is "entirely without measure and without definition." Thus the fourth dimension is impossible.

What Ptolemy actually proved was that it is impossible to visualize the fourth dimension with our three-dimensional brains. (In fact, today we know that many objects in mathematics cannot be visualized but can be shown to exist.) Ptolemy may go down in history as the man who opposed two great ideas in science: the sun-centered solar system and the fourth dimension.

Over the centuries, in fact, some mathematicians went out of their way to denounce the fourth dimension. In 1685, the mathematician John Wallis polemicized against the concept, calling it a "Monster in Nature, less possible than a Chimera or Centaure. . . . Length, Breadth, and Thickness, take up the whole of Space. Nor can Fansie imagine how there should be a Fourth Local Dimension beyond these Three."[3] For several thousand years, mathematicians would repeat this simple but fatal mistake, that the fourth dimension cannot exist because we cannot picture it in our minds.

The Unity of All Physical Law

The decisive break with Euclidean geometry came when Gauss asked his student Riemann to prepare an oral presentation on the "foundation of geometry." Gauss was keenly interested in seeing if his student could develop an alternative to Euclidean geometry. (Decades before, Gauss had privately expressed deep and extensive reservations about Euclidean geometry. He even spoke to his colleagues of hypothetical "bookworms" that might live entirely on a two-dimensional surface. He spoke of generalizing this to the geometry of higher-dimensional space. However, being a deeply conservative man, he never published any of his work on higher dimensions because of the outrage it would create among the narrow-minded, conservative old guard. He derisively called them "Boeotians" after a mentally retarded Greek tribe.[4])

Riemann, however, was terrified. This timid man, terrified of public speaking, was being asked by his mentor to prepare a lecture before the entire faculty on the most difficult mathematical problem of the century.

Over the next several months, Riemann began painfully developing the theory of higher dimensions, straining his health to the point of a nervous breakdown. His stamina further deteriorated because of his dismal financial situation. He was forced to take low-paying tutoring jobs to provide for his family. Furthermore, he was becoming sidetracked trying to explain problems of physics. In particular, he was helping another professor, Wilhelm Weber, conduct experiments in a fascinating new field of research, electricity.

Electricity, of course, had been known to the ancients in the form of lightning and sparks. But in the early nineteenth century, this phenomenon became the central focus of physics research. In particular, the discovery that passing a current of wire across a compass needle can make the needle spin riveted the attention of the physics community. Conversely, moving a bar magnet across a wire can induce an electric current in the wire. (This is called Faraday's Law, and today all electric generators and transformers—and hence much of the foundation of modern technology—are based on this principle.)

To Riemann, this phenomenon indicated that electricity and magnetism are somehow manifestations of the same force. Riemann was excited by the new discoveries and was convinced that he could give a mathematical explanation that would unify electricity and magnetism. He immersed himself in Weber's laboratory, convinced that the new mathematics would yield a comprehensive understanding of these forces.

Now, burdened with having to prepare a major public lecture on the "foundation of geometry," to support his family, and to conduct scientific experiments, his health finally collapsed and he suffered a nervous breakdown in 1854. Later, he wrote to his father, "I became so absorbed in my investigation of the unity of all physical laws that when the subject of the trial lecture was given me, I could not tear myself away from my research. Then, partly as a result of brooding on it, partly from staying indoors too much in this vile weather, I fell ill."[5] This letter is significant, for it clearly shows that, even during months of illness,

Riemann firmly believed that he would discover the "unity of all physical laws" and that mathematics would eventually pave the way for this unification.

Force = Geometry

Eventually, despite his frequent illnesses, Riemann developed a startling new picture of the meaning of a "force." Ever since Newton, scientists had considered a force to be an instantaneous interaction between two distant bodies. Physicists called it action-at-a-distance, which meant that a body could influence the motions of distant bodies instantaneously. Newtonian mechanics undoubtedly could describe the motions of the planets. However, over the centuries, critics argued that action-at-a-distance was unnatural, because it meant that one body could change the direction of another without even touching it.

Riemann developed a radically new physical picture. Like Gauss's "bookworms," Riemann imagined a race of two-dimensional creatures living on a sheet of paper. But the decisive break he made was to put these bookworms on a *crumpled* sheet of paper.[6] What would these bookworms think about their world? Riemann realized that they would conclude that their world was still perfectly flat. Because their bodies would also be crumpled, these bookworms would never notice that their world was distorted. However, Riemann argued that if these bookworms tried to move across the crumpled sheet of paper, they would feel a mysterious, unseen "force" that prevented them from moving in a straight line. They would be pushed left and right every time their bodies moved over a wrinkle on the sheet.

Thus Riemann made the first momentous break with Newton in 200 years, banishing the action-at-a-distance principle. To Riemann, *"force" was a consequence of geometry.*

Riemann then replaced the two-dimensional sheet with our three-dimensional world crumpled in the fourth dimension. It would not be obvious to us that our universe was warped. However, we would immediately realize that something was amiss when we tried to walk in a straight line. We would walk like a drunkard, as though an unseen force were tugging at us, pushing us left and right.

Riemann concluded that electricity, magnetism, and gravity are caused by the crumpling of our three-dimensional universe in the unseen fourth dimension. Thus a "force" has no independent life of its own; it is only the apparent effect caused by the distortion of geometry.

By introducing the fourth spatial dimension, Riemann accidentally stumbled on what would become one of the dominant themes in modern theoretical physics, that the laws of nature appear simple when expressed in higher-dimensional space. He then set about developing a mathematical language in which this idea could be expressed.

Riemann's Metric Tensor: A New Pythagorean Theorem

Riemann spent several months recovering from his nervous breakdown. Finally, when he delivered his oral presentation in 1854, the reception was enthusiastic. In retrospect, this was, without question, one of the most important public lectures in the history of mathematics. Word spread quickly throughout Europe that Riemann had decisively broken out of the confines of Euclidean geometry that had ruled mathematics for 2 millennia. News of the lecture soon spread throughout all the centers of learning in Europe, and his contributions to mathematics were being hailed throughout the academic world. His talk was translated into several languages and created quite a sensation in mathematics. There was no turning back to the work of Euclid.

Like many of the greatest works in physics and mathematics, the essential kernel underlying Riemann's great paper is simple to understand. Riemann began with the famous Pythagorean Theorem, one of the Greeks' greatest discoveries in mathematics. The theorem establishes the relationship between the lengths of the three sides of a right triangle: It states that the sum of the squares of the smaller sides equals the square of the longest side, the hypotenuse; that is, if a and b are the lengths of the two short sides, and c is the length of the hypotenuse, then $a^2 + b^2 = c^2$. (The Pythagorean Theorem, of course, is the foundation of all architecture; every structure built on this planet is based on it.)

For three-dimensional space, the theorum can easily be generalized. It states that the sum of the squares of three adjacent sides of a cube is equal to the square of the diagonal; so if a, b, and c represent the sides of a cube, and d is its diagonal length, then $a^2 + b^2 + c^2 = d^2$ (Figure 2.1).

It is now simple to generalize this to the case of N-dimensions. Imagine an N-dimensional cube. If $a,b,c, \ldots$ are the lengths of the sides of a "hypercube," and z is the length of the diagonal, then $a^2 + b^2 + c^2 + d^2 + \ldots = z^2$. Remarkably, even though our brains cannot visualize an N-dimensional cube, it is easy to write down the formula for its sides. (This is a common feature of working in hyperspace. Mathematically

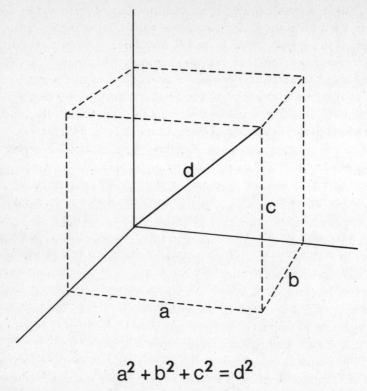

$$a^2 + b^2 + c^2 = d^2$$

Figure 2.1. The length of a diagonal of a cube is given by a three-dimensional version of the Pythagorean Theorem: $a^2 + b^2 + c^2 = d^2$. *By simply adding more terms to the Pythagorean Theorem, this equation easily generalizes to the diagonal of a hypercube in* N *dimensions. Thus although higher dimensions cannot be visualized, it is easy to represent* N *dimensions mathematically.*

manipulating *N*-dimensional space is no more difficult than manipulating three-dimensional space. It is nothing short of amazing that on a plain sheet of paper, you can mathematically describe the properties of higher-dimensional objects that cannot be visualized by our brains.)

Riemann then generalized these equations for spaces of arbitrary dimension. These spaces can be either flat or curved. If flat, then the usual axioms of Euclid apply: The shortest distance between two points is a straight line, parallel lines never meet, and the sum of the interior angles of a triangle add to 180 degrees. But Riemann also found that surfaces can have "positive curvature," as in the surface of a sphere, where parallel lines always meet and where the sum of the angles of a triangle can exceed 180 degrees. Surfaces can also have "negative curvature," as in a saddle-shaped or a trumpet-shaped surface. On these surfaces, the sum of the interior angles of a triangle add to less than 180 degrees. Given a line and a point off that line, there are an infinite number of parallel lines one can draw through that point (Figure 2.2).

Riemann's aim was to introduce a new object in mathematics that would enable him to describe all surfaces, no matter how complicated. This inevitably led him to reintroduce Faraday's concept of the field.

Faraday's field, we recall, was like a farmer's field, which occupies a region of two-dimensional space. Faraday's field occupies a region of three-dimensional space; at any point in space, we assign a collection of numbers that describes the magnetic or electric force at that point. Riemann's idea was to introduce a collection of numbers at every point in space that would describe how much it was bent or curved.

For example, for an ordinary two-dimensional surface, Riemann introduced a collection of three numbers at every point that completely describe the bending of that surface. Riemann found that in four spatial dimensions, one needs a collection of ten numbers at each point to describe its properties. No matter how crumpled or distorted the space, this collection of ten numbers at each point is sufficient to encode all the information about that space. Let us label these ten numbers by the symbols g_{11}, g_{12}, g_{13}, ... (When analyzing a four-dimensional space, the lower index can range from one to four.) Then Riemann's collection of ten numbers can be symmetrically arranged as in Figure 2.3.[7] (It appears as though there are 16 components. However, $g_{12} = g_{21}$, $g_{13} = g_{31}$, and so on, so there are actually only ten independent components.) Today, this collection of numbers is called the Riemann *metric tensor*. Roughly speaking, the greater the value of the metric tensor, the greater the crumpling of the sheet. No matter how crumpled the sheet of paper, the metric tensor gives us a simple means of measuring its curvature at any point. If we flattened the crumpled sheet completely, then we would retrieve the formula of Pythagoras.

Riemann's metric tensor allowed him to erect a powerful apparatus for describing spaces of any dimension with arbitrary curvature. To his surprise, he found that all these spaces are well defined and self-consistent. Previously, it was thought that terrible contradictions would arise when investigating the forbidden world of higher dimensions. To his surprise, Riemann found none. In fact, it was almost trivial to extend his work to N-dimensional space. The metric tensor would now resemble the squares of a checker board that was $N \times N$ in size. This will have profound physical implications when we discuss the unification of all forces in the next several chapters.

Zero
curvature

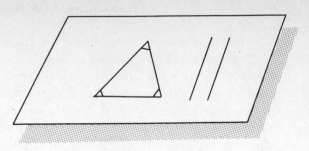

Positive
curvature

Negative
curvature

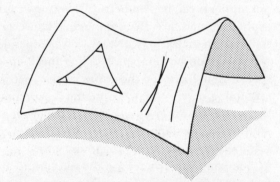

Figure 2.2. A plane has zero curvature. In Euclidean geometry, the interior angles of a triangle sum to 180 degrees, and parallel lines never meet. In non-Euclidean geometry, a sphere has positive curvature. A triangle's interior angles sum to greater than 180 degrees and parallel lines always meet. (Parallel lines include arcs whose centers coincide with the center of the sphere. This rules out latitudinal lines.) A saddle has negative curvature. The interior angles sum to less than 180 degrees. There are an infinite number of lines parallel to a given line that go through a fixed point.

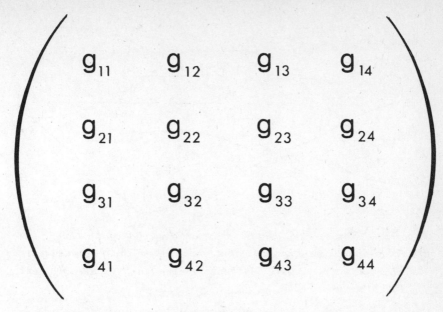

$$\begin{pmatrix} g_{11} & g_{12} & g_{13} & g_{14} \\ g_{21} & g_{22} & g_{23} & g_{24} \\ g_{31} & g_{32} & g_{33} & g_{34} \\ g_{41} & g_{42} & g_{43} & g_{44} \end{pmatrix}$$

Figure 2.3. Riemann's metric tensor contains all the information necessary to describe mathematically a curved space in N dimensions. It takes 16 numbers to describe the metric tensor for each point in four-dimensional space. These numbers can be arranged in a square array (six of these numbers are actually redundant; so the metric tensor has ten independent numbers).

(The secret of unification, we will see, lies in expanding Riemann's metric to N-dimensional space and then chopping it up into rectangular pieces. Each rectangular piece corresponds to a different force. In this way, we can describe the various forces of nature by slotting them into the metric tensor like pieces of a puzzle. This is the mathematical expression of the principle that higher-dimensional space unifies the laws of nature, that there is "enough room" to unite them in N-dimensional space. More precisely, there is "enough room" in Riemann's metric to unite the forces of nature.)

Riemann anticipated another development in physics; he was one of the first to discuss multiply connected spaces, or wormholes. To visualize this concept, take two sheets of paper and place one on top of the other. Make a short cut on each sheet with scissors. Then glue the two sheets together along the two cuts (Figure 2.4). (This is topologically the same as Figure 1.1, except that the neck of the wormhole has length zero.)

If a bug lives on the top sheet, he may one day accidentally walk into the cut and find himself on the bottom sheet. He will be puzzled because everything is in the wrong place. After much experimentation, the bug

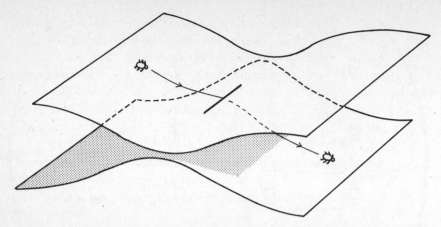

Figure 2.4. Riemann's cut, with two sheets are connected together along a line. If we walk around the cut, we stay within the same space. But if we walk through the cut, we pass from one sheet to the next. This is a multiply connected surface.

will find that he can re-emerge in his usual world by re-entering the cut. If he walks around the cut, then his world looks normal; but when he tries to take a short-cut through the cut, he has a problem.

Riemann's cuts are an example of a wormhole (except that it has zero length) connecting two spaces. Riemann's cuts were used with great effect by the mathematician Lewis Carroll in his book *Through the Looking-Glass*. Riemann's cut, connecting England with Wonderland, is the looking glass. Today, Riemann's cuts survive in two forms. First, they are cited in every graduate mathematics course in the world when applied to the theory of electrostatics or conformal mapping. Second, Riemann's cuts can be found in episodes of ''The Twilight Zone.'' (It should be stressed that Riemann himself did not view his cuts as a mode of travel between universes.)

Riemann's Legacy

Riemann persisted with his work in physics. In 1858, he even announced that he had finally succeeded in a unified description of light and electricity. He wrote, ''I am fully convinced that my theory is the correct one, and that in a few years it will be recognized as such.''[8] Although his metric tensor gave him a powerful way to describe any curved space in any dimension, he did not know the precise equations that the metric tensor obeyed; that is, he did not know what made the sheet crumple.

Unfortunately, Riemann's efforts to solve this problem were continually thwarted by grinding poverty. His successes did not translate into money. He suffered another nervous breakdown in 1857. After many years, he was finally appointed to Gauss's coveted position at Göttingen, but it was too late. A life of poverty had broken his health, and like many of the greatest mathematicians throughout history, he died prematurely of consumption at the age of 39, before he could complete his geometric theory of gravity and electricity and magnetism.

In summary, Riemann did much more than lay the foundation of the mathematics of hyperspace. In retrospect, we see that Riemann anticipated some of the major themes in modern physics. Specifically,

1. He used higher-dimensional space to simplify the laws of nature; that is, to him, electricity and magnetism as well as gravity were just effects caused by the crumpling or warping of hyperspace.
2. He anticipated the concept of wormholes. Riemann's cuts are the simplest examples of multiply connected spaces.
3. He expressed gravity as a field. The metric tensor, because it describes the force of gravity (via curvature) at every point in space, is precisely Faraday's field concept when applied to gravity.

Riemann was unable to complete his work on force fields because he lacked the field equations that electricity and magnetism and gravity obey. In other words, he did not know precisely how the universe would be crumpled in order to yield the force of gravity. He tried to discover the field equations for electricity and magnetism, but he died before he could finish that project. At his death, he still had no way of calculating how much crumpling would be necessary to describe the forces. These crucial developments would be left to Maxwell and Einstein.

Living in a Space Warp

The spell was finally broken.

Riemann, in his short life, lifted the spell cast by Euclid more than 2,000 years before. Riemann's metric tensor was the weapon with which young mathematicians could defy the Boeotians, who howled at any mention of higher dimensions. Those who followed in Riemann's footsteps found it easier to speak of unseen worlds.

Soon, research bloomed all over Europe. Prominent scientists began

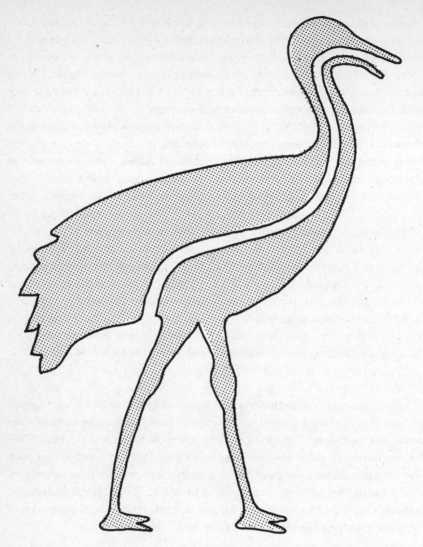

Figure 2.5. A two-dimensional being cannot eat. Its digestive tract necessarily divides it into two distinct pieces, and the being falls apart.

popularizing the idea for the general public. Hermann von Helmholtz, perhaps the most famous German physicist of his generation, was deeply affected by Riemann's work and wrote and spoke extensively to the general public about the mathematics of intelligent beings living on a ball or sphere.

According to Helmholtz, these creatures, with reasoning powers similar to our own, would independently discover that all of Euclid's postulates and theorems were useless. On a sphere, for example, the sums of the interior angles of a triangle do not add up to 180 degrees. The "bookworms" first talked about by Gauss now found themselves inhabiting Helmholtz's two-dimensional spheres. Helmholtz wrote that "geometrical axioms must vary according to the kind of space inhabited by beings whose powers of reasoning are quite in conformity with ours."[9] However, in his *Popular Lectures of Scientific Subjects* (1881), Helmholtz warned his readers that it is impossible for us to visualize the fourth dimension. In fact, he said "such a 'representation' is as impossible as the 'representation' of colours would be to one born blind."[10]

Some scientists, marveling at the elegance of Riemann's work, tried to find physical applications for such a powerful apparatus.[11] While some scientists were exploring the applications of higher dimension, other scientists asked more practical, mundane questions, such as: How does a two-dimensional being eat? In order for Gauss's two-dimensional people to eat, their mouths would have to face to the side. But if we now draw their digestive tract, we notice that this passageway completely bisects their bodies (Figure 2.5). Thus if they eat, their bodies will split into two pieces. In fact, any tube that connects two openings in their bodies will separate them into two unattached pieces. This presents us with a difficult choice. Either these people eat like we do and their bodies break apart, or they obey different laws of biology.

Unfortunately, the advanced mathematics of Riemann outstripped the relatively backward understanding of physics in the nineteenth century. There was no physical principle to guide further research. We would have to wait another century for the physicists to catch up with the mathematicians. But this did not stop nineteenth-century scientists from speculating endlessly about what beings from the fourth dimension would look like. Soon, they realized that such a fourth-dimensional being would have almost God-like powers.

To Be a God

Imagine being able to walk through walls.

You wouldn't have to bother with opening doors; you could pass right through them. You wouldn't have to go around buildings; you could enter them through their walls and pillars and out through the back wall. You wouldn't have to detour around mountains; you could step right into them. When hungry, you could simply reach through the refrigerator door without opening it. You could never be accidentally locked outside your car; you could simply step through the car door.

Imagine being able to disappear or reappear at will. Instead of driving to school or work, you would just vanish and rematerialize in your classroom or office. You wouldn't need an airplane to visit far-away places, you could just vanish and rematerialize where you wanted. You would never be stuck in city traffic during rush hours; you and your car would simply disappear and rematerialize at your destination.

Imagine having x-ray eyes. You would be able to see accidents happening from a distance. After vanishing and rematerializing at the site of any accident, you could see exactly where the victims were, even if they were buried under debris.

Imagine being able to reach into an object without opening it. You could extract the sections from an orange without peeling or cutting it. You would be hailed as a master surgeon, with the ability to repair the internal organs of patients without ever cutting the skin, thereby greatly reducing pain and the risk of infection. You would simply reach into the person's body, passing directly through the skin, and perform the delicate operation.

Imagine what a criminal could do with these powers. He could enter the most heavily guarded bank. He could see through the massive doors of the vault for the valuables and cash and reach inside and pull them out. He could then stroll outside as the bullets from the guards passed right through him. With these powers, no prison could hold a criminal.

No secrets could be kept from us. No treasures could be hidden from us. No obstructions could stop us. We would truly be miracle workers, performing feats beyond the comprehension of mortals. We would also be omnipotent.

What being could possess such God-like power? The answer: a being from a higher-dimensional world. Of course, these feats are beyond the capability of any three-dimensional person. For us, walls are solid and prison bars are unbreakable. Attempting to walk through walls will only give us a painful, bloody nose. But for a four-dimensional being, these feats would be child's play.

To understand how these miraculous feats can be performed, consider again Gauss's mythical two-dimensional beings, living on a two-dimensional table top. To jail a criminal, the Flatlanders simply draw a circle around him. No matter which way the criminal moves, he hits the impenetrable circle. However, it is a trivial task for us to spring the prisoner from jail. We just reach down, grab the Flatlander, peel him off the two-dimensional world, and redeposit him elsewhere on his world (Fig-

Figure 2.6. In Flatland, a "jail" is a circle drawn around a person. Escape from this circle is impossible in two dimensions. However, a three-dimensional person can yank a Flatlander out of jail into the third dimension. To a jailer, it appears as though the prisoner has mysteriously vanished into thin air.

ure 2.6). This feat, which is quite ordinary in three dimensions, appears fantastic in two dimensions.

To his jailer, the prisoner has suddenly disappeared from an escape-proof prison, vanishing into thin air. Then just as suddenly, he reappears somewhere else. If you explain to the jailer that the prisoner was moved "up" and off Flatland, he would not understand what you were saying. The word *up* does not exist in the Flatlander's vocabulary, nor can he visualize the concept.

The other feats can be similarly explained. For example, notice that the internal organs (like the stomach or heart) of a Flatlander are completely visible to us, in the same way that we can see the internal structure of cells on a microscope slide. It's now trivial to reach inside a Flatlander and perform surgery without cutting the skin. We can also peel the Flat-

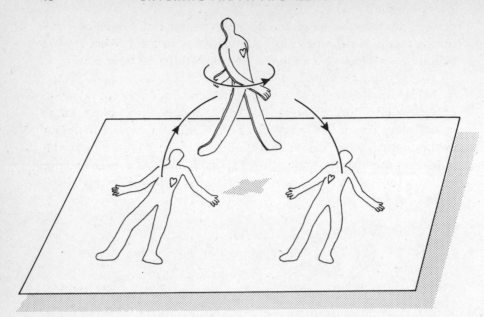

Figure 2.7. If we peel a Flatlander from his world and flip him over in three dimensions, his heart now appears on the right-hand side. All his internal organs have been reversed. This transformation is a medical impossibility to someone who lives strictly in Flatland.

lander off his world, flip him around, and put him back down. Notice that his left and right organs are now reversed, so that his heart is on the right side (Figure 2.7).

Viewing Flatland, notice also that we are omnipotent. Even if the Flatlander hides inside a house or under the ground, we can see him perfectly. He would regard our powers as magical; we, however, would know that not magic, but simply a more advantageous perspective, is at work. (Although such feats of "magic" are, in principle, possible within the realm of hyperspace physics, we should caution, once again, that the technology necessary to manipulate space–time far exceeds anything possible on the earth, at least for hundreds of years. The ability to manipulate space–time may be within the domain of only some extraterrestrial life in the universe far in advance of anything found on the earth, with the technology to master energy on a scale a quadrillion times larger than our most powerful machines.)

Although Riemann's famous lecture was popularized by the work of Helmholtz and many others, the lay public could make little sense of this or the eating habits of two-dimensional creatures. For the average person, the question was more direct: What kind of beings can walk through walls, see through steel, and perform miracles? What kind of beings are omnipotent and obey a set of laws different from ours?

Why ghosts, of course!

In the absence of any physical principle motivating the introduction of higher dimensions, the theory of the fourth dimension suddenly took an unexpected turn. We will now begin a strange but important detour in the history of hyperspace, examining its unexpected but profound impact on the arts and philosophy. This tour through popular culture will show how the mystics gave us clever ways in which to "visualize" higher-dimensional space.

Ghosts from the Fourth Dimension

The fourth dimension penetrated the public's consciousness in 1877, when a scandalous trial in London gave it an international notoriety.

The London newspapers widely publicized the sensational claims and bizarre trial of psychic Henry Slade. The raucous proceedings drew in some of the most prominent physicists of the day. As a result of all the publicity, talk of the fourth dimension left the blackboards of abstract mathematicians and burst into polite society, turning up in dinner-table conversations throughout London. The "notorious fourth dimension" was now the talk of the town.

It all began, innocently enough, when Slade, a psychic from the United States, visited London and held seances with prominent townspeople. He was subsequently arrested for fraud and charged with "using subtle crafts and devices, by palmistry and otherwise," to deceive his clients.[12] Normally, this trial might have gone unnoticed. But London society was scandalized and amused when eminent physicists came to his defense, claiming that his psychic feats actually proved that he could summon spirits living in the fourth dimension. This scandal was fueled by the fact that Slade's defenders were not ordinary British scientists, but rather some of the greatest physicists in the world. Many went on to win the Nobel Prize in physics.

Playing a leading role in stirring up this scandal was Johann Zollner, a professor of physics and astronomy at the University of Leipzig. It was Zollner who marshaled a galaxy of leading physicists to come to Slade's defense.

That mystics could perform parlor tricks for the royal court and proper society, of course, was nothing new. For centuries, they had claimed that they could summon spirits to read the writing within closed envelopes, pull objects from closed bottles, reseal broken match sticks, and intertwine rings. The strange twist to this trial was that leading scientists claimed these feats were possible by manipulating objects in the fourth dimension. In the process, they gave the public its first understanding of how to perform these miraculous feats via the fourth dimension.

Zollner enlisted the help of internationally prominent physicists who participated in the Society for Psychical Research and who even rose to lead the organization, including some of the most distinguished names of nineteenth-century physics: William Crookes, inventor of the cathode ray tube, which today is used in every television set and computer monitor in the world;[13] Wilhelm Weber, Gauss's collaborator and the mentor of Riemann (today, the international unit of magnetism is officially named the "weber" after him); J. J. Thompson, who won the Nobel Prize in 1906 for the discovery of the electron; and Lord Rayleigh, recognized by historians as one of the greatest classical physicists of the late nineteenth century and winner of the Nobel Prize in physics in 1904.

Crookes, Weber, and Zollner, in particular, took a special interest in the work of Slade, who was eventually convicted of fraud by the court. However, he insisted that he could prove his innocence by duplicating his feats before a scientific body. Intrigued, Zollner took up the challenge. A number of controlled experiments were conducted in 1877 to test Slade's ability to send objects through the fourth dimension. Several distinguished scientists were invited by Zollner to evaluate Slade's abilities.

First, Slade was given two separate, unbroken wooden rings. Could he push one wooden ring past the other, so that they were intertwined without breaking? If Slade succeeded, Zollner wrote, it would "represent a miracle, that is, a phenomenon which our conceptions heretofore of physical and organic processes would be absolutely incompetent to explain."[14]

Second, he was given the shell of a sea snail, which twisted either to the right or to the left. Could Slade transform a right-handed shell into a left-handed shell and vice versa?

Third, he was given a closed loop of rope made of dried animal gut. Could he make a knot in the circular rope without cutting it?

Slade was also given variations of these tests. For example, a rope was tied into a right-handed knot and its ends were sealed with wax and impressed with Zollner's personal seal. Slade was asked to untie the knot, without breaking the wax seal, and retie the rope in a left-handed knot.

Since knots can always be untied in the fourth dimension, this feat should be easy for a fourth-dimensional person. Slade was also asked to remove the contents of a sealed bottle without breaking the bottle.

Could Slade demonstrate this astounding ability?

Magic in the Fourth Dimension

Today we realize that the manipulation of higher-dimensional space, as claimed by Slade, would require a technology far in advance of anything possible on this planet for the conceivable future. However, what is interesting about this notorious case is that Zollner correctly concluded that Slade's feats of wizardry could be explained if one could somehow move objects through the fourth dimension. Thus for pedagogical reasons, the experiments of Zollner are compelling and worth discussing.

For example, in three dimensions, separate rings cannot be pushed through each other until they intertwine without breaking them. Similarly, closed, circular pieces of rope cannot be twisted into knots without cutting them. Any Boy or Girl Scout who has struggled with knots for his or her merit badges knows that knots in a circular loop of rope cannot be removed. However, in higher dimensions, knots are easily unraveled and rings can be intertwined. This is because there is "more room" in which to move ropes past each other and rings into each other. If the fourth dimension existed, ropes and rings could be lifted off our universe, intertwined, and then returned to our world. In fact, in the fourth dimension, knots can never remain tied. They can always be unraveled without cutting the rope. This feat is impossible in three dimensions, but trivial in the fourth. The third dimension, as it turns out, is the only dimension in which knots stay knotted. (The proof of this rather unexpected result is given in the notes.[15])

Similarly, in three dimensions it is impossible to convert a rigid left-handed object into a right-handed one. Humans are born with hearts on their left side, and no surgeon, no matter now skilled, can reverse human internal organs. This is possible (as first pointed out by mathematician August Möbius in 1827) only if we lift the body out of our universe, rotate it in the fourth dimension, and then reinsert it back into our universe. Two of these tricks are depicted in Figure 2.8; they can be performed only if objects can be moved in the fourth dimension.

Polarizing the Scientific Community

Zollner sparked a storm of controversy when, publishing in both the *Quarterly Journal of Science* and *Transcendental Physics*, he claimed that

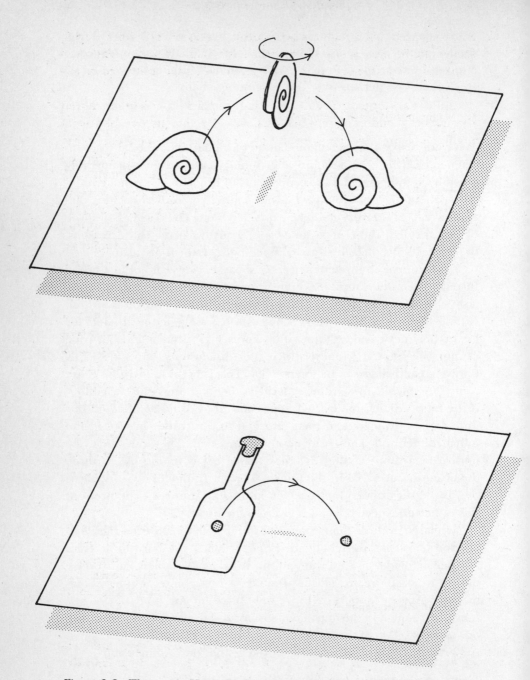

Figure 2.8. The mystic Henry Slade claimed to be able to change right-handed snail shells into left-handed ones, and to remove objects from sealed bottles. These feats are impossible in three dimensions, but are trivial if one can move objects through the fourth dimension.

Slade amazed his audiences with these "miraculous" feats during seances in the presence of distinguished scientists. (However, Slade also flunked some of the tests that were conducted under controlled conditions.)

Zollner's spirited defense of Slade's feats was sensationalized throughout London society. (In fact, this was actually one of several highly publicized incidents involving spiritualists and mediums in the late nineteenth century. Victorian England was apparently fascinated with the occult.) Scientists, as well as the general public, quickly took sides in the matter. Supporting Zollner's claims was his circle of reputable scientists, including Weber and Crookes. These were not average scientists, but masters of the art of science and seasoned observers of experiment. They had spent a lifetime working with natural phenomena, and now before their eyes, Slade was performing feats that were possible only if spirits lived in the fourth dimension.

But detractors of Zollner pointed out that scientists, because they are trained to trust their senses, are the worst possible people to evaluate a magician. A magician is trained specifically to distract, deceive, and confuse those very senses. A scientist may carefully observe the magician's right hand, but it is the left hand that secretly performs the trick. Critics also pointed out that only another magician is clever enough to detect the sleight-of-hand tricks of a fellow magician. Only a thief can catch a thief.

One particularly savage piece of criticism, published in the science quarterly magazine *Bedrock*, was made against two other prominent physicists, Sir W. F. Barrett and Sir Oliver Lodge, and their work on telepathy. The article was merciless:

> It is not necessary either to regard the phenomena of so-called telepathy as inexplicable or to regard the mental condition of Sir W. F. Barrett and Sir Oliver Lodge as indistinguishable from idiocy. There is a third possibility. *The will to believe* has made them ready to accept evidence obtained under conditions which they would recognize to be unsound if they had been trained in experimental psychology.

Over a century later, precisely the same arguments, pro and con, would be used in the debate over the feats of the Israeli psychic Uri Geller, who convinced two reputable scientists at the Stanford Research Institute in California that he could bend keys by mental power alone and perform other miracles. (Commenting on this, some scientists have repeated a saying that dates back to the Romans: "Populus vult decipi, ergo decipiatur" [People want to be deceived, therefore let them be deceived].)

The passions raging within the British scientific community touched off a lively debate that quickly spread across the English Channel. Unfortunately, in the decades following Riemann's death, scientists lost sight of his original goal, to simplify the laws of nature through higher dimensions. As a consequence, the theory of higher dimensions wandered into many interesting but questionable directions. This is an important lesson. Without a clear physical motivation or a guiding physical picture, pure mathematical concepts sometimes drift into speculation.

These decades were not a complete loss, however, because mathematicians and mystics like Charles Hinton would invent ingenious ways in which to "see" the fourth dimension. Eventually, the pervasive influence of the fourth dimension would come full circle and cross-pollinate the world of physics once again.

3

The Man Who "Saw" the Fourth Dimension

[T]he fourth dimension had become almost a household word by 1910. . . . Ranging from an ideal Platonic or Kantian reality—or even Heaven—the answer to all of the problems puzzling contemporary science, the fourth dimension could be all things to all people.

Linda Dalrymple Henderson

WITH the passions aroused by the trial of the "notorious Mr. Slade," it was perhaps inevitable that the controversy would eventually spawn a best-selling novel.

In 1884, after a decade of acrimonious debate, clergyman Edwin Abbot, headmaster of the City of London School, wrote the surprisingly successful and enduring novel *Flatland: A Romance of Many Dimensions by a Square.** Because of the intense public fascination with higher dimen-

---

*It wasn't surprising that a clergyman wrote the novel, since theologians of the Church of England were among the first to jump into the fray created by the sensationalized trial. For uncounted centuries, clergymen had skillfully dodged such perennial questions as Where are heaven and hell? and Where do angels live? Now, they found a convenient resting place for these heavenly bodies: the fourth dimension. The Christian spiritualist A. T. Schofield, in his 1888 book *Another World*, argued at length that God and the spirits resided in the fourth dimension.[1] Not to be outdone, in 1893 the theologian Arthur Willink wrote *The World of the Unseen*, in which he claimed that it was unworthy of God to reside in the lowly fourth dimension. Willink claimed that the only domain magnificent enough for God was infinite-dimensional space.[2]

55

sions, the book was an instant success in England, with nine successive reprintings by the year 1915, and editions too numerous to count today.

What was surprising about the novel *Flatland* was that Abbott, for the first time, used the controversy surrounding the fourth dimension as a vehicle for biting social criticism and satire. Abbot took a playful swipe at the rigid, pious individuals who refused to admit the possibility of other worlds. The "bookworms" of Gauss became the Flatlanders. The Boeotians whom Gauss so feared became the High Priests, who would persecute—with the vigor and impartiality of the Spanish Inquisition—anyone who dared mention the unseen third dimension.

Abbot's *Flatland* is a thinly disguised criticism of the subtle bigotry and suffocating prejudice prevalent in Victorian England. The hero of the novel is Mr. Square, a conservative gentleman who lives in a socially stratified, two-dimensional land where everyone is a geometric object. Women, occupying the lowest rank in the social hierarchy, are mere lines, the nobility are polygons, while the High Priests are circles. The more sides people have, the higher their social rank.

Discussion of the third dimension is strictly forbidden. Anyone mentioning it is sentenced to severe punishment. Mr. Square is a smug, self-righteous person who would never think of challenging the Establishment for its injustices. One day, however, his life is permanently turned upside down when he is visited by a mysterious Lord Sphere, a three-dimensional sphere. Lord Sphere appears to Mr. Square as a circle that can magically change size (Figure 3.1)

Lord Sphere patiently tries to explain that he comes from another world called Spaceland, where all objects have three dimensions. However, Mr. Square remains unconvinced; he stubbornly resists the idea that a third dimension can exist. Frustrated, Lord Sphere decides to resort to deeds, not mere words. He then peels Mr. Square off the two-dimensional Flatland and hurls him into Spaceland. It is a fantastic, almost mystical experience that changes Mr. Square's life.

As the flat Mr. Square floats in the third dimension like a sheet of paper drifting in the wind, he can visualize only two-dimensional slices of Spaceland. Mr. Square, seeing only the cross sections of three-dimensional objects, views a fantastic world where objects change shape and even appear and disappear into thin air. However, when he tries to tell his fellow Flatlanders of the marvels he saw in his visit to the third dimension, the High Priests consider him a blabbering, seditious maniac. Mr. Square becomes a threat to the High Priests because he dares to challenge their authority and their sacred belief that only two dimensions can possibly exist.

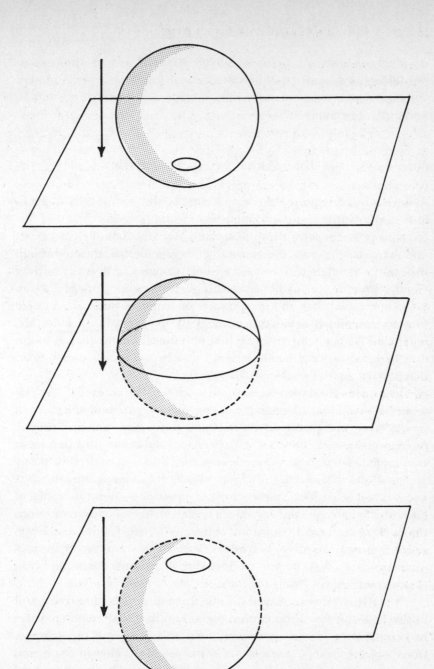

Figure 3.1. In Flatland, Mr. Square encounters Lord Sphere. As Lord Sphere passes through Flatland, he appears to be a circle that becomes successivley larger and then smaller. Thus Flatlanders cannot visualize three-dimensional beings, but can understand their cross sections.

The book ends on a pessimistic note. Although he is convinced that he did, indeed, visit the third-dimensional world of Spaceland, Mr. Square is sent to jail and condemned to spend the rest of his days in solitary confinement.

A Dinner Party in the Fourth Dimension

Abbot's novel is important because it was the first widely read popularization of a visit to a higher-dimensional world. His description of Mr. Square's psychedelic trip into Spaceland is mathematically correct. In popular accounts and the movies, interdimensional travel through hyperspace is often pictured with blinking lights and dark, swirling clouds. However, the mathematics of higher-dimensional travel is much more interesting than the imagination of fiction writers. To visualize what an interdimensional trip would look like, imagine peeling Mr. Square off Flatland and throwing him into the air. As he floats through our three-dimensional world, let's say that he comes across a human being. What do we look like to Mr. Square?

Because his two-dimensional eyes can see only flat slices of our world, a human would look like a singularly ugly and frightening object. First, he might see two leather circles hovering in front of him (our shoes). As he drifts upward, these two circles change color and turn into cloth (our pants). Then these two circles coalesce into one circle (our waist) and split into three circles of cloth and change color again (our shirt and our arms). As he continues to float upward, these three circles of cloth merge into one smaller circle of flesh (our neck and head). Finally, this circle of flesh turns into a mass of hair, and then abruptly disappears as Mr. Square floats above our heads. To Mr. Square, these mysterious "humans" are a nightmarish, maddeningly confusing collection of constantly changing circles made of leather, cloth, flesh, and hair.

Similarly, if we were peeled off our three-dimensional universe and hurled into the fourth dimension, we would find that common sense becomes useless. As we drift through the fourth dimension, blobs appear from nowhere in front of our eyes. They constantly change in color, size, and composition, defying all the rules of logic of our three-dimensional world. And they disappear into thin air, to be replaced by other hovering blobs.

If we were invited to a dinner party in the fourth dimension, how would we tell the creatures apart? We would have to recognize them by the differences in how these blobs change. Each person in higher dimensions would have his or her own characteristic sequences of changing blobs. Over a period of time, we would learn to tell these creatures apart by recognizing their distinctive patterns of changing blobs and colors. Attending dinner parties in hyperspace might be a trying experience.

Class Struggle in the Fourth Dimension

The concept of the fourth dimension had so pervasively infected the intellectual climate by the late nineteenth century that even playwrights poked fun at it. In 1891, Oscar Wilde wrote a spoof on these ghost stories, "The Canterville Ghost," which lampoons the exploits of a certain gullible "Psychical Society" (a thinly veiled reference to Crookes's Society for Psychical Research). Wilde wrote of a long-suffering ghost who encounters the newly arrived American tenants of Canterville. Wilde wrote, "There was evidently no time to be lost, so hastily adopting the Fourth Dimension of Space as a means of escape, he [the ghost] vanished through the wainscoting and the house became quiet."

A more serious contribution to the literature of the fourth dimension was the work of H. G. Wells. Although he is principally remembered for his works in science fiction, he was a dominant figure in the intellectual life of London society, noted for his literary criticism, reviews, and piercing wit. In his 1894 novel, *The Time Machine,* he combined several mathematical, philosophical, and political themes. He popularized a new idea in science—that the fourth dimension might also be viewed as time, not necessarily space:*

Clearly . . . any real body must have extension in *four* directions: it must have Length, Breadth, Thickness, and—Duration. But through a natural infirmity of the flesh . . . we incline to overlook this fact. There are really four dimensions, three which we call the three lanes of Space, and a Fourth, Time. There is, however, a tendency to draw an unreal distinction between the former three dimensions and the latter, because it happens that our consciousness moves intermittently in one direction along the latter from the beginning to the end of our lives.[3]

*Wells was not the first to speculate that time could be viewed as a new type of fourth dimension, different from a spatial one. Jean d'Alembert had considered time as the fourth dimension in his 1754 article "Dimension."

Like *Flatland* before it, what makes *The Time Machine* so enduring, even a century after its conception, is its sharp political and social critique. England in the year 802,701, Wells's protagonist finds, is not the gleaming citadel of modern scientific marvels that the positivists foretold. Instead, the future England is a land where the class struggle went awry. The working class was cruelly forced to live underground, until the workers mutated into a new, brutish species of human, the Morlocks, while the ruling class, with its unbridled debauchery, deteriorated and evolved into the useless race of elflike creatures, the Eloi.

Wells, a prominent Fabian socialist, was using the fourth dimension to reveal the ultimate irony of the class struggle. The social contract between the poor and the rich had gone completely mad. The useless Eloi are fed and clothed by the hard-working Morlocks, but the workers get the final revenge: The Morlocks eat the Eloi. The fourth dimension, in other words, became a foil for a Marxist critique of modern society, but with a novel twist: The working class will not break the chains of the rich, as Marx predicted. They will eat the rich.

In a short story, "The Plattner Story," Wells even toyed with the paradox of handedness. Gottfried Plattner, a science teacher, is performing an elaborate chemical experiment, but his experiment blows up and sends him into another universe. When he returns from the netherworld to the real world, he discovers that his body has been altered in a curious fashion: His heart is now on his right side, and he is now left handed. When they examine him, his doctors are stunned to find that Plattner's entire body has been reversed, a biological impossibility in our three-dimensional world: "[T]he curious inversion of Plattner's right and left sides is proof that he has moved out of our space into what is called the Fourth Dimension, and that he has returned again to our world." However, Plattner resists the idea of a postmortem dissection after his death, thereby postponing "perhaps forever, the positive proof that his entire body had had its left and right sides transposed."

Wells was well aware that there are two ways to visualize how left-handed objects can be transformed into right-handed objects. A Flatlander, for example, can be lifted out of his world, flipped over, and then placed back in Flatland, thereby reversing his organs. Or the Flatlander may live on a Möbius strip, created by twisting a strip of paper 180 degrees and then gluing the ends together. If a Flatlander walks completely around the Möbius strip and returns, he finds that his organs have been reversed (Figure 3.2). Möbius strips have other remarkable properties that have fascinated scientists over the past century. For example, if you walk completely around the surface, you will find that it has

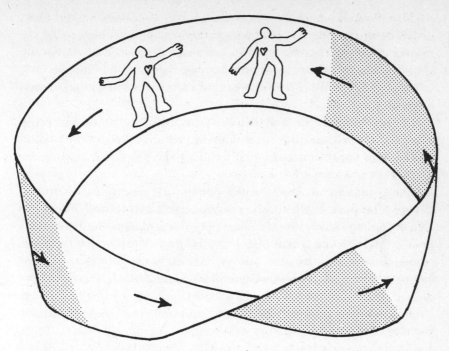

Figure 3.2. A Möbius strip is a strip with only one side. Its outside and inside are identical. If a Flatlander wanders around a Mobius strip, his internal organs will be reversed.

only one side. Also, if you cut it in half along the center strip, it remains in one piece. This has given rise to the mathematicians' limerick:

> A mathematician confided
> That a Möbius band is one-sided
> And you'll get quite a laugh
> If you cut it in half,
> For it stays in one piece when divided.

In his classic *The Invisible Man*, Wells speculated that a man might even become invisible by some trick involving "a formula, a geometrical expression involving four dimensions." Wells knew that a Flatlander disappears if he is peeled off his two-dimensional universe; similarly, a man could become invisible if he could somehow leap into the fourth dimension.

In the short story "The Remarkable Case of Davidson's Eyes," Wells explored the idea that a "kink in space" might enable an individual to see across vast distances. Davidson, the hero of the story, one day finds he has the disturbing power of being able to see events transpiring on a distant South Sea island. This "kink in space" is a space warp whereby light from the South Seas goes through hyperspace and enters his eyes in England. Thus Wells used Riemann's wormholes as a literary device in his fiction.

In *The Wonderful Visit,* Wells explored the possibility that heaven exists in a parallel world or dimension. The plot revolves around the predicament of an angel who accidentally falls from heaven and lands in an English country village.

The popularity of Wells's work opened up a new genre of fiction. George McDonald, a friend of mathematician Lewis Carroll, also speculated about the possibility of heaven being located in the fourth dimension. In McDonald's fantasy *Lilith,* written in 1895, the hero creates a dimensional window between our universe and other worlds by manipulating mirror reflections. And in the 1901 story *The Inheritors* by Joseph Conrad and Ford Madox Ford, a race of supermen from the fourth dimension enters into our world. Cruel and unfeeling, these supermen begin to take over the world.

The Fourth Dimension as Art

The years 1890 to 1910 may be considered the Golden Years of the Fourth Dimension. It was a time during which the ideas originated by Gauss and Riemann permeated literary circles, the avant garde, and the thoughts of the general public, affecting trends in art, literature, and philosophy. The new branch of philosophy, called Theosophy, was deeply influenced by higher dimensions.

On the one hand, serious scientists regretted this development because the rigorous results of Riemann were now being dragged through tabloid headlines. On the other hand, the popularization of the fourth dimension had a positive side. Not only did it make the advances in mathematics available to the general public, but it also served as a metaphor that could enrich and cross-fertilize cultural currents.

Art historian Linda Dalrymple Henderson, writing in *The Fourth Dimension and Non-Euclidean Geometry in Modern Art,* elaborates on this and argues that the fourth dimension crucially influenced the development of Cubism and Expressionism in the art world. She writes that "it was among the Cubists that the first and most coherent art theory based on the new geometries was developed."[4] To the avant garde, the fourth

Figure 3.3. One scene in the Bayeux Tapestry depicts frightened English troops pointing to an apparition in the sky (Halley's comet). The figures are flat, as in most of the art done in the Middle Ages. This signified that God was omnipotent. Pictures were thus drawn two dimensionally. (Giraudon/Art Resource)

dimension symbolized the revolt against the excesses of capitalism. They saw its oppressive positivism and vulgar materialism as stifling creative expression. The Cubists, for example, rebelled against the insufferable arrogance of the zealots of science whom they perceived as dehumanizing the creative process.

The avant garde seized on the fourth dimension as their vehicle. On the one hand, the fourth dimension pushed the boundaries of modern science to their limit. It was more scientific than the scientists. On the other hand, it was mysterious. And flaunting the fourth dimension tweaked the noses of the stiff, know-it-all positivists. In particular, this took the form of an artistic revolt against the laws of perspective.

In the Middle Ages, religious art was distinctive for its deliberate lack of perspective. Serfs, peasants, and kings were depicted as though they were flat, much in the way children draw people. These paintings largely reflected the church's view that God was omnipotent and could therefore see all parts of our world equally. Art had to reflect his point of view, so the world was painted two dimensionally. For example, the famous Bayeux Tapestry (Figure 3.3) depicts the superstitious soldiers of King Harold II of England pointing in frightened wonder at an omi-

Figure 3.4. During the Renaissance, painters discovered the third dimension. Pictures were painted with perspective and were viewed from the vantage point of a single eye, not God's eye. Note that all the lines in Leonardo da Vinci's fresco The Last Supper *converge to a point at the horizon. (Bettmann Archive)*

nous comet soaring overhead in April 1066, convinced that it is an omen of impending defeat. (Six centuries later, the same comet would be christened Halley's comet.) Harold subsequently lost the crucial Battle of Hastings to William the Conqueror, who was crowned the king of England, and a new chapter in English history began. However, the Bayeux Tapestry, like other medieval works of art, depicts Harold's soldiers' arms and faces as flat, as though a plane of glass had been placed over their bodies, compressing them against the tapestry.

Renaissance art was a revolt against this flat God-centered perspective, and man-centered art began to flourish, with sweeping landscapes and realistic, three-dimensional people painted from the point of view of a person's eye. In Leonardo da Vinci's powerful studies on perspective, we see the lines in his sketches vanishing into a single point on the horizon. Renaissance art reflected the way the eye viewed the world, from the singular point of view of the observer. In Michelangelo's frescoes or in da Vinci's sketch book, we see bold, imposing figures jumping out of the second dimension. In other words, Renaissance art discovered the third dimension (Figure 3.4).

With the beginning of the machine age and capitalism, the artistic world revolted against the cold materialism that seemed to dominate industrial society. To the Cubists, positivism was a straitjacket that confined us to what could be measured in the laboratory, suppressing the fruits of our imagination. They asked: Why must art be clinically "realistic"? This Cubist "revolt against perspective" seized the fourth dimension because it touched the third dimension from all possible perspectives. Simply put, Cubist art embraced the fourth dimension.

Picasso's paintings are a splendid example, showing a clear rejection of the perspective, with women's faces viewed simultaneously from several angles. Instead of a single point of view, Picasso's paintings show multiple perspectives, as though they were painted by someone from the fourth dimension, able to see all perspectives simultaneously (Figure 3.5).

Picasso was once accosted on a train by a stranger who recognized him. The stranger complained: Why couldn't he draw pictures of people the way they actually were? Why did he have to distort the way people looked? Picasso then asked the man to show him pictures of his family. After gazing at the snapshot, Picasso replied, "Oh, is your wife really that small and flat?" To Picasso, any picture, no matter how "realistic," depended on the perspective of the observer.

Abstract painters tried not only to visualize people's faces as though painted by a four-dimensional person, but also to treat time as the fourth dimension. In Marcel Duchamp's painting *Nude Descending a Staircase,* we see a blurred representation of a woman, with an infinite number of her images superimposed over time as she walks down the stairs. This is how a four-dimensional person would see people, viewing all time sequences at once, if time were the fourth dimension.

In 1937, art critic Meyer Schapiro summarized the influence of these new geometries on the art world when he wrote, "Just as the discovery of non-Euclidean geometry gave a powerful impetus to the view that mathematics was independent of existence, so abstract painting cut at the roots of the classic ideas of artistic imitation." Or, as art historian Linda Henderson has said, "the fourth dimension and non-Euclidean geometry emerge as among the most important themes unifying much of modern art and theory."[5]

Bolsheviks and the Fourth Dimension

The fourth dimension also crossed over into Czarist Russia via the writings of the mystic P. D. Ouspensky, who introduced Russian intellectuals to its mysteries. His influence was so pronounced that even Fyodor Dos-

Figure 3.5. Cubism was heavily influenced by the fourth dimension. For example, it tried to view reality through the eyes of a fourth-dimensional person. Such a being, looking at a human face, would see all angles simultaneously. Hence, both eyes would be seen at once by a fourth-dimensional being, as in Picasso's painting Portrait of Dora Maar. *(Giraudon/Art Resource. © 1993. Ars, New York/ Spadem, Paris)*

toyevsky, in *The Brothers Karamazov,* had his protagonist Ivan Karamazov speculate on the existence of higher dimensions and non-Euclidean geometries during a discussion on the existence of God.

Because of the historic events unfolding in Russia, the fourth dimension was to play a curious role in the Bolshevik Revolution. Today, this strange interlude in the history of science is important because Vladimir Lenin would join the debate over the fourth dimension, which would eventually exert a powerful influence on the science of the former Soviet Union for the next 70 years.[6] (Russian physicists, of course, have played key roles in developing the present-day ten-dimensional theory.)

After the Czar brutally crushed the 1905 revolution, a faction called the Otzovists, or "God-builders," developed within the Bolshevik party. They argued that the peasants weren't ready for socialism; to prepare them, Bolsheviks should appeal to them through religion and spiritualism. To bolster their heretical views, the God-builders quoted from the work of the German physicist and philosopher Ernst Mach, who had written eloquently about the fourth dimension and the recent discovery of a new, unearthly property of matter called radioactivity. The God-builders pointed out that the discovery of radioactivity by the French scientist Henri Becquerel in 1896 and the discovery of radium by Marie Curie in 1896 had ignited a furious philosophical debate in French and German literary circles. It appeared that matter could slowly disintegrate and that energy (in the form of radiation) could reappear.

Without question, the new experiments on radiation showed that the foundation of Newtonian physics was crumbling. Matter, thought by the Greeks to be eternal and immutable, was now disintegrating before our very eyes. Uranium and radium, confounding accepted belief, were mutating in the laboratory. To some, Mach was the prophet who would lead them out of the wilderness. However, he pointed in the wrong direction, rejecting materialism and declaring that space and time were products of our sensations. In vain, he wrote, "I hope that nobody will defend ghost-stories with the help of what I have said and written on this subject."[7]

A split developed within the Bolsheviks. Their leader, Vladimir Lenin, was horrified. Are ghosts and demons compatible with socialism? In exile in Geneva in 1908, he wrote a mammoth philosophical tome, *Materialism and Empirio-Criticism,* defending dialectical materialism from the onslaught of mysticism and metaphysics. To Lenin, the mysterious disappearance of matter and energy did not prove the existence of spirits. He argued that this meant instead that a *new dialectic* was emerging, which would embrace both matter and energy. No longer could they be viewed as separate entities, as Newton had done. They must now be viewed as two poles of a dialectical unity. A new conservation principle was needed. (Unknown to Lenin, Einstein had proposed the correct principle 3 years earlier, in 1905.) Furthermore, Lenin questioned Mach's easy embrace of the fourth dimension. First, Lenin praised Mach, who "has raised the very important and useful question of a space of n dimensions as a conceivable space." Then he took Mach to task for failing to emphasize that only the three dimensions of space could be verified experimentally. Mathematics may explore the fourth dimension and the world of what is possible, and this is good, wrote Lenin, but the Czar can be overthrown only in the third dimension![8]

Fighting on the battleground of the fourth dimension and the new theory of radiation, Lenin needed years to root out Otzovism from the Bolshevik party. Nevertheless, he won the battle shortly before the outbreak of the 1917 October Revolution.

Bigamists and the Fourth Dimension

Eventually, the ideas of the fourth dimension crossed the Atlantic and came to America. Their messenger was a colorful English mathematician named Charles Howard Hinton. While Albert Einstein was toiling at his desk job in the Swiss patent office in 1905, discovering the laws of relativity, Hinton was working at the United States Patent Office in Washington, D.C. Although they probably never met, their paths would cross in several interesting ways.

Hinton spent his entire adult life obsessed with the notion of popularizing and visualizing the fourth dimension. He would go down in the history of science as the man who "saw" the fourth dimension.

Hinton was the son of James Hinton, a renowned British ear surgeon of liberal persuasion. Over the years, the charismatic elder Hinton evolved into a religious philosopher, an outspoken advocate of free love and open polygamy, and finally the leader of an influential cult in England. He was surrounded by a fiercely loyal and devoted circle of free-thinking followers. One of his best-known remarks was "Christ was the Savior of men, but I am the savior of women, and I don't envy Him a bit!"[9]

His son Charles, however, seemed doomed to lead a respectable, boring life as a mathematician. He was fascinated not by polygamy, but by polygons! Having graduated from Oxford in 1877, he became a respectable master at the Uppingham School while working on his master's degree in mathematics. At Oxford, Hinton became intrigued with trying to visualize the fourth dimension. As a mathematician, he knew that one cannot visualize a four-dimensional object in its entirety. However, it is possible, he reasoned, to visualize the cross section or the unraveling of a four-dimensional object.

Hinton published his notions in the popular press. He wrote the influential article "What is the Fourth Dimension?" for the *Dublin University Magazine* and the *Cheltenham Ladies' College Magazine,* reprinted in 1884 with the catchy subtitle "Ghosts Explained."

Hinton's life as a comfortable academic, however, took a sharp turn for the worse in 1885 when he was arrested for bigamy and put on trial. Earlier, Hinton had married Mary Everest Boole, the daughter of a member of his father's circle, and widow of the great mathematician George Boole (founder of Boolean algebra). However, he was also the father of twins born to a certain Maude Weldon.

The headmaster at Uppingham, noticing Hinton in the presence of his wife, Mary, and his mistress, Maude, had assumed that Maude was Hinton's sister. All was going well for Hinton, until he made the mistake of marrying Maude as well. When the headmaster learned that Hinton was a bigamist, it set off a scandal. He was promptly fired from his job at Uppingham and placed on trial for bigamy. He was imprisoned for 3 days, but Mary Hinton declined to press charges and together they left England for the United States.

Hinton was hired as an instructor in the mathematics department at Princeton University, where his obsession with the fourth dimension was temporarily sidetracked when he invented the baseball machine. The Princeton baseball team benefited from Hinton's machine, which could fire baseballs at 70 miles per hour. The descendants of Hinton's creation can now be found on every major baseball field in the world.

Hinton was eventually fired from Princeton, but managed to get a job at the United States Naval Observatory through the influence of its director, a devout advocate of the fourth dimension. Then, in 1902, he took a job at the Patent Office in Washington.

Hinton's Cubes

Hinton spent years developing ingenious methods by which the average person and a growing legion of followers, not only professional mathematicians, could "see" four-dimensional objects. Eventually, he perfected special cubes that, if one tried hard enough, could allow one to visualize hypercubes, or cubes in four dimensions. These would eventually be called Hinton's cubes. Hinton even coined the official name for an unraveled hypercube, a tesseract, which found its way into the English language.

Hinton's cubes were widely advertised in women's magazines and were even used in seances, where they soon became objects of mystical importance. By mediating on Hinton's cubes, it was claimed by members of high society, you could catch glimpses of the fourth dimension and hence the nether world of ghosts and the dearly departed. His disciples spent hours contemplating and meditating on these cubes, until they attained the ability to mentally rearrange and reassemble these cubes via the fourth dimension into a hypercube. Those who could perform this mental feat, it was said, would attain the highest state of nirvana.

As an analogy, take a three-dimensional cube. Although a Flatlander cannot visualize a cube in its entirety, it is possible for us to unravel the cube in three dimensions, so that we have a series of six squares making a cross. Of course, a Flatlander cannot reassemble the squares to make a cube. In the second dimension, the joints between each square are rigid and cannot be moved. However, these joints are easy to bend in the third dimension. A Flatlander witnessing this event would see the squares disappear, leaving only one square in his universe (Figure 3.6).

Likewise, a hypercube in four dimensions cannot be visualized. But one can unravel a hypercube into its lower components, which are ordinary three-dimensional cubes. These cubes, in turn, can be arranged in a three-dimensional cross—a tesseract. It is impossible for us to visualize how to wrap up these cubes to form a hypercube. However, a higher-dimensional person can "lift" each cube off our universe and then wrap up the cube to form a hypercube. (Our three-dimensional eyes, witnessing this spectacular event, would only see the other cubes disappear, leaving only one cube in our universe.) So pervasive was Hinton's influence that Salvadore Dali used Hinton's tesseract in his famous painting *Christus Hypercubus,* on display at the Metropolitan Museum of Art in New York, which depicts Christ being crucified on a four-dimensional cross (Figure 3.7).

Hinton also knew of a second way to visualize higher-dimensional objects: by looking at the shadows they cast in lower dimensions. For example, a Flatlander can visualize a cube by looking at its two-dimensional shadow. A cube looks like two squares joined together. Similarly, a hypercube's shadow cast on the third dimension becomes a cube within a cube (Figure 3.8).

In addition to visualizing unravelings of hypercubes and examining their shadows, Hinton was aware of a third way to conceptualize the

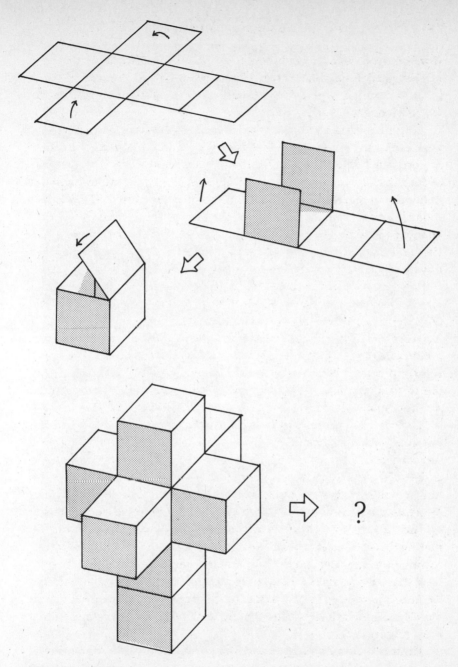

Figure 3.6. Flatlanders cannot visualize a cube, but they can conceptualize a three-dimensional cube by unraveling it. To a Flatlander, a cube, when unfolded, resembles a cross, consisting of six squares. Similarly, we cannot visualize a four-dimensional hypercube, but if we unfold it we have a series of cubes arranged in a crosslike tesseract. Although the cubes of a tesseract appear immobile, a four-dimensional person can "wrap up" the cubes into a hypercube.

71

Figure 3.7. In Christus Hypercubus, *Salvador Dali depicted Christ as being crucified on a tesseract, an unraveled hypercube. (The Metropolitan Museum of Art. Gift of Chester Dale, Collection, 1955.* © *1993. Ars, New York/Demart Pro Arte, Geneva)*

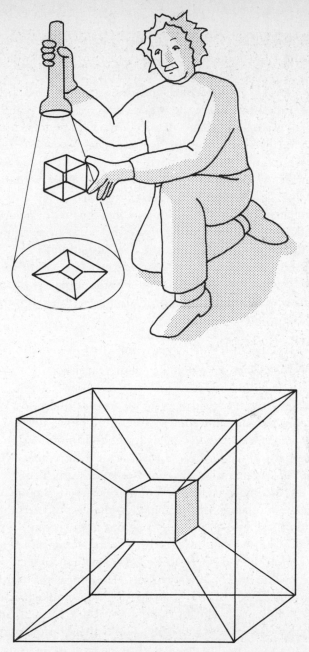

Figure 3.8. A Flatlander can visualize a cube by examining its shadow, which appears as a square within a square. If the cube is rotated, the squares execute motions that appear impossible to a Flatlander. Similarly, the shadow of a hypercube is a cube within a cube. If the hypercube is rotated in four dimensions, the cubes execute motions that appear impossible to our three-dimensional brains.

fourth dimension: by cross sections. For example, when Mr. Square is sent into the third dimension, his eyes can see only two-dimensional cross sections of the third dimension. Thus he can see only circles appear, get larger, change color, and then suddenly disappear. If Mr. Square moved past an apple, he would see a red circle materialize out of nowhere, gradually expand, then contract, then turn into a small brown circle (the stem), and finally disappear. Likewise, Hinton knew that if we were hurled into the fourth dimension, we would see strange objects suddenly appear out of nowhere, get larger, change color, change shape, get smaller, and finally disappear.

In summary, Hinton's contribution may be his popularization of higher-dimensional figures using three methods: by examining their shadows, their cross sections, and their unravellings. Even today, these three methods are the chief ways in which professional mathematicians and physicists conceptualize higher-dimensional objects in their work. The scientists whose diagrams appear in today's physics journals owe a small debt of gratitude to Hinton's work.

The Contest on the Fourth Dimension

In his articles, Hinton had answers for all possible questions. When people asked him to name the fourth dimension, he would reply that the words *ana* and *kata* described moving in the fourth dimension and were the counterparts of the terms *up* and *down,* or *left* and *right.* When asked where the fourth dimension was, he also had a ready answer.

For the moment, consider the motion of cigarette smoke in a closed room. Because the atoms of the smoke, by the laws of thermodynamics, spread and diffuse into all possible locations in the room, we can determine if there are any regions of ordinary three-dimensional space that the smoke molecules miss. However, experimental observations show that there are no such hidden regions. Therefore, the fourth spatial dimension is possible only if it is smaller than the smoke particles. Thus if the fourth dimension actually exists, it must be incredibly small, even smaller than an atom. This is the philosophy that Hinton adopted, that all objects in our three-dimensional universe exist in the fourth dimension, but that the fourth dimension is so small that it evades any experimental observation. (We will find that physicists today adopt essentially the same philosophy as Hinton and conclude that the higher dimensions are too small to be experimentally seen. When asked, "What is light?" he also had a ready answer. Following Riemann, Hinton believed that light is a vibration of the unseen fourth dimension, which is essentially the viewpoint taken today by many theoretical physicists.)

In the United States, Hinton single-handedly sparked an enormous public interest in the fourth dimension. Popular magazines like *Harper's Weekly, McClure's, Current Literature, Popular Science Monthly,* and *Science* all devoted pages to the blossoming interest in the fourth dimension. But what probably ensured Hinton's fame in America was the famous contest sponsored by *Scientific American* in 1909. This unusual contest offered a $500 prize (a considerable amount of money in 1909) to "the best popular explanation of the Fourth Dimension." The magazine's editors were pleasantly surprised by the deluge of letters that poured into their offices, including entries from as far away as Turkey, Austria, Holland, India, Australia, France, and Germany.

The object of the contest was to "set forth in an essay not longer than twenty-five hundred words the meaning of the term so that the ordinary lay reader could understand it." It drew a large number of serious essays. Some lamented the fact that people like Zollner and Slade had besmirched the reputation of the fourth dimension by confusing it with spiritualism. However, many of the essays recognized Hinton's pioneering work on the fourth dimension. (Surprisingly, not one essay mentioned the work of Einstein. In 1909, it was still far from clear that Einstein had uncovered the secret of space and time. In fact, the idea of time as the fourth dimension did not appear in a single essay.)

Without experimental verification, the *Scientific American* contest could not, of course, resolve the question of the existence of higher dimensions. However, the contest did address the question of what higher-dimensional objects might look like.

Monsters from the Fourth Dimension

What would it be like to meet a creature from a higher dimension?

Perhaps the best way to explain the wonder and excitement of a hypothetical visit to other dimensions is through science fiction, where writers have tried to grapple with this question.

In "The Monster from Nowhere," writer Nelson Bond tried to imagine what would happen if an explorer in the jungles of Latin America encountered a beast from a higher dimension.

Our hero is Burch Patterson, adventurer, bon vivant, and soldier of fortune, who hits on the idea of capturing wild animals in the towering mountains of Peru. The expedition will be paid for by various zoos, which put up the money for the trip in return for whatever animals Patterson can find. With much hoopla and fanfare, the press covers the progress of the expedition as it journeys into unexplored territory. But after a few weeks, the expedition loses contact with the outside world and mysteriously disappears without a trace. After a long and futile search, the authorities reluctantly give the explorers up for dead.

Two years later, Burch Patterson abruptly reappears. He meets secretly with reporters and tells them an astonishing story of tragedy and heroism. Just before the expedition disappeared, it encountered a fantastic animal in the Maratan Plateau of upper Peru, an unearthly bloblike creature that was constantly changing shape in the most bizarre fashion. These black blobs hovered in midair, disappearing and reappearing and changing shape and size. The blobs then unexpectedly attacked the expedition, killing most of the men. The blobs hoisted some of the remaining men off the ground; they screamed and then disappeared into thin air.

Only Burch escaped the rout. Dazed and frightened, he nonetheless studied these blobs from a distance and gradually formed a theory about what they were and how to capture them. He had read *Flatland* years before, and imagined that anyone sticking his fingers into and out of Flatland would startle the two-dimensional inhabitants. The Flatlanders would see pulsating rings of flesh hovering in midair (our fingers poking through Flatland), constantly changing size. Likewise, reasoned Patterson, any higher-dimensional creature sticking his foot or arms through our universe would appear as three-dimensional, pulsating blobs of flesh, appearing out of nowhere and constantly changing shape and size. That would also explain why his team members had disappeared into thin air: They had been dragged into a higher-dimensional universe.

But one question still plagued him: How do you capture a higher-dimensional being? If a Flatlander, seeing our finger poke its way through his two-dimensional universe, tried to capture our finger, he would be at a loss. If he tried to lasso our finger, we could simply remove our finger and disappear. Similarly, Patterson reasoned, he could put a net around one of these blobs, but then the higher-dimensional creature could simply pull his "finger" or "leg" out of our universe, and the net would collapse.

Suddenly, the answer came to him: If a Flatlander were to try to capture our finger as it poked its way into Flatland, the Flatlander could stick a needle *through our finger*, painfully impaling it to the two-dimensional universe. Thus Patterson's strategy was to drive a spike through one of the blobs and impale the creature in our universe!

After months of observing the creature, Patterson identified what looked like the creature's "foot" and drove a spike right through it. It took him 2 years to capture the creature and ship the writhing, struggling blob back to New Jersey.

Finally, Patterson announces a major press conference where he will unveil a fantastic creature caught in Peru. Journalists and scientists alike gasp in horror when the creature is unveiled, writhing and struggling against a large steel rod. Like a scene from *King Kong,* one newspaperman, against the rules, takes flash pictures of the creature. The flash enrages the creature, which then struggles so hard against the rod that its flesh begins to tear. Suddenly, the monster is free, and pandemonium breaks out. People are torn to shreds, and Patterson and others are grabbed by the creature and then disappear into the fourth dimension.

In the aftermath of the tragedy, one of the survivors of the massacre decides to burn all evidence of the creature. Better to leave this mystery forever unsolved.

Building a Four-Dimensional House

In the previous section, the question of what happens when we encounter a higher-dimensional being was explored. But what happens in the reverse situation, when we visit a higher-dimensional universe? As we have seen, a Flatlander cannot possibly visualize a three-dimensional universe in its entirety. However, there are, as Hinton showed, several ways in which the Flatlander can comprehend revealing fragments of higher-dimensional universes.

In his classic short story ". . . And He Built a Crooked House . . . ," Robert Heinlein explored the many possibilities of living in an unraveled hypercube.

Quintus Teal is a brash, flamboyant architect whose ambition is to build a house in a truly revolutionary shape: a tesseract, a hypercube that has been unraveled in the third dimension. He cons his friends Mr. and Mrs. Bailey into buying the house.

Built in Los Angeles, the tesseract is a series of eight ultramodern cubes stacked on top of one another in the shape of a cross. Unfortunately, just as Teal is about to show off his new creation to the Baileys, an earthquake strikes southern California, and the house collapses into itself. The cubes begin to topple, but strangely only a single cube is left standing. The other cubes have mysteriously disappeared. When Teal and the Baileys cautiously enter the house, now just a single cube, they are amazed that the other missing rooms are clearly visible through the windows of the first floor. But that is impossible. The house is now only a single cube. How can the interior of a single cube be connected to a series of other cubes that cannot be seen from the outside?

They climb the stairs and find the master bedroom above the entry-way. Instead of finding the third floor, however, they find themselves back on the ground floor. Thinking the house is haunted, the frightened Baileys race to the front door. Instead of leading to the outside, the front door just leads to another room. Mrs. Bailey faints.

As they explore the house, they find that each room is connected to an impossible series of other rooms. In the original house, each cube had windows to view the outside. Now, all windows face other rooms. There is no outside!

Scared out of their wits, they slowly try all the doors of the house, only to wind up in other rooms. Finally, in the study they decide to open the four Venetian blinds and look outside. When they open the first Venetian blind, they find that they are peering down at the Empire State Building. Apparently, that window opened up to a "window" in space just above the spire of the tower. When they open the second Venetian blind, they find themselves staring at a vast ocean, except it is upside down. Opening the third Venetian blind, they find themselves looking at Nothing. Not empty space. Not inky blackness. Just Nothing. Finally, opening up the last Venetian blind, they find themselves gazing at a bleak desert landscape, probably a scene from Mars.

After a harrowing tour through the rooms of the house, with each room impossibly connected to the other rooms, Teal finally figures it all out. The earthquake, he reasons, must have collapsed the joints of various cubes and folded the house in the fourth dimension.[10]

On the outside, Teal's house originally looked like an ordinary sequence of cubes. The house did not collapse because the joints between the cubes were rigid and stable in three dimensions. However, viewed from the fourth dimension, Teal's house is an unraveled hypercube that can be reassembled or folded back into a hypercube. Thus when the house was shaken by the earthquake, it somehow folded up in four dimensions, leaving only a single cube dangling in our third dimension. Anyone walking into the single remaining cube would view a series of rooms connected in a seemingly impossible fashion. By racing through the various rooms, Teal has moved through the fourth dimension without noticing it.

Although our protagonists seem doomed to spend their lives fruit-lessly wandering in circles inside a hypercube, another violent earthquake shakes the tesseract. Holding their breath, Teal and the terrified Baileys leap out the nearest window. When they land, they find themselves in Joshua Tree National Monument, miles from Los Angeles. Hours later, hitching a ride back to the city, they return to the house, only to find that the last remaining cube has vanished. Where did the tesseract go? It is probably drifting somewhere in the fourth dimension.

The Useless Fourth Dimension

In retrospect, Riemann's famous lecture was popularized to a wide audience via mystics, philosophers, and artists, but did little to further our understanding of nature. From the perspective of modern physics, we can also see why the years 1860 to 1905 did not produce any fundamental breakthroughs in our understanding of hyperspace.

First, there was no attempt to use hyperspace to simplify the laws of nature. Without Riemann's original guiding principle—that the laws of nature become simple in higher dimensions—scientists during this period were groping in the dark. Riemann's seminal idea of using geometry—that is, crumpled hyperspace—to explain the essence of a "force" was forgotten during those years.

Second, there was no attempt to exploit Faraday's field concept or Riemann's metric tensor to find the field equations obeyed by hyperspace. The mathematical apparatus developed by Riemann became a province of pure mathematics, contrary to Riemann's original intentions. Without field theory, you cannot make any predictions with hyperspace.

Thus by the turn of the century, the cynics claimed (with justification) that there was no experimental confirmation of the fourth dimension. Worse, they claimed, there was no physical motivation for introducing the fourth dimension, other than to titillate the general public with ghost stories. This deplorable situation would soon change, however. Within a few decades, the theory of the fourth dimension (of time) would forever change the course of human history. It would give us the atomic bomb and the theory of Creation itself. And the man who would do it would be an obscure physicist named Albert Einstein.

4

The Secret of Light: Vibrations in the Fifth Dimension

If [relativity] should prove to be correct, as I expect it will, he
will be considered the Copernicus of the twentieth century.
                                        Max Planck on Albert Einstein

THE life of Albert Einstein appeared to be one long series of failures
and disappointments. Even his mother was distressed at how slowly
he learned to talk. His elementary-school teachers thought him a foolish
dreamer. They complained that he was constantly disrupting classroom
discipline with his silly questions. One teacher even told the boy bluntly
that he would prefer that Einstein drop out of his class.

He had few friends in school. Losing interest in his courses, he
dropped out of high school. Without a high-school diploma, he had to
take special exams to enter college, but he did not pass them and had
to take them a second time. He even failed the exam for the Swiss military because he had flat feet.

After graduation, he could not get a job. He was an unemployed
physicist who was passed over for a teaching position at the university
and was rejected for jobs everywhere he applied. He earned barely 3
francs an hour—a pittance—by tutoring students. He told his friend
Maurice Solovine that "an easier way of earning a living would be to
play the violin in public places."

Einstein was a man who rejected the things most men chase after, such as power and money. However, he once noted pessimistically, "By the mere existence of his stomach, everyone is condemned to participate in that chase." Finally, through the influence of a friend, he landed a lowly job as a clerk at the Swiss patent office in Bern, earning just enough money so his parents would not have to support him. On his meager salary, he supported his young wife and their newborn baby.

Lacking financial resources or connections with the scientific establishment, Einstein began to work in solitude at the patent office. In between patent applications, his mind drifted to problems that had intrigued him as a youth. He then undertook a task that would eventually change the course of human history. His tool was the fourth dimension.

Children's Questions

Wherein lies the essence of Einstein's genius? In *The Ascent of Man,* Jacob Bronowski wrote: "The genius of men like Newton and Einstein lies in that: they ask transparent, innocent questions which turn out to have catastrophic answers. Einstein was a man who could ask immensely simple questions."[1] As a child, Einstein asked himself the simple question: What would a light beam look like if you could catch up with one? Would you see a stationary wave, frozen in time? This question set him on a 50-year journey through the mysteries of space and time.

Imagine trying to overtake a train in a speeding car. If we hit the gas pedal, our car races neck-and-neck with the train. We can peer inside the train, which now appears to be at rest. We can see the seats and the people, who are acting as though the train weren't moving. Similarly, Einstein as a child imagined traveling alongside a light beam. He thought that the light beam should resemble a series of stationary waves, frozen in time; that is, the light beam should appear motionless.

When Einstein was 16 years old, he spotted the flaw in this argument. He recalled later,

After ten years of reflection such a principle resulted from a paradox upon which I had already hit at the age of sixteen: If I pursue a beam of light with the velocity c (velocity of light in a vacuum) I should observe such a beam of light as a spatially oscillatory electromagnetic field at rest. However, there seems to be no such thing, whether on the basis of experience or according to Maxwell's equations.[2]

In college, Einstein confirmed his suspicions. He learned that light can be expressed in terms of Faraday's electric and magnetic fields, and that these fields obey the field equations found by James Clerk Maxwell. As he suspected, he found that stationary, frozen waves are not allowed by Maxwell's field equations. In fact, Einstein showed that a light beam travels at the *same* velocity *c*, no matter how hard you try to catch up with it.

At first, this seemed absurd. This meant that we could never overtake the train (light beam). Worse, no matter how fast we drove our car, the train would always seem to be traveling ahead of us at the *same* velocity. In other words, a light beam is like the "ghost ship" that old sailors love to spin tall tales about. It is a phantom vessel that can never be caught. No matter how fast we sail, the ghost ship always eludes us, taunting us.

In 1905, with plenty of time on his hands at the patent office, Einstein carefully analyzed the field equations of Maxwell and was led to postulate the principle of *special relativity:* The speed of light is the same in all constantly moving frames. This innocent-sounding principle is one of the greatest achievements of the human spirit. Some have said that it ranks with Newton's law of gravitation as one of the greatest scientific creations of the human mind in the 2 million years our species has been evolving on this planet. From it, we can logically unlock the secret of the vast energies released by the stars and galaxies.

To see how this simple statement can lead to such profound conclusions, let us return to the analogy of the car trying to overtake the train. Let us say that a pedestrian on the sidewalk clocks our car traveling at 99 miles per hour, and the train traveling at 100 miles per hour. Naturally, from our point of view in the car, we see the train moving ahead of us at 1 mile per hour. This is because velocities can be added and subtracted, just like ordinary numbers.

Now let us replace the train by a light beam, but keep the velocity of light at just 100 miles per hour. The pedestrian still clocks our car traveling at 99 miles per hour in hot pursuit of the light beam traveling at 100 miles per hour. According to the pedestrian, we should be closing in on the light beam. However, according to relativity, we in the car actually see the light beam not traveling ahead of us at 1 mile per hour, as expected, but speeding ahead of us at 100 miles per hour. Remarkably, we see the light beam racing ahead of us as though we were at rest. Not believing our own eyes, we slam on the gas pedal until the pedestrian clocks our car racing ahead at 99.99999 miles per hour. Surely, we think, we must be about to overtake the light beam. However, when we look out the window, we see the light beam still speeding ahead of us at 100 miles per hour.

Uneasily, we reach several bizarre, disturbing conclusions. First, no matter how much we gun the engines of our car, the pedestrian tells us that we can approach but never exceed 100 miles per hour. This seems to be the top velocity of the car. Second, no matter how close we come to 100 miles per hour, we still see the light beam speeding ahead of us at 100 miles per hour, as though we weren't moving at all.

But this is absurd. How can both people in the speeding car and the stationary person measure the velocity of the light beam to be the same? Ordinarily, this is impossible. It appears to be nature's colossal joke.

There is only one way out of this paradox. Inexorably, we are led to the astonishing conclusion that shook Einstein to the core when he first conceived of it. The only solution to this puzzle is that *time slows down* for us in the car. If the pedestrian takes a telescope and peers into our car, he sees everyone in the car moving exceptionally slowly. However, we in the car never notice that time is slowing down because our brains, too, have slowed down, and everything seems normal to us. Furthermore, he sees that the car has become flattened in the direction of motion. The car has shrunk like an accordion. However, we never feel this effect because our bodies, too, have shrunk.

Space and time play tricks on us. In actual experiments, scientists have shown that the speed of light is always *c*, no matter how fast we travel. This is because the faster we travel, the slower our clocks tick and the shorter our rulers become. In fact, our clocks slow down and our rulers shrink just enough so that whenever we measure the speed of light, it comes out the same.

But why can't we see or feel this effect? Since our brains are thinking more slowly, and our bodies are also getting thinner as we approach the speed of light, we are blissfully unaware that we are turning into slow-witted pancakes.

These relativistic effects, of course, are too small to be seen in everyday life because the speed of light is so great. Being a New Yorker, however, I am constantly reminded of these fantastic distortions of space and time whenever I ride the subway. When I am on the subway platform with nothing to do except wait for the next subway train, I sometimes let my imagination drift and wonder what it would be like if the speed of light were only, say, 30 miles per hour, the speed of a subway train. Then when the train finally roars into the station, it appears squashed, like an accordion. The train, I imagine, would be a flattened slab of metal 1 foot thick, barreling down the tracks. And everyone inside the subway cars would be as thin as paper. They would also be virtually frozen in time, as though they were motionless statues. However, as the train comes to a grinding halt, it suddenly expands, until this slab of metal gradually fills the entire station.

As absurd as these distortions might appear, the passengers inside the train would be totally oblivious to these changes. Their bodies and space itself would be compressed along the direction of motion of the train; everything would appear to have its normal shape. Furthermore, their brains would have slowed down, so that everyone inside the train would act normally. Then when the subway train finally comes to a halt, they are totally unaware that their train, to someone on the platform, appears to miraculously expand until it fills up the entire platform. When the passengers depart from the train, they are totally oblivious to the profound changes demanded by special relativity.*

The Fourth Dimension and High-School Reunions

There have been, of course, hundreds of popular accounts of Einstein's theory, stressing different aspects of his work. However, few accounts capture the essence behind the theory of special relativity, which is that time is the fourth dimension and that the laws of nature are simplified and unified in higher dimensions. Introducing time as the fourth dimension overthrew the concept of time dating all the way back to Aristotle. Space and time would now be forever dialectically linked by special relativity. (Zollner and Hinton had assumed that the next dimension to be discovered would be the fourth spatial dimension. In this respect, they were wrong and H. G. Wells was correct. The next dimension to be discovered would be time, a fourth temporal dimension. Progress in understanding the fourth spatial dimension would have to wait several more decades.)

To see how higher dimensions simplify the laws of nature, we recall that any object has length, width, and depth. Since we have the freedom

---

*Similarly, passengers riding in the train would think that the train was at rest and that the subway station was coming toward the train. They would see the platform and everyone standing on it compressed like an accordian. Then this leads us to a contradiction, that people on the train and in the station each think that the other has been compressed. The resolution of this paradox is a bit delicate.[3]

to rotate an object by 90 degrees, we can turn its length into width and its width into depth. By a simple rotation, we can interchange any of the three spatial dimensions. Now if time is the fourth dimension, then it is possible to make "rotations" that convert space into time and vice versa. These four-dimensional "rotations" are precisely the distortions of space and time demanded by special relativity. In other words, space and time have mixed in an essential way, governed by relativity. The meaning of time as being the fourth dimension is that time and space can rotate into each other in a mathematically precise way. From now on, they must be treated as two aspects of the same quantity: space–time. Thus adding a higher dimension helped to unify the laws of nature.

Newton, writing 300 years ago, thought that time beat at the same rate everywhere in the universe. Whether we sat on the earth, on Mars, or on a distant star, clocks were expected to tick at the same rate. There was thought to be an absolute, uniform rhythm to the passage of time throughout the entire universe. Rotations between time and space were inconceivable. Time and space were two distinct quantities with no relationship between them. Unifying them into a single quantity was unthinkable. However, according to special relativity, time can beat at different rates, depending on how fast one is moving. Time being the fourth dimension means that time is intrinsically linked with movement in space. How fast a clock ticks depends on how fast it is moving in space. Elaborate experiments done with atomic clocks sent into orbit around the earth have confirmed that a clock on the earth and a clock rocketing in outer space tick at different rates.

I was graphically reminded of the relativity principle when I was invited to my twentieth high-school reunion. Although I hadn't seen most of my classmates since graduation, I assumed that all of them would show the same telltale signs of aging. As expected, most of us at the reunion were relieved to find that the aging process was universal: It seemed that all of us sported graying temples, expanding waistlines, and a few wrinkles. Although we were separated across space and time by several thousand miles and 20 years, each of us had assumed that time had beat uniformly for all. We automatically assumed that each of us would age at the same rate.

Then my mind wandered, and I imagined what would happen if a classmate walked into the reunion hall looking *exactly* as he had on graduation day. At first, he would probably draw stares from his classmates. Was this the same person we knew 20 years ago? When people realized that he was, a panic would surge through the hall.

We would be jolted by this encounter because we tacitly assume that clocks beat the same everywhere, even if they are separated by vast distances. However, if time is the fourth dimension, then space and time can rotate into each other and clocks can beat at different rates, depending on how fast they move. This classmate, for example, may have entered a rocket traveling at near-light speeds. For us, the rocket trip may have lasted for 20 years. However, for him, because time slowed down in the speeding rocket, he aged only a few moments from graduation day. To him, he just entered the rocket, sped into outer space for a few minutes, and then landed back on earth in time for his twentieth high-school reunion after a short, pleasant journey, still looking youthful amid a field of graying hair.

I am also reminded that the fourth dimension simplifies the laws of nature whenever I think back to my first encounter with Maxwell's field equations. Every undergraduate student learning the theory of electricity and magnetism toils for several years to master these eight abstract equations, which are exceptionally ugly and very opaque. Maxwell's eight equations are clumsy and difficult to memorize because time and space are treated separately. (To this day, I have to look them up in a book to make sure that I get all the signs and symbols correct.) I still remember the relief I felt when I learned that these equations collapse into one trivial-looking equation when time is treated as the fourth dimension. In one masterful stroke, the fourth dimension simplifies these equations in a beautiful, transparent fashion.[4] Written in this way, the equations possess a higher *symmetry;* that is, space and time can turn into each other. Like a beautiful snowflake that remains the same when we rotate it around its axis, Maxwell's field equations, written in relativistic form, remain the same when we rotate space into time.

Remarkably, this one simple equation, written in a relativistic fashion, contains the same physical content as the eight equations originally written down by Maxwell over 100 years ago. This one equation, in turn, governs the properties of dynamos, radar, radio, television, lasers, household appliances, and the cornucopia of consumer electronics that appear in everyone's living room. This was one of my first exposures to the concept of *beauty* in physics—that is, that the symmetry of four-dimensional space can explain a vast ocean of physical knowledge that would fill an engineering library.

Once again, this demonstrates one of the main themes of this book, that the addition of higher dimensions helps to simplify and unify the laws of nature.

Matter as Condensed Energy

This discussion of unifying the laws of nature, so far, has been rather abstract, and would have remained so had Einstein not taken the next fateful step. He realized that if space and time can be unified into a single entity, called space–time, then perhaps matter and energy can also be united into a dialectical relationship. If rulers can shrink and clocks slow down, he reasoned, then everything that we measure with rulers and clocks must also change. However, almost everything in a physicist's laboratory is measured by rulers and clocks. This meant that physicists had to recalibrate all the laboratory quantities they once took for granted to be constant.

Specifically, energy is a quantity that depends on how we measure distances and time intervals. A speeding test car slamming into a brick wall obviously has energy. If the speeding car approaches the speed of light, however, its properties become distorted. It shrinks like an accordion and clocks in it slow down.

More important, Einstein found that the mass of the car also increases as it speeds up. But where did this excess mass come from? Einstein concluded that it came from the energy.

This had disturbing consequences. Two of the great discoveries of nineteenth-century physics were the conservation of mass and the conservation of energy; that is, the total mass and total energy of a closed system, taken separately, do not change. For example, if the speeding car hits the brick wall, the energy of the car does not vanish, but is converted into the sound energy of the crash, the kinetic energy of the flying brick fragments, heat energy, and so on. The total energy (and total mass) before and after the crash is the same.

However, Einstein now said that the energy of the car could be converted into mass—a new conservation principle that said that the sum total of the mass added to energy must always remain the same. Matter does not suddenly disappear, nor does energy spring out of nothing. In this regard, the God-builders were wrong and Lenin was right. Matter disappears only to unleash enormous quantities of energy, or vice versa.

When Einstein was 26 years old, he calculated precisely how energy must change if the relativity principle was correct, and he discovered the relation $E = mc^2$. Since the speed of light squared (c^2) is an astronomically large number, a small amount of matter can release a vast amount of energy. Locked within the smallest particles of matter is a storehouse of energy, more than 1 million times the energy released in a chemical explosion. Matter, in some sense, can be seen as an almost inexhaustible storehouse of energy; that is, matter is condensed energy.

In this respect, we see the profound difference between the work of the mathematician (Charles Hinton) and that of the physicist (Albert Einstein). Hinton spent most of his adult years trying to visualize higher *spatial* dimensions. He had no interest in finding a physical interpretation for the fourth dimension. Einstein saw, however, that the fourth dimension can be taken as a *temporal* one. He was guided by a conviction and physical intuition that higher dimensions have a purpose: to unify the principles of nature. By adding higher dimensions, he could unite physical concepts that, in a three-dimensional world, have no connection, such as matter and energy.

From then on, the concept of matter and energy would be taken as a single unit: matter–energy. The direct impact of Einstein's work on the fourth dimension was, of course, the hydrogen bomb, which has proved to be the most powerful creation of twentieth-century science.

"The Happiest Thought of My Life"

Einstein, however, wasn't satisfied. His special theory of relativity alone would have guaranteed him a place among the giants of physics. But there was something missing.

Einstein's key insight was to use the fourth dimension to unite the laws of nature by introducing two new concepts: space–time and matter–energy. Although he had unlocked some of the deepest secrets of nature, he realized there were several gaping holes in his theory. What was the relationship between these two new concepts? More specifically, what about accelerations, which are ignored in special relativity? And what about gravitation?

His friend Max Planck, the founder of the quantum theory, advised the young Einstein that the problem of gravitation was too difficult. Planck told him that he was too ambitious: "As an older friend I must advise you against it for in the first place you will not succeed; and even if you succeed, no one will believe you."[5] Einstein, however, plunged ahead to unravel the mystery of gravitation. Once again, the key to his momentous discovery was to ask questions that only children ask.

When children ride in an elevator, they sometimes nervously ask, "What happens if the rope breaks?" The answer is that you become weightless and float inside the elevator, as though in outer space, because both you and the elevator are falling at the same rate. Even though both you and the elevator are accelerating in the earth's gravitational field, the acceleration is the same for both, and hence it appears that you are weightless in the elevator (at least until you reach the bottom of the shaft).

In 1907, Einstein realized that a person floating in the elevator might think that someone had mysteriously turned off gravity. Einstein once recalled, "I was sitting in a chair in the patent office at Bern when all of a sudden a thought occurred to me: 'If a person falls freely he will not feel his own weight.' I was startled. This simple thought made a deep impression on me. It impelled me toward a theory of gravitation."[6] Einstein would call it "the happiest thought of my life."

Reversing the situation, he knew that someone in an accelerating rocket will feel a force pushing him into his seat, as though there were a gravitational pull on him. (In fact, the force of acceleration felt by our astronauts is routinely measured in g's—that is, multiples of the force of the earth's gravitation.) The conclusion he reached was that someone accelerating in a speeding rocket may think that these forces were caused by gravity.

From this children's question, Einstein grasped the fundamental nature of gravitation: *The laws of nature in an accelerating frame are equivalent to the laws in a gravitational field.* This simple statement, called the *equivalence principle,* may not mean much to the average person, but once again, in the hands of Einstein, it became the foundation of a theory of the cosmos.

(The equivalence principle also gives simple answers to complex physics questions. For example, if we are holding a helium balloon while riding in a car, and the car suddenly swerves to the left, our bodies will be jolted to the right, but which way will the balloon move? Common sense tells us that the balloon, like our bodies, will move to the right. However, the correct resolution of this subtle question has stumped even experienced physicists. The answer is to use the equivalence principle. Imagine a gravitational field pulling on the car from the right. Gravity will make us lurch us to the right, so the helium balloon, which is lighter than air and always floats "up," opposite the pull of gravity, must float to the left, into the direction of the swerve, defying common sense.)

Einstein exploited the equivalence principle to solve the long-standing problem of whether a light beam is affected by gravity. Ordinarily, this is a highly nontrivial question. Through the equivalence principle, however, the answer becomes obvious. If we shine a flashlight inside an accelerating rocket, the light beam will bend downward toward the floor (because the rocket has accelerated beneath the light beam during the time it takes for the light beam to move across the room). Therefore, argued Einstein, a gravitational field will also bend the path of light.

Einstein knew that a fundamental principle of physics is that a light beam will take the path requiring the least amount of time between two points. (This is called Fermat's least-time principle.) Ordinarily, the path with the smallest time between two points is a straight line, so light beams are straight. (Even when light bends upon entering glass, it still obeys the least-time principle. This is because light slows down in glass, and the path with the least time through a combination of air and glass is now a bent line. This is called refraction, which is the principle behind microscopes and telescopes.)*

However, if light takes the path with the least time between two points, and light beams bend under the influence of gravity, then the shortest distance between two points is a curved line. Einstein was shocked by this conclusion: If light could be observed traveling in a curved line, it would mean that *space itself is curved.*

Space Warps

At the core of Einstein's belief was the idea that "force" could be explained using pure geometry. For example, think of riding on a merry-go-round. Everyone knows that if we change horses on a merry-go-round, we feel a "force" tugging at us as we walk across the platform. Because the outer rim of the merry-go-round moves faster than the center, the outer rim of the merry-go-round must shrink, according to special relativity. However, if the platform of the merry-go-round now has a shrunken rim or circumference, the platform as a whole must be curved. To someone on the platform, light no longer travels in a straight line, as though a "force" were pulling it toward the rim. The usual theorems of geometry no longer hold. Thus the "force" we feel while walking between horses on a merry-go-round can be explained as the curving of space itself.

Einstein independently discovered Riemann's original program, to give a purely geometric explanation of the concept of "force." We recall

---

*For example, imagine being a lifeguard on a beach, at some distance from the water; out of the corner of your eye, you spy someone drowning in the ocean far off at an angle. Assume that you can run very slowly in the soft sand, but can swim swiftly in the water. A straight path to the victim will spend too much time on the sand. The path with the least time is a bent line, one that reduces the time spent running on the sand and maximizes the time spent swimming in the water.

that Riemann used the analogy of Flatlanders living on a crumpled sheet of paper. To us, it is obvious that Flatlanders moving over a wrinkled surface will be incapable of walking in a straight line. Whichever way they walk, they will experience a "force" that tugs at them from left and right. To Riemann, the bending or warping of space causes the appearance of a force. Thus forces do not really exist; what is actually happening is that space itself is being bent out of shape.

The problem with Riemann's approach, however, was that he had no idea specifically how gravity or electricity and magnetism caused the warping of space. His approach was purely mathematical, without any concrete physical picture of precisely how the bending of space was accomplished. Here Einstein succeeded where Riemann failed.

Imagine, for example, a rock placed on a stretched bedsheet. Obviously the rock will sink into the sheet, creating a smooth depression. A small marble shot onto the bedsheet will then follow a circular or an elliptical path around the rock. Someone looking from a distance at the marble orbiting around the rock may say that there is an "instantaneous force" emanating from the rock that alters the path of the marble. However, on close inspection it is easy to see what is really happening: The rock has warped the bedsheet, and hence the path of the marble.

By analogy, if the planets orbit around the sun, it is because they are moving in space that has been curved by the presence of the sun. Thus the reason we are standing on the earth, rather than being hurled into the vacuum of outer space, is that the earth is constantly warping the space around us (Figure 4.1).

Einstein noticed that the presence of the sun warps the path of light from the distant stars. This simple physical picture therefore gave a way in which the theory could be tested experimentally. First, we measure the position of the stars at night, when the sun is absent. Then, during an eclipse of the sun, we measure the position of the stars, when the sun is present (but doesn't overwhelm the light from the stars). According to Einstein, the apparent relative position of the stars should change when the sun is present, because the sun's gravitational field will have bent the path of the light of those stars on its way to the earth. By comparing the photographs of the stars at night and the stars during an eclipse, one should be able to test this theory.

This picture can be summarized by what is called Mach's principle, the guide Einstein used to create his general theory of relativity. We recall that the warping of the bedsheet was determined by the presence of the rock. Einstein summarized this analogy by stating: The presence of matter–energy determines the curvature of the space–time surrounding it. This is the essence of the physical principle that Riemann failed

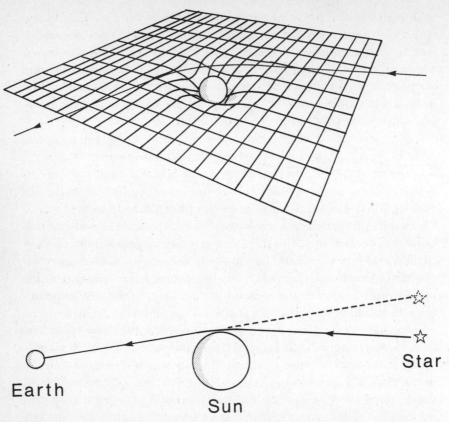

Figure 4.1. To Einstein, "gravity" was an illusion caused by the bending of space. He predicted that starlight moving around the sun would be bent, and hence the relative positions of the stars should appear distored in the presence of the sun. This has been verified by repeated experiments.

to discover, that the bending of space is directly related to the amount of energy and matter contained within that space.

This, in turn, can be summarized by Einstein's famous equation,[7] which essentially states:

$$\text{Matter–energy} \rightarrow \text{curvature of space–time}$$

where the arrow means "determines." This deceptively short equation is one of the greatest triumphs of the human mind. From it emerge the principles behind the motions of stars and galaxies, black holes, the Big Bang, and perhaps the fate of the universe itself.

Nevertheless, Einstein was still missing a piece of the puzzle. He had discovered the correct physical principle, but lacked a rigorous mathematical formalism powerful enough to express this principle. He lacked a version of Faraday's fields for gravity. Ironically, Riemann had the mathematical apparatus, but not the guiding physical principle. Einstein, by contrast, discovered the physical principle, but lacked the mathematical apparatus.

Field Theory of Gravity

Because Einstein formulated this physical principle without knowing of Riemann, he did not have the mathematical language or skill with which to express his principle. He spent 3 long, frustrating years, from 1912 to 1915, in a desperate search for a mathematical formalism powerful enough to express the principle. Einstein wrote a desperate letter to his close friend, mathematician Marcel Grossman, pleading, "Grossman, you must help me or else I'll go crazy!"[8]

Fortunately, Grossman, when combing through the library for clues to Einstein's problem, accidentally stumbled on the work of Riemann. Grossman showed Einstein the work of Riemann and his metric tensor, which had been ignored by physicists for 60 years. Einstein would later recall that Grossman "checked through the literature and soon discovered that the mathematical problem had already been solved by Riemann, Ricci, and Levi-Civita. . . . Riemann's achievement was the greatest one."

To his shock, Einstein found Riemann's celebrated 1854 lecture to be the key to the problem. He found that he could incorporate the entire body of Riemann's work in the reformulation of his principle. Almost line for line, the great work of Riemann found its true home in Einstein's principle. This was Einstein's proudest piece of work, even more than his celebrated equation $E = mc^2$. The physical reinterpretation of Riemann's famous 1854 lecture is now called *general relativity,* and Einstein's field equations rank among the most profound ideas in scientific history.

Riemann's great contribution, we recall, was that he introduced the concept of the metric tensor, a field that is defined at all points in space. The metric tensor is not a single number. At each point in space, it consists of a collection of ten numbers. Einstein's strategy was to follow Maxwell and write down the field theory of gravity. The object of his search for a field to describe gravity was found practically on the first page of Riemann's lecture. In fact, Riemann's metric tensor was precisely the Faraday field for gravity!

When Einstein's equations are fully expressed in terms of Riemann's metric tensor, they assume an elegance never before seen in physics. Nobel laureate Subrahmanyan Chandrasekhar once called it "the most beautiful theory there ever was." (In fact, Einstein's theory is so simple yet so powerful that physicists are sometimes puzzled as to why it works so well. MIT physicist Victor Weisskopf once said, "It's like the peasant who asks the engineer how the steam engine works. The engineer explains to the peasant exactly where the steam goes and how it moves through the engine and so on. And then the peasant says: 'Yes, I understand all that, but where is the horse?' That's how I feel about general relativity. I know all the details, I understand where the steam goes, but I'm still not sure I know where the horse is."[9])

In retrospect, we now see how close Riemann came to discovering the theory of gravity 60 years before Einstein. The entire mathematical apparatus was in place in 1854. His equations were powerful enough to describe the most complicated twisting of space–time in any dimension. However, he lacked the physical picture (that matter–energy determines the curvature of space–time) and the keen physical insight that Einstein provided.

Living in Curved Space

I once attended a hockey game in Boston. All the action, of course, was concentrated on the hockey players as they glided on the ice rink. Because the puck was being rapidly battered back and forth between the various players, it reminded me of how atoms exchange electrons when they form chemical elements or molecules. I noticed that the skating rink, of course, did not participate in the game. It only marked the various boundaries; it was a passive arena on which the hockey players scored points.

Next, I imagined what it must be like if the skating rink actively participated in the game: What would happen if the players were forced to play on an ice rink whose surface was curved, with rolling hills and steep valleys?

The hockey game would suddenly became more interesting. The players would have to skate along a curved surface. The rink's curvature would distort their motion, acting like a "force" pulling the players one way or another. The puck would move in a curved line like a snake, making the game much more difficult.

Then I imagined taking this one step further; I imagined that the players were forced to play on a skating rink shaped like a cylinder. If the players could generate enough speed, they could skate upside down and move entirely around the cylinder. New strategies could be devised, such as ambushing an opposing player by skating upside down around the cylinder and catching him unawares. Once the ice rink was bent in the shape of a circle, space would become the decisive factor in explaining the motion of matter on its surface.

Another, more relevant example for our universe might be living in a curved space given by a hypersphere, a sphere in four dimensions.[10] If we look ahead, light will circle completely around the small perimeter of the hypersphere and return to our eyes. Thus we will see someone standing in front of us, with his back facing us, a person who is wearing the same clothes as we are. We look disapprovingly at the unruly, unkempt mass of hair on this person's head, and then remember that we forgot to comb our hair that day.

Is this person a fake image created by mirrors? To find out, we stretch out our hand and put it on his shoulder. We find that the person in front of us is a real person, not just a fake. If we look into the distance, in fact, we see an infinite number of identical people, each facing forward, each with his hand on the shoulder of the person in front.

But what is most shocking is that we feel someone's hand sneaking up from behind, which then grabs our shoulder. Alarmed, we look back, and see another infinite sequence of identical people behind us, with their faces turned the other way.

What's really happening? We, of course, are the only person living in this hypersphere. The person in front of us is really ourself. We are staring at the back of our own head. By placing our hand in front of us, we are really stretching our hand around the hypersphere, until we place our hand on our own shoulder.

The counterintuitive stunts that are possible in a hypersphere are physically interesting because many cosmologists believe that our universe is actually a large hypersphere. There are also other equally strange topologies, like hyperdoughnuts and Möbius strips. Although they may ultimately have no practical application, they help to illustrate many of the features of living in hyperspace.

For example, let us assume that we are living on a hyperdoughnut. If we look to our left and right, we see, much to our surprise, a person on either side. Light circles completely around the larger perimeter of the doughnut, and returns to its starting point. Thus if we turn our heads and look to the left, we see the right side of someone's body. By turning our heads the other way, we see someone's left side. No matter how fast we turn our heads, the people ahead of us and to our sides turn their heads just as fast, and we can never see their faces.

Now imagine stretching our arms to either side. Both the person on the left and the one on the right will also stretch their arms. In fact, if you are close enough, you can grab the left and right hands of the persons to either side. If you look carefully in either direction, you can see an infinitely long, straight line of people all holding hands. If you look ahead, there is another infinite sequence of people standing before you, arranged in a straight line, all holding hands.

What's actually happening? In reality our arms are long enough to reach around the doughnut, until the arms have touched. Thus we have actually grabbed our own hands (Figure 4.2)!

Now we find ourselves tiring of this charade. These people seem to be taunting us; they are copy-cats, doing exactly what we do. We get annoyed—so we get a gun and point it at the person in front of us. Just before we pull the trigger, we ask ourselves: Is this person a fake mirror image? If so, then the bullet will go right through him. But if not, then the bullet will go completely around the universe and hit us in the back. Maybe firing a gun in this universe is not such a good idea!

For an even more bizarre universe, imagine living on a Möbius strip, which is like a long strip of paper twisted 180 degrees and then reglued back together into a circular strip. When a right-handed Flatlander moves completely around the Möbius strip, he finds that he has become left-handed. Orientations are reversed when traveling around the universe. This is like H. G. Wells's "The Plattner Story," in which the hero returns to earth after an accident to find that his body is completely reversed; for example, his heart is on his right side.

If we lived on a hyper-Möbius strip, and we peered in front of us, we would see the back of someone's head. At first, we wouldn't think it could be our head, because the part of the hair would be on the wrong side. If we reached out and placed our right hand on his shoulder, then he would lift up his left hand and place it on the shoulder of the person ahead of him. In fact, we would see an infinite chain of people with hands on each other's shoulders, except the hands would alternate from the left to the right shoulders.

If we left some of our friends at one spot and walked completely around this universe, we would find that we had returned to our original spot. But our friends would be shocked to find that our body was

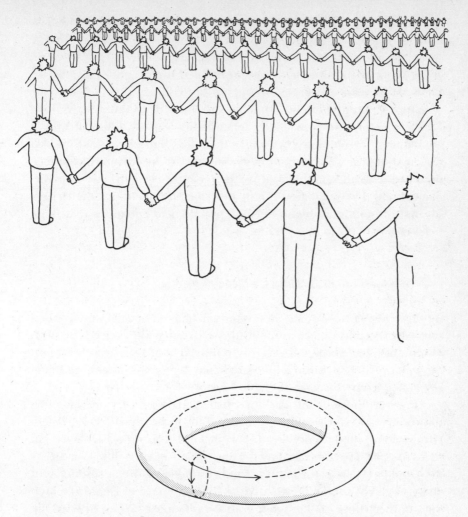

Figure 4.2. If we lived in a hyperdoughnut, we would see an infinite succession of ourselves repeated in front of us, to the back of us, and to our sides. This is because there are two ways that light can travel around the doughnut. If we hold hands with the people to our sides, we are actually holding our own hands; that is, our arms are actually encircling the doughnut.

reversed. The part in our hair and the rings on our fingers would be on the wrong side, and our internal organs would have been reversed. Our friends would be amazed at the reversal of our body, and would ask if we felt well. In fact, we would feel completely normal; to us, it would be our friends who had been completely turned around! An argument would now ensue over who was really reversed.

These and other interesting possibilities open up when we live in a universe where space and time are curved. No longer a passive arena, space becomes an active player in the drama unfolding in our universe.

In summary, we see that Einstein fulfilled the program initiated by Riemann 60 years earlier, to use higher dimensions to simplify the laws of nature. Einstein, however, went beyond Riemann in several ways. Like Riemann before him, Einstein independently realized that "force" is a consequence of geometry, but unlike Riemann, Einstein was able to find the physical principle behind this geometry, that the curvature of space–time is due to the presence of matter–energy. Einstein, also like Riemann, knew that gravitation can be described by a field, the metric tensor, but Einstein was able to find the precise field equations that these fields obey.

A Universe Made of Marble

By the mid-1920s, with the development of both special and general relativity, Einstein's place in the history of science was assured. In 1921, astronomers had verified that starlight indeed bends as it travels around the sun, precisely as Einstein had predicted. By then, Einstein was being celebrated as the successor to Isaac Newton.

However, Einstein still was not satisfied. He would try one last time to produce another world-class theory. But on his third try, he failed. His third and final theory was to have been the crowning achievement of his lifetime. He was searching for the "theory of everything," a theory that would explain all the familiar forces found in nature, including light and gravity. He coined this theory the *unified field theory*. Alas, his search for a unified theory of light and gravity was fruitless. When he died, he left only the unfinished ideas of various manuscripts on his desk.

Ironically, the source of Einstein's frustration was the structure of his own equation. For 30 years, he was disturbed by a fundamental flaw in this formulation. On one side of the equation was the curvature of space–time, which he likened to "marble" because of its beautiful geometric structure. To Einstein, the curvature of space–time was like the epitome of Greek architecture, beautiful and serene. However, he hated the other side of this equation, describing matter–energy, which he considered to be ugly and which he compared to "wood." While the "marble" of space–time was clean and elegant, the "wood" of matter–energy was a horrible jumble of confused, seemingly random forms, from subatomic particles, atoms, polymers, and crystals to rocks, trees, planets, and stars. But in the 1920s and 1930s, when Einstein was actively working on the unified field theory, the true nature of matter remained an unsolved mystery.

Einstein's grand strategy was to turn wood into marble—that is, to give a completely geometric origin to matter. But without more physical clues and a deeper physical understanding of the wood, this was impossible. By analogy, think of a magnificent, gnarled tree growing in the middle of a park. Architects have surrounded this grizzled tree with a plaza made of beautiful pieces of the purest marble. The architects have carefully assembled the marble pieces to resemble a dazzling floral pattern with vines and roots emanating from the tree. To paraphrase Mach's principle: The presence of the tree determines the pattern of the marble surrounding it. But Einstein hated this dichotomy between wood, which seemed to be ugly and complicated, and marble, which was simple and pure. His dream was *to turn the tree into marble;* he would have liked to have a plaza completely made of marble, with a beautiful, symmetrical marble statue of a tree at its center.

In retrospect, we can probably spot Einstein's error. We recall that the laws of nature simplify and unify in higher dimensions. Einstein correctly applied this principle twice, in special and general relativity. However, on his third try, he abandoned this fundamental principle. Very little was known about the structure of atomic and nuclear matter in his time; consequently, it was not clear how to use higher-dimensional space as a unifying principle.

Einstein blindly tried a number of purely mathematical approaches. He apparently thought that "matter" could be viewed as kinks, vibrations, or distortions of space–time. In this picture, matter was a concentrated distortion of space. In other words, everything we see around us, from the trees and clouds to the stars in the heavens, was probably an illusion, some form of crumpling of hyperspace. However, without any more solid leads or experimental data, this idea led to a blind alley.

It would be left to an obscure mathematician to take the next step, which would lead us to the fifth dimension.

The Birth of Kaluza–Klein Theory

In April 1919, Einstein received a letter that left him speechless.

It was from an unknown mathematician, Theodr Kaluza, at the University of Königsberg in Germany, in what is Kaliningrad in the former Soviet Union. In a short article, only a few pages long, this obscure mathematician was proposing a solution to one of the greatest problems of the century. In just a few lines, Kaluza was uniting Einstein's theory of gravity with Maxwell's theory of light by introducing the *fifth* dimension (that is, four dimensions of space and one dimension of time).

In essence, he was resurrecting the old "fourth dimension" of Hinton and Zollner and incorporating it into Einstein's theory in a fresh fashion as the fifth dimension. Like Riemann before him, Kaluza assumed that light is a disturbance caused by the rippling of this higher dimension. The key difference separating this new work from Riemann's, Hinton's, and Zollner's was that Kaluza was proposing a genuine field theory.

In this short note, Kaluza began, innocently enough, by writing down Einstein's field equations for gravity in five dimensions, not the usual four. (Riemann's metric tensor, we recall, can be formulated in any number of dimensions.) Then he proceeded to show that these five-dimensional equations contained within them Einstein's earlier four-dimensional theory (which was to be expected) with an additional piece. But what shocked Einstein was that this additional piece was precisely Maxwell's theory of light. In other words, this unknown scientist was proposing to combine, in one stroke, the two greatest field theories known to science, Maxwell's and Einstein's, by mixing them in the fifth dimension. This was a theory made of pure marble—that is, pure geometry.

Kaluza had found the first important clue in turning wood into marble. In the analogy of the park, we recall that the marble plaza is two dimensional. Kaluza's observation was that we could build a "tree" of marble if we could move the pieces of marble *up* into the third dimension.

To the average layman, light and gravity have nothing in common. After all, light is a familiar force that comes in a spectacular variety of colors and forms, while gravity is invisible and more distant. On the earth, it is the electromagnetic force, not gravity, that has helped us tame nature; it is the electromagnetic force that powers our machines, electrifies our cities, lights our neon signs, and brightens our television sets. Gravity, by contrast, operates on a larger scale; it is the force that guides the planets and keeps the sun from exploding. It is a cosmic force that permeates the universe and binds the solar system. (Along with Weber and Riemann, one of the first scientists to search actively for a link between light and gravity in the laboratory was Faraday himself. The actual experimental apparatus used by Faraday to measure the link between these two forces can still be found in the Royal Institution in Piccadilly, London. Although he failed experimentally to find any connection at all between the two forces, Faraday was confident of the power of unification. He wrote, "If the hope [of unification] should prove well founded, how great and mighty and sublime in its hitherto unchangeable character is the force I am trying to deal with, and how large may be the new domain of knowledge that may be opened to the mind of man."[11])

Even mathematically, light and gravity are like oil and water. Maxwell's field theory of light requires four fields, while Einstein's metric theory of gravity requires ten. Yet Kaluza's paper was so elegant and compelling that Einstein could not reject it.

At first, it seemed like a cheap mathematical trick simply to expand the number or dimensions of space and time from four to five. This was because, as we recall, there was no experimental evidence for the fourth spatial dimension. What astonished Einstein was that once the five-dimensional field theory was broken down to a four-dimensional field theory, both Maxwell's and Einstein's equations remained. In other words, Kaluza succeeded in joining the two pieces of the jigsaw puzzle because both of them were part of a larger whole, a five-dimensional space.

"Light" was emerging as the warping of the geometry of higher-dimensional space. This was the theory that seemed to fulfill Riemann's old dream of explaining forces as the crumpling of a sheet of paper. In his article, Kaluza claimed that his theory, which synthesized the two most important theories up to that time, possessed "virtually unsurpassed formal unity." He furthermore insisted that the sheer simplicity and beauty of his theory could not "amount to the mere alluring play of a capricious accident."[12] What shook Einstein was the audacity and simplicity of the article. Like all great ideas, Kaluza's essential argument was elegant and compact.

The analogy with piecing together the parts of a jigsaw puzzle is a meaningful one. Recall that the basis of Riemann's and Einstein's work is the metric tensor—that is, a collection of ten numbers defined at each point in space. This was a natural generalization of Faraday's field concept. In Figure 2.2, we saw how these ten numbers can be arranged as in the pieces of a checker board with dimensions 4×4. We can denote these ten numbers as g_{11}, g_{12}, ... Furthermore, the field of Maxwell is a collection of four numbers defined at each point in space. These four numbers can be represented by the symbols A_1, A_2, A_3, A_4.

To understand Kaluza's trick, let us now begin with Riemann's theory in five dimensions. Then the metric tensor can be arranged in a 5×5 checkerboard. Now, by definition, we will rename the components of

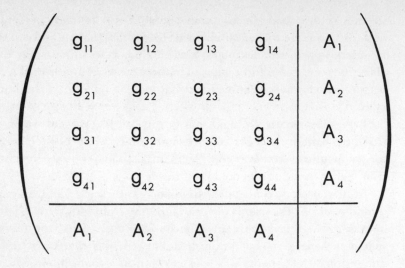

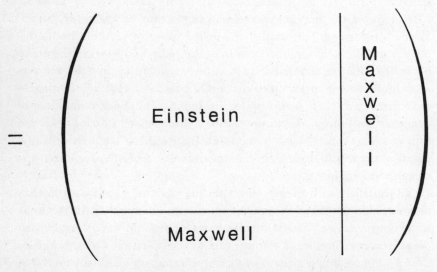

Figure 4.3. Kaluza's brilliant idea was to write down the Riemann metric in five dimensions. The fifth column and row are identified as the electromagnetic field of Maxwell, while the remaining 4 × 4 block is the old four-dimensional metric of Einstein. In one stroke, Kaluza unified the theory of gravity with light simply by adding another dimension.

Kaluza's field, so that some of them become Einstein's original field and some of them become Maxwell's field (Figure 4.3). This is the essence of Kaluza's trick, which caught Einstein totally by surprise. By simply adding Maxwell's field to Einstein's, Kaluza was able to reassemble both of them into a five-dimensional field.

Notice that there is "enough room" within the 15 components of Riemann's five-dimensional gravity to fit both the ten components of Einstein's field and the four components of Maxwell's field! Thus Kaluza's brilliant idea can be crudely summarized as

$$15 = 10 + 4 + 1$$

(the leftover component is a scalar particle, which is unimportant for our discussion). When carefully analyzing the full five-dimensional theory, we find that Maxwell's field is nicely included within the Riemann metric tensor, just as Kaluza claimed. This innocent-looking equation thus summarized one of the seminal ideas of the century.

In summary, the five-dimensional metric tensor included both Maxwell's field and Einstein's metric tensor. It seemed incredible to Einstein that such a simple idea could explain the two most fundamental forces of nature: gravity and light.

Was it just a parlor trick? Or numerology? Or black magic? Einstein was deeply shaken by Kaluza's letter and, in fact, refused to respond to the article. He mulled over the letter for 2 years, an unusually long time for someone to hold up publication of an important article. Finally, convinced that this article was potentially important, he submitted it for publication in the *Sitzungsberichte Preussische Akademie der Wissenschaften*. It bore the imposing title "On the Unity Problem of Physics."

In the history of physics, no one had found any use for the fourth spatial dimension. Ever since Riemann, it was known that the mathematics of higher dimensions was one of breathtaking beauty, but without physical application. For the first time, someone had found a use for the fourth spatial dimension: to unite the laws of physics! In some sense, Kaluza was proposing that the four dimensions of Einstein were "too small" to accommodate both the electromagnetic and gravitational forces.

We can also see historically that Kaluza's work was not totally unexpected. Most historians of science, when they mention Kaluza's work at all, say that the idea of a fifth dimension was a bolt out of the blue, totally unexpected and original. Given the continuity of physics research, these historians are startled to find a new avenue of science opening up without any historical precedent. But their amazement is probably due to their unfamiliarity with the nonscientific work of the mystics, literati, and avante garde. A closer look at the cultural and historical setting shows that Kaluza's work was not such an unexpected development. As we have seen, because of Hinton, Zollner, and others, the possible existence of higher dimensions was perhaps the single most popular quasi-scientific idea circulating within the arts. From this larger cultural point of view, it was only a matter of time before some physicist took seriously Hinton's widely known idea that light is a vibration of the fourth spatial dimension. In this sense, the work of Riemann pollinated the world of arts and letters via Hinton and Zollner, and then probably cross-pollinated back into the world of science through the work of Kaluza. (In support of this thesis, it was recently revealed by Freund that Kaluza was actually not the first one to propose a five-dimensional theory of gravity. Gunnar Nordstrom, a rival of Einstein, actually published the first five-dimensional field theory, but it was too primitive to include both Einstein's and Maxwell's theories. The fact that both Kaluza and Nordstrom independently tried to exploit the fifth dimension indicates that the concepts widely circulating within popular culture affected their thinking.[13])

The Fifth Dimension

Every physicist receives quite a jolt when confronting the fifth dimension for the first time. Peter Freund remembers clearly the precise moment when he first encountered the fifth and higher dimensions. It was an event that left a deep impression on his thinking.

It was 1953 in Romania, the country of Freund's birth. Joseph Stalin had just died, an important event that led to a considerable relaxation of tensions. Freund was a precocious college freshman that year, and he attended a talk by George Vranceanu. He vividly remembers hearing Vranceanu discuss the important question: Why should light and gravity be so disparate? Then the lecturer mentioned an old theory that could contain both the theory of light and Einstein's equations of gravity. The secret was to use Kaluza–Klein theory, which was formulated in five dimensions.

Freund was shocked. Here was a brilliant idea that took him completely by surprise. Although only a freshman, he had the audacity to pose the obvious question: How does this Kaluza–Klein theory explain the other forces? He asked, "Even if you achieve a unification of light and gravity, you will not achieve anything because there is still the nuclear force.'' He realized that the nuclear force was outside Kaluza–Klein theory. (In fact, the hydrogen bomb, which hung like a sword over everyone on the planet at the height of the Cold War, was based on unleashing the nuclear force, not electromagnetism or gravity.)

The lecturer had no answer. In his youthful enthusiasm, Freund blurted out, ''What about adding more dimensions?''

''But how many more dimensions?'' asked the lecturer.

Freund was caught off guard. He did not want to give a low number of dimensions, only to be scooped by someone else. So he proposed a number that no one could possibly top: an infinite number of dimensions![14] (Unfortunately for this precocious physicist, an infinite number of dimensions does not seem to be physically possible.)

Life on a Cylinder

After the initial shock of confronting the fifth dimension, most physicists invariably begin to ask questions. In fact, Kaluza's theory raised more questions than it answered. The obvious question to ask Kaluza was: Where is the fifth dimension? Since all earthly experiments showed conclusively that we live in a universe with three dimensions of space and one of time, the embarrassing question still remained.

Kaluza had a clever response. His solution was essentially the same as that proposed by Hinton years before, that the higher dimension, which was not observable by experiment, was different from the other dimensions. It had, in fact, collapsed down to a circle so small that even atoms could not fit inside it. Thus the fifth dimension was not a mathematical trick introduced to manipulate electromagnetism and gravity, but a physical dimension that provided the glue to unite these two fundamental forces into one force, but was just too small to measure.

Anyone walking in the direction of the fifth dimension would eventually find himself back where he started. This is because the fifth dimension is topologically identical to a circle, and the universe is topologically identical to a cylinder.

Freund explains it this way:

Think of some imaginary people living in Lineland, which consists of a single line. Throughout their history, they believed that their world was just a single line. Then, a scientist in Lineland proposed that their world was not just a one-dimensional line, but a two-dimensional world. When asked where this mysterious and unobservable second dimension was, he would reply that the second dimension was curled up into a small ball. Thus, the line people actually live on the surface of a long, but very thin, cylinder. The radius of the cylinder is too small to be measured; it is so small, in fact, that it appears that the world is just a line.[15]

If the radius of the cylinder were larger, the line people could move off their universe and move perpendicular to their line world. In other words, they could perform interdimensional travel. As they moved perpendicular to Lineland, they would encounter an infinite number of parallel line worlds that coexisted with their universe. As they moved farther into the second dimension, they would eventually return to their own line world.

Now think of Flatlanders living on a plane. Likewise, a scientist on Flatland may make the outrageous claim that traveling through the third dimension is possible. In principle, a Flatlander could rise off the surface of Flatland. As this Flatlander slowly floated upward in the third dimension, his "eyes" would see an incredible sequence of different parallel universes, each coexisting with his universe. Because his eyes would be able to see only parallel to the surface of Flatland, he would see different Flatland universes appearing before him. If the Flatlander drifted too far above the plane, eventually he would return to his original Flatland universe.

Now, imagine that our present three-dimensional world actually has another dimension that has curled up into a circle. For the sake of argument, assume that the fifth dimension is 10 feet long. By leaping into the fifth dimension, we simply disappear instantly from our present universe. Once we move in the fifth dimension, we find that, after moving 10 feet, we are back where we started from. But why did the fifth dimension curl up into a circle in the first place? In 1926, the mathematician Oskar Klein made several improvements on the theory, stating that perhaps the quantum theory could explain why the fifth dimension rolled up. On this basis, he calculated that the size of the fifth dimension should be 10^{-33} centimeters (the Planck length), which is much too small for any earthly experiment to detect its presence. (This is the same argument used today to justify the ten-dimensional theory.)

On the one hand, this meant that the theory was in agreement with experiment because the fifth dimension was too small to be measured. On the other hand, it also meant that the fifth dimension was so fantastically small that one could never build machines powerful enough to prove the theory was really correct. (The quantum physicist Wolfgang

Pauli, in his usual caustic way, would dismiss theories he didn't like by saying, "It isn't even wrong." In other words, they were so half-baked that one could not even determine if they were correct. Given the fact that Kaluza's theory could not be tested, one could also say that it wasn't even wrong.)

The Death of Kaluza–Klein Theory

As promising as Kaluza–Klein theory was for giving a purely geometric foundation to the forces of nature, by the 1930s the theory was dead. On the one hand, physicists weren't convinced that the fifth dimension really existed. Klein's conjecture that the fifth dimension was curled up into a tiny circle the size of the Planck length was untestable. The energy necessary to probe this tiny distance can be computed, and it is called the *Planck energy*, or 10^{19} billion electron volts. This fabulous energy is almost beyond comprehension. It is 100 billion billion times the energy locked in a proton, an energy beyond anything we will be able to produce within the next several centuries.

On the other hand, physicists left this area of research in droves because of the discovery of a new theory that was revolutionizing the world of science. The tidal wave unleashed by this theory of the subatomic world completely swamped research in Kaluza–Klein theory. The new theory was called quantum mechanics, and it sounded the death knell for Kaluza–Klein theory for the next 60 years. Worse, quantum mechanics challenged the smooth, geometric interpretation of forces, replacing it with discrete packets of energy.

Was the program initiated by Riemann and Einstein completely wrong?

PART II

Unification in
Ten Dimensions

5

Quantum Heresy

Anyone who is not shocked by the quantum theory does not understand it.

Niels Bohr

A Universe Made of Wood

IN 1925, a new theory burst into existence. With dizzying, almost meteoric speed, this theory overthrew long-cherished notions about matter that had been held since the time of the Greeks. Almost effortlessly, it vanquished scores of long-standing fundamental problems that had stumped physicists for centuries. What is matter made of? What holds it together? Why does it come in an infinite variety of forms, such as gases, metals, rocks, liquids, crystals, ceramics, glasses, lightning bolts, stars, and so on?

The new theory was christened *quantum mechanics,* and gave us the first comprehensive formulation with which to pry open the secrets of the atom. The subatomic world, once a forbidden realm for physicists, now began to spill its secrets into the open.

To understand the speed with which this revolution demolished its rivals, we note that in the early 1920s some scientists still held serious reservations about the existence of "atoms." What couldn't be seen or measured directly in the laboratory, they scoffed, didn't exist. But by 1925 and 1926, Erwin Schrödinger, Werner Heisenberg, and others had

111

developed an almost complete mathematical description of the hydrogen atom. With devastating precision, they could now explain nearly all the properties of the hydrogen atom from pure mathematics. By 1930, quantum physicists such as Paul A. M. Dirac were declaring that *all of chemistry* could be derived from first principles. They even made the brash claim that, given enough time on a calculating machine, they could predict all the chemical properties of matter found in the universe. To them, chemistry would no longer be a fundamental science. From now on, it would be "applied physics."

Not only did its dazzling rise include a definitive explanation of the bizarre properties of the atomic world; but quantum mechanics also eclipsed Einstein's work for many decades: One of the first casualties of the quantum revolution was Einstein's geometric theory of the universe. In the halls of the Institute for Advanced Study, young physicists began to whisper that Einstein was over the hill, that the quantum revolution had bypassed him completely. The younger generation rushed to read the latest papers written about quantum theory, not those about the theory of relativity. Even the director of the institute, J. Robert Oppenheimer, confided privately to his close friends that Einstein's work was hopelessly behind the times. Even Einstein began to think of himself as an "old relic."

Einstein's dream, we recall, was to create a universe made of "marble"—that is, pure geometry. Einstein was repelled by the relative ugliness of matter, with its confusing, anarchistic jumble of forms, which he called "wood." Einstein's goal was to banish this blemish from his theories forever, to turn wood into marble. His ultimate hope was to create a theory of the universe based entirely on marble. To his horror, Einstein realized that the quantum theory was a theory made *entirely of wood!* Ironically, it now appeared that he had made a monumental blunder, that the universe apparently preferred wood to marble.

In the analogy between wood and marble, we recall that Einstein wanted to convert the tree in the marble plaza to a marble statue, creating a park completely made of marble. The quantum physicists, however, approached the problem from the opposite perspective. Their dream was to take a sledge hammer and pulverize all the marble. After removing the shattered marble pieces, they would cover the park completely with wood.

Quantum theory, in fact, turned Einstein on his head. In almost every sense of the word, quantum theory is the opposite of Einstein's theory. Einstein's general relativity is a theory of the cosmos, a theory of stars and galaxies held together via the smooth fabric of space and time. Quantum theory, by contrast, is a theory of the microcosm, where sub-atomic particles are held together by particlelike forces dancing on the sterile stage of space–time, which is viewed as an empty arena, devoid of any content. Thus the two theories are hostile opposites. In fact, the tidal wave generated by the quantum revolution swamped all attempts at a geometric understanding of forces for over a half-century.

Throughout this book, we have developed the theme that the laws of physics appear simple and unified in higher dimensions. However, with the appearance of the quantum heresy after 1925, we see the first serious challenge to this theme. In fact, for the next 60 years, until the mid-1980s, the ideology of the quantum heretics would dominate the world of physics, almost burying the geometric ideas of Riemann and Einstein under an avalanche of undeniable successes and stunning experimental victories.

Fairly rapidly, quantum theory began to give us a comprehensive framework in which to describe the visible universe: The material universe consists of atoms and its constituents. There are about 100 different types of atoms, or elements, out of which we can build all the known forms of matter found on earth and even in outer space. Atoms, in turn, consist of electrons orbiting around nuclei, which in turn are composed of neutrons and protons. In essence, the key differences between Einstein's beautiful geometric theory and quantum theory can now be summarized as follows.

1. Forces are created by the exchange of discrete packets of energy, called *quanta*.

In contrast to Einstein's geometric picture of a "force," in quantum theory light was to be chopped up into tiny pieces. These packets of light were named *photons,* and they behave very much like point particles. When two electrons bump into each other, they repel each other not because of the curvature of space, but because they exchange a packet of energy, the photon.

The energy of these photons is measured in units of something called *Planck's constant* ($\hbar \sim 10^{-27}$ erg sec). The almost infinitesimal size of Planck's constant means that quantum theory gives tiny corrections to Newton's laws. These are called *quantum corrections*, and can be neglected when describing our familiar, macroscopic world. That is why we can, for the most part, forget about quantum theory when describing every-day phenomena. However, when dealing with the microscopic subatomic world, these quantum corrections begin to dominate any physical process, accounting for the bizarre, counterintuitive properties of subatomic particles.

2. Different forces are caused by the exchange of different quanta.

The weak force, for example, is caused by the exchange of a different type of quantum, called a W particle (W stands for "weak"). Similarly, the strong force holding the protons and neutrons together within the nucleus of the atom is caused by the exchange of subatomic particles called π mesons. Both W bosons and π mesons have been seen experimentally in the debris of atom smashers, thereby verifying the fundamental correctness of this approach. And finally, the subnuclear force holding the protons and neutrons and even the π mesons together are called gluons.

In this way, we have a new "unifying principle" for the laws of physics. We can unite the laws of electromagnetism, the weak force, and the strong force by postulating a variety of different quanta that mediate them. Three of the four forces (excluding gravity) are therefore united by quantum theory, giving us unification without geometry, which appears to contradict the theme of this book and everything we have considered so far.

3. We can never know simultaneously the velocity and position of a subatomic particle.

This is the Heisenberg Uncertainty Principle, which is by far the most controversial aspect of the theory, but one that has resisted every challenge in the laboratory for half a century. There is no known experimental deviation to this rule.

The Uncertainty Principle means that we can never be sure where an electron is or what its velocity is. The best we can do is to calculate the probability that the electron will appear at a certain place with a certain velocity. The situation is not as hopeless as one might suspect, because we can calculate with mathematical rigor the probability of finding that electron. Although the electron is a point particle, it is accompanied by a wave that obeys a well-defined equation, the Schrödinger wave equation. Roughly speaking, the larger the wave, the greater the probability of finding the electron at that point.

Thus quantum theory merges concepts of both particle and wave into a nice dialectic: The fundamental physical objects of nature are particles, but the probability of finding a particle at any given place in space and time is given by a probability wave. This wave, in turn, obeys a well-defined mathematical equation given by Schrödinger.

What is so crazy about the quantum theory is that it reduces everything to these baffling probabilities. We can predict with great precision *how many* electrons in a beam will scatter when moving through a screen with holes in it. However, we can never know precisely *which* electron will scatter in which direction. This is not a matter of having crude instruments; according to Heisenberg, it is a law of nature.

This formulation, of course, had unsettling philosophical implications. The Newtonian vision held that the universe was a gigantic clock, wound at the beginning of time and ticking ever since because it obeyed Newton's three laws of motion; this picture of the universe was now replaced by uncertainty and chance. Quantum theory demolished, once and for all, the Newtonian dream of mathematically predicting the motion of all the particles in the universe.

If quantum theory violates our common sense, it is only because nature does not seem to care much about our common sense. As alien and disturbing as these ideas may seem, they can be readily verified in the laboratory. This is illustrated by the celebrated double-slit experiment. Let us say we fire a beam of electrons at a screen with two small slits. Behind the screen, there is sensitive photographic paper. According to nineteenth-century classical physics, there should be two tiny spots burned into the photographic paper by the beam of electrons behind each hole. However, when the experiment is actually performed in the laboratory, we find an interference pattern (a series of bright and dark lines) on the photographic paper, which is commonly associated with wavelike, not particlelike, behavior (Figure 5.1). (The simplest way of creating an interference pattern is to take a quiet bath and then rhythmically splash waves on the water's surface. The spiderweblike pattern of waves criss-crossing the surface of the water is an interference pattern caused by the collision of many wave fronts.) The pattern on the photographic sheet corresponds to a wave that has penetrated both holes simultaneously and then interfered with itself behind the screen. Since the interference pattern is created by the collective motion of many individual electrons, and since the wave has gone through both holes simultaneously, naively we come to the absurd conclusion that electrons can somehow enter *both holes* simultaneously. But how can an electron be in two places at the same time? According to quantum theory, the electron is indeed a point particle that went through one or the other hole, but the wave function of the electron spread out over space, went

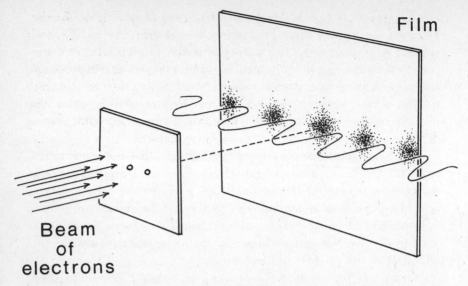

Figure 5.1. A beam of electrons is shot through two small holes and exposes some film. We expect to see two dots on the film. Instead, we find an undulating interference pattern. How can this be? According to quantum theory, the electron is indeed a pointlike particle and cannot go through both holes, but the Schrödinger wave associated with each electron can pass through both holes and interfere with itself.

through both holes, and then interacted with itself. As unsettling as this idea is, it has been verified repeatedly by experiment. As physicist Sir James Jeans once said, "It is probably as meaningless to discuss how much room an electron takes up as it is to discuss how much room a fear, an anxiety, or an uncertainty takes up."[1] (A bumper sticker I once saw in Germany summed this up succinctly. It read, "Heisenberg may have slept here.")

4. There is a finite probability that particles may "tunnel" through or make a quantum leap through impenetrable barriers.

This is one of more stunning predictions of quantum theory. On the atomic level, this prediction has had nothing less than spectacular success. "Tunneling," or quantum leaps through barriers, has survived every experimental challenge. In fact, a world without tunneling is now unimaginable.

One simple experiment that demonstrates the correctness of quantum tunneling starts by placing an electron in a box. Normally, the electron does not have enough energy to penetrate the walls of the box. If classical physics is correct, then the electron would never leave the box. However, according to quantum theory, the electron's probability wave will spread through the box and seep into the outside world. The seepage through the wall can be calculated precisely with the Schrödinger wave equation; that is, there is a small probability that the electron's position is somewhere *outside* the box. Another way of saying this is that there is a finite but small probability that the electron will tunnel its way through the barrier (the wall of the box) and emerge from the box. In the laboratory, when one measures the rate at which electrons tunnel through these barriers, the numbers agree precisely with the quantum theory.

This quantum tunneling is the secret behind the tunnel diode, which is a purely quantum-mechanical device. Normally, electricity might not have enough energy to penetrate past the tunnel diode. However, the wave function of these electrons can penetrate through barriers in the diode, so there is a non-negligible probability that electricity will emerge on the other side of the barrier by tunneling through it. When you listen to the beautiful sounds of stereo music, remember that you are listening to the rhythms of trillions of electrons obeying this and other bizarre laws of quantum mechanics.

But if quantum mechanics were incorrect, then all of electronics, including television sets, computers, radios, stereo, and so on, would cease to function. (In fact, if quantum theory were incorrect, the atoms in our bodies would collapse, and we would instantly disintegrate. According to Maxwell's equations, the electrons spinning in an atom should lose their energy within a microsecond and plunge into the nucleus. This sudden collapse is prevented by quantum theory. Thus the fact that we exist is living proof of the correctness of quantum mechanics.)

This also means that there is a finite, calculable probability that "impossible" events will occur. For example, I can calculate the probability that I will unexpectedly disappear and tunnel through the earth and reappear in Hawaii. (The time we would have to wait for such an event to occur, it should be pointed out, is longer than the lifetime of the universe. So we cannot use quantum mechanics to tunnel to vacation spots around the world.)

The Yang–Mills Field, Successor to Maxwell

Quantum physics, after an initial flush of success in the 1930s and 1940s unprecedented in the history of science, began to run out of steam by the 1960s. Powerful atom smashers built to break up the nucleus of the atom found hundreds of mysterious particles among the debris. Physicists, in fact, were deluged by mountains of experimental data spewing from these particle accelerators.

While Einstein guessed the entire framework of general relativity with only physical intuition, particle physicists were drowning in a mass of experimental data in the 1960s. As Enrico Fermi, one of the builders of the atomic bomb, confessed, "If I could remember the names of all these particles, I would have become a botanist."[2] As hundreds of "elementary" particles were discovered in the debris of smashed atoms, particle physicists would propose innumerable schemes to explain them, all without luck. So great were the number of incorrect schemes that it was sometimes said that the half-life of a theory of subatomic physics is only 2 years.

Looking back at all the blind alleys and false starts in particle physics during that period, one is reminded of the story of the scientist and the flea.

A scientist once trained a flea to jump whenever he rang a bell. Using a microscope, he then anesthetized one of the flea's legs and rang the bell again. The flea still jumped.

The scientist then anesthetized another leg and then rang the bell. The flea still jumped.

Eventually, the scientist anesthetized more and more legs, each time ringing the bell, and each time recording that the flea jumped.

Finally, the flea had only one leg left. When the scientist anesthetized the last leg and rang the bell, he found to his surprise that the flea no longer jumped.

Then the scientist solemnly declared his conclusion, based on irrefutable scientific data: Fleas hear through their legs!

Although high-energy physicists have often felt like the scientist in that story, over the decades a consistent quantum theory of matter has slowly emerged. In 1971, the key development that propelled a unified description of three of the quantum forces (excluding gravity) and changed the landscape of theoretical physics was made by a Dutch graduate student, Gerard 't Hooft, who was still in his twenties.

Based on the analogy with photons, the quanta of light, physicists believed that the weak and strong forces were caused by the exchange of a quantum of energy, called the Yang–Mills field. Discovered by C. N. Yang and his student R. L. Mills in 1954, the Yang–Mills field is a generalization of the Maxwell field introduced a century earlier to describe light, except that the Yang–Mills field has many more components and can have an electrical charge (the photon carries no electrical charge). For the weak interactions, the quantum corresponding to the Yang–Mills field is the *W* particle, which can have charge +1, 0, and −1. For the strong interactions, the quantum corresponding to the Yang–Mills field, the "glue" that holds the protons and neutrons together, was christened the gluon.

Although this general picture was compelling, the problem that bedeviled physicists in the 1950s and 1960s was that the Yang–Mills field is not "renormalizable"; that is, it does not yield finite, meaningful quantities when applied to simple interactions. This rendered quantum theory useless in describing the weak and strong interactions. Quantum physics had hit a brick wall.

This problem arose because physicists, when they calculate what happens when two particles bump into each other, use something called *perturbation theory*, which is a fancy way of saying they use clever approximations. For example, in Figure 5.2(a), we see what happens when an electron bumps into another weakly interacting particle, the elusive neutrino. As a first guess, this interaction can be described by a diagram (called a *Feynman diagram*) showing that a quantum of the weak interactions, the *W* particle, is exchanged between the electron and the neutrino. To a first approximation, this gives us a crude but reasonable fit to the experimental data.

But according to quantum theory, we must also add small quantum corrections to our first guess. To make our calculation rigorous, we must also add in the Feynman diagrams for *all* possible graphs, including ones that have "loops" in them, as in Figure 5.2(b). Ideally, these quantum corrections should be tiny. After all, as we mentioned earlier, quantum theory was meant to give tiny quantum corrections to Newtonian physics. But much to the horror of physicists, these quantum corrections, or "loop graphs," instead of being small, were infinite. No matter how physicists tinkered with their equations or tried to disguise these infinite quantities, these divergences were persistently found in any calculation of quantum corrections.

Furthermore, the Yang–Mills field had a formidable reputation of being devilishly hard to calculate with, compared with the simpler Maxwell field. There was a mythology surrounding the Yang–Mills field that held that it was simply too complicated for practical calculations. Perhaps it was fortunate that 't Hooft was only a graduate student and wasn't influenced by the prejudices of more "seasoned" physicists. Using techniques pioneered by his thesis adviser, Martinus Veltman, 't Hooft showed that whenever we have "symmetry breaking" (which we will

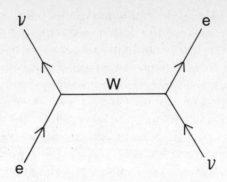

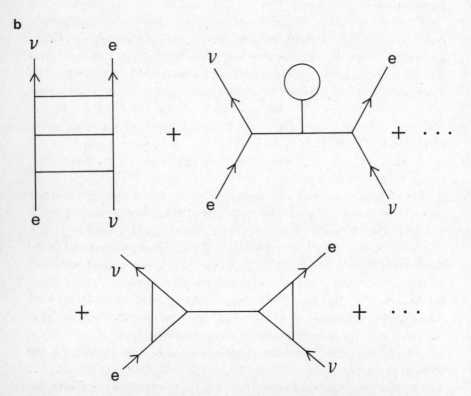

Figure 5.2. (a) In quantum theory, when subatomic particles bump into one another, they exchange packets of energy, or quanta. Electrons and neutrinos interact by exchanging a quantum of the weak force, called the W particle. (b) To calculate the complete interaction of electrons and neutrinos, we must add up an infinite series of graphs, called Feynman diagrams, where the quanta are exchanged in increasingly complicated geometric patterns. This process of adding up an infinite series of Feynman graphs is called perturbation theory.

explain later), the Yang–Mills field acquires a mass but remains a finite theory. He demonstrated that the infinities due to the loop graphs can all be canceled or shuffled around until they become harmless.

Almost 20 years after its being proposed by Yang and Mills, 't Hooft finally showed that the Yang–Mills field is a well-defined theory of particle interactions. News of 't Hooft's work spread like a flash fire. Nobel laureate Sheldon Glashow remembers that when he heard the news, he exclaimed, "Either this guy's a total idiot, or he's the biggest genius to hit physics in years!"[3] Developments came thick and fast. An earlier theory of the weak interactions, proposed in 1967 by Steven Weinberg and Abdus Salam, was rapidly shown to be the correct theory of the weak interactions. By the mid-1970s, the Yang–Mills field was applied to the strong interactions. In the 1970s came the stunning realization that the secret of all nuclear matter could be unlocked by the Yang–Mills field.

This was the missing piece in the puzzle. The secret of wood that bound matter together was the Yang–Mills field, not the geometry of Einstein. It appeared as though this, and not geometry, was the central lesson of physics.

The Standard Model

Today, the Yang–Mills field has made possible a comprehensive theory of all matter. In fact, we are so confident of this theory that we blandly call it the *Standard Model.*

The Standard Model can explain every piece of experimental data concerning subatomic particles, up to about 1 trillion electron volts in energy (the energy created by accelerating an electron by 1 trillion volts). This is about the limit of the atom smashers currently on line. Consequently, it is no exaggeration to state that the Standard Model is the most successful theory in the history of science.

According to the Standard Model, each of the forces binding the various particles is created by exchanging different kinds of quanta. Let us now discuss each force separately, and then assemble them into the Standard Model.

The Strong Force

The Standard Model states that the protons, neutrons, and other heavy particles are not fundamental particles at all, but consist of some even tinier particles, called *quarks.* These quarks, in turn, come in a wide variety: three "colors" and six "flavors." (These names have nothing to do with actual colors and flavors.) There are also the antimatter counterparts of the quarks, called antiquarks. (Antimatter is identical to matter in all respects, except that the charges are reversed and it annihilates on contact with ordinary matter.) This gives us a total of $3 \times 6 \times 2 = 36$ quarks.

The quarks, in turn, are held together by the exchange of small packets of energy, called *gluons*. Mathematically, these gluons are described by the Yang–Mills field, which "condenses" into a sticky, taffylike substance that "glues" the quarks permanently together. The gluon field is so powerful and binds the quarks so tightly together that the quarks can never be torn away from one another. This is called *quark confinement*, and may explain why free quarks have never been seen experimentally.

For example, the proton and neutron can be compared to three steel balls (quarks) held together by a Y-shaped string (gluon) in the shape of a bola. Other strongly interacting particles, such as the π meson, can be compared to a quark and an antiquark held together by a single string (Figure 5.3).

Obviously, by kicking this arrangement of steel balls, we can set this contraption vibrating. In the quantum world, only a discrete set of vibrations is allowed. Each vibration of this set of steel balls or quarks corresponds to a different type of subatomic particle. Thus this simple (but powerful) picture explains the fact that there are an infinite number of strongly interacting particles. This part of the Standard Model describing the strong force is called quantum chromodynamics (QCD)—that is, the quantum theory of the color force.

The Weak Force

In the Standard Model, the weak force governs the properties of "leptons," such as the electron, the muon, and the tau meson, and their neutrino partners. Like the other forces, the leptons interact by exchanging quanta, called W and Z bosons. These quanta are also described mathematically by the Yang–Mills field. Unlike the gluon force, the force generated by exchanging the W and Z bosons is too weak to bind the leptons into a resonance, so we do not see an infinite number of leptons emerging from our atom smashers.

The Electromagnetic Force

The Standard Model includes the theory of Maxwell interacting with the other particles. This part of the Standard Model governing the interac-

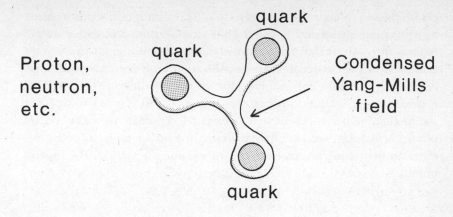

Proton,
neutron,
etc.

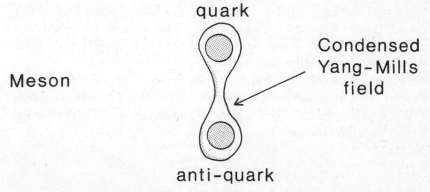

Meson

Figure 5.3. Strongly interacting particles are actually composites of even smaller particles, called quarks, which are bound together by a taffylike "glue," which is described by the Yang–Mills field. The proton and neutron are each made up of three quarks, while mesons are made up of a quark and an antiquark.

tion of electrons and light is called quantum electrodynamics (QED), which has been experimentally verified to be correct to within one part in 10 million, technically making it the most accurate theory in history.

In sum, the fruition of 50 years of research, and several hundred million dollars in government funds, has given us the following picture of subatomic matter: *All matter consists of quarks and leptons, which interact by exchanging different types of quanta, described by the Maxwell and Yang–Mills fields*. In one sentence, we have captured the essence of the past century of frustrating investigation into the subatomic realm. From this simple picture one can derive, from pure mathematics alone, all the myriad and baffling properties of matter. (Although it all seems so easy now, Nobel laureate Steven Weinberg, one of the creators of the Standard Model, once reflected on how tortuous the 50-year journey to discover the model had been. He wrote, "There's a long tradition of theoretical physics, which by no means affected everyone but certainly affected me, that said the strong interactions [were] too complicated for the human mind."[4])

Symmetry in Physics

The details of the Standard Model are actually rather boring and unimportant. The most interesting feature of the Standard Model is that it is based on symmetry. What has propelled this investigation into matter (wood) is that we can see the unmistakable sign of symmetry within each of these interactions. Quarks and leptons are not random, but occur in definite patterns in the Standard Model.

Symmetry, of course, is not strictly the province of physicists. Artists, writers, poets, and mathematicians have long admired the beauty that is to be found in symmetry. To the poet William Blake, symmetry possessed mystical, even fearful qualities, as expressed in the poem "Tyger! Tyger! burning bright":

> Tyger! Tyger! burning bright
> In the forests of the night
> What immortal hand or eye
> Could frame thy fearful symmetry?[5]

To mathematician Lewis Carroll, symmetry represented a familiar, almost playful concept. In the "The Hunting of the Snark," he captured the essence of symmetry when he wrote:

> You boil it in sawdust:
> You salt it in glue:
> You condense it with locusts in tape:
> Still keeping one principal object in view—
> To preserve its symmetrical shape.

In other words, symmetry is the preservation of the shape of an object even after we deform or rotate it. Several kinds of symmetries occur repeatedly in nature. The first is the symmetry of rotations and reflections. For example, a snowflake remains the same if we rotate it by 60 degrees. The symmetry of a kaleidoscope, a flower, or a starfish is of this type. We call these space–time symmetries, which are created by rotating the object through a dimension of space or time. The symmetry of special relativity is of this type, since it describes rotations between space and time.

Another type of symmetry is created by reshuffling a series of objects. Think of a shell game, where a huckster shuffles three shells with a pea hidden beneath one of them. What makes the game difficult is that there are many ways in which the shells can be arranged. In fact, there are six different ways in which three shells can be shuffled. Since the pea is hidden, these six configurations are identical to the observer. Mathematicians like to give names to these various symmetries. The name for the symmetries of a shell game is called S_3, which describes the number of ways that three identical objects may be interchanged.

If we replace the shells with quarks, then the equations of particle physics must remain the same if we shuffle the quarks among themselves. If we shuffle three colored quarks and the equations remain the same, then we say that the equations possess something called SU(3) symmetry. The 3 represents the fact that we have three types of colors, and the SU stands for a specific mathematical property of the symmetry.* We say that there are three quarks in a *multiplet*. The quarks in a multiplet can be shuffled among one another without changing the physics of the theory.

Similarly, the weak force governs the properties of two particles, the electron and the neutrino. The symmetry that interchanges these particles, yet leaves the equation the same, is called SU(2). This means that a multiplet of the weak force contains an electron and a neutrino, which can be rotated into each other. Finally, the electromagnetic force has U(1) symmetry, which rotates the components of the Maxwell field into itself.

Each of these symmetries is simple and elegant. However, the most controversial aspect of the Standard Model is that it "unifies" the three fundamental forces by simply splicing all three theories into one large symmetry, SU(3) × SU(2) × U(1), which is just the product of the

---

*SU stands for "special unitary" matrices—that is, matrices that have unit determinant and are unitary.

symmetries of the individual forces. (This can be compared to assembling a jigsaw puzzle. If we have three jigsaw pieces that don't quite fit, we can always take Scotch tape and splice them together by hand. This is how the Standard Model is formed, by taping three distinct multiplets together. This may not be aesthetically pleasing, but at least the three jigsaw puzzles now hang together by tape.)

Ideally, one might have expected that "the ultimate theory" would have all the particles inside just a single multiplet. Unfortunately, the Standard Model has three distinct multiplets, which cannot be rotated among one another.

Beyond the Standard Model

Promoters of the Standard Model can say truthfully that it fits all known experimental data. They can correctly point out that there are no experimental results that contradict the Standard Model. Nonetheless, nobody, not even its most fervent advocates, believes it is the final theory of matter. There are several deep reasons why it cannot be the final theory.

First, the Standard Model does not describe gravity, so it is necessarily incomplete. When attempts are made to splice Einstein's theory with the Standard Model, the resulting theory gives nonsensical answers. When we calculate, say, the probability of an electron being deflected by a gravitational field, the hybrid theory gives us an infinite probability, which makes no sense. Physicists say that quantum gravity is *nonrenormalizable,* meaning that it cannot yield sensible, finite numbers to describe simple physical processes.

Second, and perhaps most important, it is very ugly because it crudely splices three very different interactions together. Personally, I think that the Standard Model can be compared to crossing three entirely dissimilar types of animals, such as a mule, an elephant, and a whale. In fact, it is so ugly and contrived that even its creators are a bit embarrassed. They are the first to apologize for its shortcomings and admit that it cannot be the final theory.

This ugliness is obvious when we write down the details of the quarks and leptons. To describe how ugly the theory is, let us list the various particles and forces within the Standard Model:

1. Thirty-six quarks, coming in six "flavors" and three "colors," and their antimatter counterparts to describe the strong interactions

2. Eight Yang–Mills fields to describe the gluons, which bind the quarks
3. Four Yang–Mills fields to describe the weak and electromagnetic forces
4. Six types of leptons to describe the weak interactions (including the electron, muon, tau lepton, and their respective neutrino counterparts)
5. A large number of mysterious "Higgs" particles necessary to fudge the masses and the constants describing the particles
6. At least 19 arbitrary constants that describe the masses of the particles and the strengths of the various interactions. These 19 constants must be put in by hand; they are not determined by the theory in any way

Worse, this long list of particles can be broken down into three "families" of quarks and leptons, which are practically indistinguishable from one another. In fact, these three families of particles appear to be exact copies of one another, giving a threefold redundancy in the number of supposedly "elementary" particles (Figure 5.4). (It is disturbing to realize that we now have vastly more "elementary" particles than the total number of subatomic particles that were discovered by the 1940s. It makes one wonder how elementary these elementary particles really are.)

The ugliness of the Standard Model can be contrasted to the simplicity of Einstein's equations, in which everything was deduced from first principles. To understand the aesthetic contrast between the Standard Model and Einstein's theory of general relativity, we must realize that when physicists speak of "beauty" in their theories, they really mean that their theory possesses at least two essential features:

1. A unifying symmetry
2. The ability to explain vast amounts of experimental data with the most economical mathematical expressions

The Standard Model fails on both counts. Its symmetry, as we have seen, is actually formed by splicing three smaller symmetries, one for each of the three forces. Second, the theory is unwieldy and awkward in form. It is certainly not economical by any means. For example, Einstein's equations, written out in their entirety, are only about an inch long and wouldn't even fill up one line of this book. From this one line of equations, we can go beyond Newton's laws and derive the warping

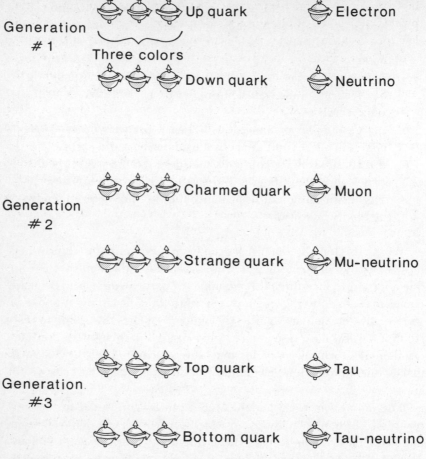

Figure 5.4. In the Standard Model, the first generation of particles consists of the "up" and "down" quark (in three colors, with their associated antiparticles) and the electron and neutrino. The embarrassing feature of the Standard Model is that there are three generation of such particles, each generation being nearly an exact copy of the previous generation. It's hard to believe that nature would be so redundant as to create, at a fundamental level, three identical copies of particles.

of space, the Big Bang, and other astronomically important phenomena. However, just to write down the Standard Model in its entirety would require two-thirds of this page and would look like a blizzard of complex symbols.

Nature, scientists like to believe, prefers economy in its creations and always seems to avoid unnecessary redundancies in creating physical, biological, and chemical structures. When nature creates panda bears, protein molecules, or black holes, it is sparing in its design. Or, as Nobel laureate C. N. Yang once said, "Nature seems to take advantage of the simple mathematical representations of the symmetry laws. When one pauses to consider the elegance and the beautiful perfection of the mathematical reasoning involved and contrast it with the complex and far-reaching physical consequences, a deep sense of respect for the power of the symmetry laws never fails to develop."[6] However, at the most fundamental level, we now find a gross violation of this rule. The existence of three identical families, each one with an odd assortment of particles, is one of the most disturbing features of the Standard Model, and raises a persistent problem for physicists: Should the Standard Model, the most spectacularly successful theory in the history of science, be thrown out just because it is ugly?

Is Beauty Necessary?

I once attended a concert in Boston, where people were visibly moved by the power and intensity of Beethoven's Ninth Symphony. After the concert, with the rich melodies still fresh in my mind, I happened to walk past the empty orchestra pit, where I noticed some people staring in wonder at the sheet music left by the musicians.

To the untrained eye, I thought, the musical score of even the most moving musical piece must appear to be a raw mass of unintelligible squiggles, bearing more resemblance to a chaotic jumble of scratches than a beautiful work of art. However, to the ear of a trained musician, this mass of bars, clefs, keys, sharps, flats, and notes comes alive and resonates in the mind. A musician can "hear" beautiful harmonies and rich resonances by simply looking at a musical score. A sheet of music, therefore, is more than just the sum of its lines.

Similarly, it would be a disservice to define a poem as "a short collection of words organized according to some principle." Not only is the definition sterile, but it is ultimately inaccurate because it fails to take into account the subtle interaction between the poem and the emotions that it evokes in the reader. Poems, because they crystallize and convey the essence of the feelings and images of the author, have a reality much greater than the words printed on a sheet of paper. A few short words of a haiku poem, for example, may transport the reader into a new realm of sensations and feelings.

Like music or art, mathematical equations can have a natural progression and logic that can evoke rare passions in a scientist. Although the lay public considers mathematical equations to be rather opaque, to a scientist an equation is very much like a movement in a larger symphony.

Simplicity. Elegance. These are the qualities that have inspired some of the greatest artists to create their masterpieces, and they are precisely the same qualities that motivate scientists to search for the laws of nature. Like a work of art or a haunting poem, equations have a beauty and rhythm all their own.

Physicist Richard Feynman expressed this when he said,

> You can recognize truth by its beauty and simplicity. When you get it right, it is obvious that it is right—at least if you have any experience—because usually what happens is that more comes out than goes in. . . . The inexperienced, the crackpots, and people like that, make guesses that are simple, but you can immediately see that they are wrong, so that does not count. Others, the inexperienced students, make guesses that are very complicated, and it sort of looks as if it is all right, but I know it is not true because the truth always turns out to be simpler than you thought.[7]

The French mathematician Henri Poincaré expressed it even more frankly when he wrote, "The scientist does not study Nature because it is useful; he studies it because he delights in it, and he delights in it because it is beautiful. If Nature were not beautiful, it would not be worth knowing, and if Nature were not worth knowing, life would not be worth living." In some sense, the equations of physics are like the poems of nature. They are short and are organized according to some principle, and the most beautiful of them convey the hidden symmetries of nature.

For example, Maxwell's equations, we recall, originally consisted of eight equations. These equations are not "beautiful." They do not possess much symmetry. In their original form, they are ugly, but they are the bread and butter of every physicist or engineer who has ever earned a living working with radar, radio, microwaves, lasers, or plasmas. These eight equations are what a tort is to a lawyer or a stethoscope is to a doctor. However, when rewritten using time as the fourth dimension, this rather awkward set of eight equations collapses into a single tensor equation. This is what a physicist calls "beauty," because both criteria are now satisfied. By increasing the number of dimensions, we reveal the true, four-dimensional symmetry of the theory and can now explain vast amounts of experimental data with a single equation.

As we have repeatedly seen, the addition of higher dimensions causes the laws of nature to simplify.

One of the greatest mysteries confronting science today is the explanation of the origin of these symmetries, especially in the subatomic world. When our powerful machines blow apart the nuclei of atoms by slamming them with energies beyond 1 trillion electron volts, we find that the fragments can be arranged according to these symmetries. Something rare and precious is unquestionably happening when we probe down to subatomic distances.

The purpose of science, however, is not to marvel at the elegance of natural laws, but to explain them. The fundamental problem facing subatomic physicists is that, historically, we had no idea of *why* these symmetries were emerging in our laboratories and our blackboards.

And here is precisely why the Standard Model fails. No matter how successful the theory is, physicists universally believe that it must be replaced by a higher theory. It fails both "tests" for beauty. It neither has a single symmetry group nor describes the subatomic world economically. But more important, the Standard Model does not explain where these symmetries originally came from. They are just spliced together by fiat, without any deeper understanding of their origin.

GUTs

Physicist Ernest Rutherford, who discovered the nucleus of the atom, once said, "All science is either physics or stamp collecting."[8]

By this, he meant that science consists of two parts. The first is physics, which is based on the foundation of physical laws or principles. The second is taxonomy ("bug collecting" or stamp collecting), which is giving erudite Greek names for objects you know almost nothing about based on superficial similarities. In this sense, the Standard Model is not real physics; it is more like stamp collecting, arranging the subatomic particles according to some superficial symmetries, but without the vaguest hint of where the symmetries come from.

Similarly, when Charles Darwin named his book *On the Origin of Species,* he was going far beyond taxonomy by giving the logical explanation for the diversity of animals in nature. What is needed in physics is a counterpart of this book, to be called *On the Origin of Symmetry,* which explains the reasons why certain symmetries are found in nature.

Because the Standard Model is so contrived, over the years attempts have been made to go beyond it, with mixed success. One prominent attempt was called the Grand Unified Theory (GUT), popular in the late 1970s, which tried to unite the symmetries of the strong, weak, and electromagnetic quanta by arranging them into a much larger symmetry group [for example, SU(5), O(10), or E(6)]. Instead of naively splicing the symmetry groups of the three forces, GUTs tried to start with a larger symmetry that required fewer arbitrary constants and fewer assumptions. GUTs vastly increased the number of particles beyond the Standard Model, but the advantage was that the ugly $SU(3) \times SU(2) \times U(1)$ was now replaced by a single symmetry group. The simplest of these GUTs, called SU(5), used 24 Yang–Mills fields, but at least all these Yang–Mills fields belonged to a single symmetry, not three separate ones.

The aesthetic advantage of the GUTs was that they put the strongly interacting quarks and the weakly interacting leptons on the same footing. In SU(5), for example, a multiplet of particles consisted of three colored quarks, an electron, and a neutrino. Under an SU(5) rotation, these five particles could rotate into one another without changing the physics.

At first, GUTs were met with intense skepticism, because the energy at which the three fundamental forces were unified was around 10^{15} billion electron volts, just a bit smaller than the Planck energy. This was far beyond the energy of any atom smasher on the earth, and that was discouraging. However, physicists gradually warmed up to the idea of GUTs when it was realized that they made a clear, testable prediction: the decay of the proton.

We recall that in the Standard Model, a symmetry like SU(3) rotates three quarks into one another; that is, a multiplet consists of three quarks. This means that each of the quarks can turn into one of the other quarks under certain conditions (such as the exchange of a Yang–Mills particle). However, quarks cannot turn into electrons. The multiplets do not mix. But in SU(5) GUT, there are five particles within a multiplet that can rotate into one another: three quarks, the electron, and the neutrino. This means that one can, under certain circumstances, turn a proton (made of quarks) into an electron or a neutrino. In other words, GUTs say that the proton, which was long held to be a stable particle with an infinite lifetime, is actually unstable. In principle, it also means that all atoms in the universe will eventually disintegrate into radiation. If correct, it means that the chemical elements, which are taught in elementary chemistry classes to be stable, are actually all unstable.

This doesn't mean that we should expect the atoms in our body to disintegrate into a burst of radiation anytime soon. The time for the proton to decay into leptons was calculated to be on the order of 10^{31} years, far beyond the lifetime of the universe (15 to 20 billion years). Although this time scale was astronomically long, this didn't faze the experimentalists. Since an ordinary tank of water contains an astronomical amount of protons, there is a measurable probability that *some* proton within the tank will decay, even if the protons *on the average* decay on a cosmological time scale.

The Search for Proton Decay

Within a few years, this abstract theoretical calculation was put to the test: Several expensive, multimillion-dollar experiments were conducted by several groups of physicists around the world. The construction of detectors sensitive enough to detect proton decay involved highly expensive and sophisticated techniques. First, experimentalists needed to construct enormous vats in which to detect proton decay. Then they had to fill the vats with a hydrogen-rich fluid (such as water or cleaning fluid) that had been filtered with special techniques in order to eliminate all impurities and contaminants. Most important, they then had to bury these gigantic tanks deep in the earth to eliminate any contamination from highly penetrating cosmic rays. And finally, they had to construct thousands of highly sensitive detectors to record the faint tracks of subatomic particles emitted from proton decay.

Remarkably, by the late 1980s six gigantic detectors were in operation around the world, such as the Kamioka detector in Japan and the IMB (Irvine, Michigan, Brookhaven) detector near Cleveland, Ohio. They contained vast amounts of pure fluid (such as water) ranging in weight from 60 to 3,300 tons. (The IMB detector, for example, is the world's largest and is contained in a huge 20-meter cube hollowed out of a salt mine underneath Lake Erie. Any proton that spontaneously decayed in the purified water would produce a microscopic burst of light, which in turn would be picked up by some of the 2,048 photoelectric tubes.)

To understand how these monstrous detectors can measure the proton lifetime, by analogy think of the American population. We know that the average American can expect to live on the order of 70 years. However, we don't have to wait 70 years to find fatalities. Because there are so many Americans, in fact more than 250 million, we expect to find some American dying every few minutes. Likewise, the simplest SU(5) GUT predicted that the half-life of the proton should be about 10^{29} years; that is, after 10^{29} years, *half* of the protons in the universe will have decayed.* (By contrast, this is about 10 billion billion times longer than the life of the universe itself.) Although this seems like an enormous lifetime, these detectors should have been able to see these rare, fleeting events simply because there were so many protons in the detector. In fact, each ton of water contains over 10^{29} protons. With that many protons, a handful of protons were expected to decay every year.

However, no matter how long the experimentalists waited, they saw no clear-cut evidence of any proton decays. At present, it seems that protons must have a lifetime larger than 10^{32} years, which rules out the simpler GUTs, but still leaves open the possibility of more complicated GUTs.

Initially, a certain amount of excitement over the GUTs spilled over into the media. The quest for a unified theory of matter and the search for the decay of the proton caught the attention of science producers and writers. Public television's "Nova" devoted several shows to it, and popular books and numerous articles in science magazines were written about it. Nevertheless, the fanfare died out by the late 1980s. No matter how long physicists waited for the proton to decay, the proton simply didn't cooperate. After tens of millions of dollars were spent by various nations looking for this event, it has not yet been found. Public interest in the GUTs began to fizzle.

The proton may still decay, and GUTs may still prove to be correct, but physicists are now much more cautious about touting the GUTs as the "final theory," for several reasons. As with the Standard Model, GUTs make no mention of gravity. If we naively combine GUTs with gravity, the theory produces numbers that are infinite and hence make no sense. Like the Standard Model, GUTs are nonrenormalizable. Moreover, the theory is defined at tremendous energies, where we certainly expected gravitational effects to appear. Thus the fact that gravity is missing in the GUT theory is a serious drawback. Furthermore, it is also plagued by the mysterious presence of three identical carbon copies or families of particles. And finally, the theory could not predict the fundamental constants, such as the quark masses. GUTs lacked a larger physical principle that would fix the quark masses and the other constants from first principles. Ultimately, it appeared that GUTs were also stamp collecting.

The fundamental problem was that the Yang–Mills field was not suf-

---

Half-life is the amount of time it takes for half of a substance to disintegrate. After two half-lives, only one-quarter of the substance remains.

ficient to provide the "glue" to unite all four interactions. The world of wood, as described by the Yang–Mills field, was not powerful enough to explain the world of marble.

After half a century of dormancy, the time had come for "Einstein's revenge."

6

Einstein's Revenge

Supersymmetry is the ultimate proposal for a complete unification of all particles.

Abdus Salam

The Resurrection of Kaluza–Klein

IT'S been called "the greatest scientific problem of all time." The press has dubbed it the "Holy Grail" of physics, the quest to unite the quantum theory with gravity, thereby creating a Theory of Everything. This is the problem that has frustrated the finest minds of the twentieth century. Without question, the person who solves this problem will win the Nobel Prize.

By the 1980s, physics was reaching an impasse. Gravity alone stubbornly stood apart and aloof from the other three forces. Ironically, although the classical theory of gravity was the first to be understood through the work of Newton, the quantum theory of gravity was the last interaction to be understood by physicists.

All the giants of physics have had their crack at this problem, and all have failed. Einstein devoted the last 30 years of his life to his unified field theory. Even the great Werner Heisenberg, one of the founders of quantum theory, spent the last years of his life chasing after his version of a unified theory of fields, even publishing a book on the subject. In 1958, Heisenberg even broadcast on radio that he and his colleague

136

Wolfgang Pauli had finally succeeded in finding the unified field theory, and that only the technical details were missing. (When the press got wind of this stunning declaration, Pauli was furious that Heisenberg had prematurely made that announcement. Pauli send a letter to his collaborator, consisting of a blank sheet of paper with the caption, "This is to show the world that I can paint like Titian. Only technical details are missing."[1])

Later that year, when Wolfgang Pauli finally gave a lecture on the Heisenberg–Pauli unified field theory, many eager physicists were in the audience, anxious to hear the missing details. When he was finished, however, the talk received a mixed response. Niels Bohr finally stood up and said, "We are all agreed that your theory is crazy. The question which divides us is whether it is crazy enough."[2] In fact, so many attempts have been made at the "final synthesis" that it has created a backlash of skepticism. Nobel laureate Julian Schwinger has said, "It's nothing more than another symptom of the urge that afflicts every generation of physicist—the itch to have all the fundamental questions answered in their own lifetimes."[3]

However, by the 1980s, the "quantum theory of wood," after a half-century of almost uninterrupted success, was beginning to run out of steam. I can vividly remember the sense of frustration among jaded young physicists during this period. Everyone sensed that the Standard Model was being killed by its own success. It was so successful that every international physics conference seemed like just another rubber stamp of approval. All the talks concerned finding yet another boring experimental success for the Standard Model. At one physics conference, I glanced back at the audience and found that half of them were slowly dozing off to sleep; the speaker was droning on with chart after chart showing how the latest data could be fit according to the Standard Model.

I felt like the physicists at the turn of the century. They, too, seemed to be facing a dead end. They spent decades tediously filling up tables of figures for the spectral lines of various gases, or calculating the solutions to Maxwell's equations for increasingly complicated metal surfaces. Since the Standard Model had 19 free parameters that could be arbitrarily "tuned" to any value, like the dials on a radio, I imagined that physicists would spend decades finding the precise values of all 19 parameters.

The time had come for a revolution. What beckoned the next generation of physicists was the world of marble.

Of course, several profound problems stood in the way of a genuine quantum theory of gravity. One problem with constructing a theory of gravity is that the force is so maddeningly weak. For example, it takes the entire mass of the earth to keep pieces of paper on my desk. However, by brushing a comb through my hair, I can pick up these pieces of paper, overwhelming the force of the planet earth. The electrons in my comb are more powerful than the gravitational pull of the entire planet. Similarly, if I were to try to construct an "atom" with electrons attracted to the nucleus by the gravitational force, and not the electrical force, the atom would be the size of the universe.

Classically, we see that the gravitational force is negligible compared with the electromagnetic force, and hence is extraordinarily difficult to measure. But if we attempt to write down a quantum theory of gravity, then the tables are turned. The quantum corrections due to gravity are on the order of the Planck energy, or 10^{19} billion electron volts, far beyond anything achievable on the planet earth in this century. This perplexing situation deepens when we try to construct a complete theory of quantum gravity. We recall that when quantum physicists try to quantize a force, they break it up into tiny packets of energy, called quanta. If you blindly try to quantize the theory of gravity, you postulate that it functions by the exchange of tiny packets of gravity, called *gravitons*. The rapid exchange of gravitons between matter is what binds them together gravitationally. In this picture, what holds us to the floor, and keeps us from flying into outer space at a thousand miles per hour, is the invisible exchange of trillions of tiny graviton particles. But whenever physicists tried to perform simple calculations to calculate quantum corrections to Newton's and Einstein's laws of gravity, they found that the result is infinite, which is useless.

For example, let us examine what happens when two electrically neutral particles bump into each other. To calculate the Feynman diagrams for this theory, we have to make an approximation, so we assume that the curvature of space–time is small, and hence the Riemann metric tensor is close to 1. For a first guess, we assume that space–time is close to being flat, not curved, so we divide the components of the metric tensor as $g_{11} = 1 + h_{11}$, where 1 represents flat space in our equations and h_{11} is the graviton field. (Einstein, of course, was horrified that quantum physicists would mutilate his equations in this way by breaking up the metric tensor. This is like taking a beautiful piece of marble and hitting it with a sledge hammer in order to break it.) After this mutilation is performed, we arrive at a conventional-looking quantum theory. In Figure 6.1(a), we see that the two neutral particles exchange a quantum of gravity, labeled by the field h.

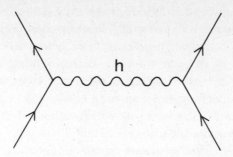

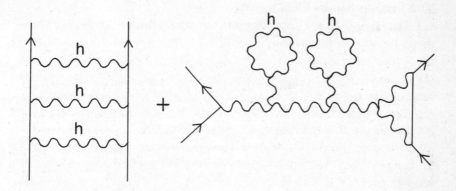

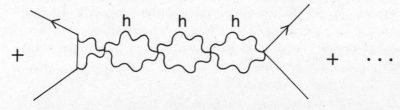

Figure 6.1. (a) In quantum theory, a quantum of the gravitational force, labeled h, is called the graviton, which is formed by breaking up Riemann's metric. In this theory, objects interact by exchanging this packet of gravity. In this way, we completely lose the beautiful geometric picture of Einstein. (b) Unfortunately, all the diagrams with loops in them are infinite, which has prevented a unification of gravity with the quantum theory for the past half-century. A quantum theory of gravity that unites it with the other forces is the Holy Grail of physics.

The problem arises when we sum over all loop diagrams: We find that they diverge, as in Figure 6.1(b). For the Yang–Mills field, we could use clever sleight-of-hand tricks to shuffle around these infinite quantities until they either cancel or are absorbed into quantities that can't be measured. However, it can be shown that the usual renormalization prescriptions fail completely when we apply them to a quantum theory of gravity. In fact, the efforts of physicists over half a century to eliminate or absorb these infinities has been in vain. In other words, the brute-force attempt to smash marble into pieces failed miserably.

Then, in the early 1980s, a curious phenomenon occurred. Kaluza–Klein theory, we recall, had been a dormant theory for 60 years. But physicists were so frustrated in their attempts to unify gravity with the other quantum forces that they began to overcome their prejudice about unseen dimensions and hyperspace. They were ready for an alternative, and that was Kaluza–Klein theory.

The late physicist Heinz Pagels summarized this excitement over the re-emergence of Kaluza–Klein theory:

> After the 1930s, the Kaluza–Klein idea fell out of favor, and for many years it lay dormant. But recently, as physicists searched out every possible avenue for the unification of gravity with other forces, it has again sprung to prominence. Today, in contrast with the 1920s, physicists are challenged to do more than unify gravity with just electromagnetism—they want to unify gravity with the weak and strong interactions as well. This requires even more dimensions, beyond the fifth.[4]

Even Nobel laureate Steven Weinberg was swept up by the enthusiasm generated by Kaluza–Klein theory. However, there were still physicists skeptical of the Kaluza–Klein renaissance. Harvard's Howard Georgi, reminding Weinberg how difficult it is to measure experimentally these compactified dimensions that have curled up, composed the following poem:

> Steve Weinberg, returning from Texas
> brings dimensions galore to perplex us
>   But the extra ones all
>   are rolled up in a ball
> so tiny it never affects us.[5]

Although Kaluza–Klein theory was still nonrenormalizable, what sparked the intense interest in the theory was that it gave the hope of a theory made of marble. Turning the ugly, confused jumble of wood into the pure, elegant marble of geometry was, of course, Einstein's dream. But in the 1930s and 1940s, almost nothing was known about the nature of wood. However, by the 1970s, the Standard Model had finally unlocked the secret of wood: that matter consists of quarks and leptons held together by the Yang–Mills field, obeying the symmetry $SU(3) \times SU(2) \times U(1)$. The problem was how to derive these particles and mysterious symmetries from marble.

At first, that seemed impossible. After all, these symmetries are the result of interchanging point particles among one another. If N quarks within a multiplet are shuffled among one another, then the symmetry is $SU(N)$. These symmetries seemed to be exclusively the symmetries of wood, not marble. What did $SU(N)$ have to do with geometry?

Turning Wood into Marble

The first small clue came in the 1960s, when physicists found, much to their delight, that there is an alternative way in which to introduce symmetries into physics. When physicists extended the old five-dimensional theory of Kaluza–Klein to N dimensions, they realized that there is the freedom to impose a symmetry on hyperspace. When the fifth dimension was curled up, they saw that the Maxwell field popped out of Riemann's metric. But when N dimensions were curled up, physicists found the celebrated Yang–Mills field, the key to the Standard Model, popping out of their equations!

To see how symmetries emerge from space, consider an ordinary beach ball. It has a symmetry: We can rotate it around its center, and the beach ball retains its shape. The symmetry of a beach ball, or a sphere, is called $O(3)$, or rotations in three dimension. Similarly, in higher dimensions, a hypersphere can also be rotated around its center and maintain its shape. The hypersphere has a symmetry called $O(N)$.

Now consider vibrating the beach ball. Ripples form on the surface of the ball. If we carefully vibrate the beach ball in a certain way, we can induce regular vibrations on it that are called *resonances*. These resonances, unlike ordinary ripples, can vibrate at only certain frequencies. In fact, if we vibrate the beach ball fast enough, we can create musical tones of a definite frequency. These vibrations, in turn, can be cataloged by the symmetry $O(3)$.

The fact that a membrane, like a beach ball, can induce resonance frequencies is a common phenomenon. The vocal chords in our throat, for example, are stretched membranes that vibrate at definite frequencies, or resonances, and can thereby produce musical tones. Another example is our hearing. Sound waves of all types impinge on our eardrums, which then resonate at definite frequencies. These vibrations are then turned into electrical signals that are sent into our brain, which interprets them as sounds. This is also the principle behind the telephone. The metallic diaphragm contained in any telephone is set into motion by electrical signals in the telephone wire. This creates mechanical vibrations or resonances in the diaphragm, which in turn create the sound waves we hear on the phone. This is also the principle behind stereo speakers as well as orchestral drums.

For a hypersphere, the effect is the same. Like a membrane, it can resonate at various frequencies, which in turn can be determined by its symmetry $O(N)$. Alternatively, mathematicians have dreamed up more sophisticated surfaces in higher dimensions that are described by complex numbers. (Complex numbers use the square root of -1, $\sqrt{-1}$.) Then it is straightforward to show that the symmetry corresponding to a complex "hypersphere" is $SU(N)$.

The key point is now this: If the wave function of a particle vibrates along this surface, it will inherit this $SU(N)$ symmetry. Thus the mysterious $SU(N)$ symmetries arising in subatomic physics can now be seen as *by-products of vibrating hyperspace!* In other words, we now have an explanation for the origin of the mysterious symmetries of wood: They are really the hidden symmetries coming from marble.

If we now take a Kaluza–Klein theory defined in $4 + N$ dimensions and then curl up N dimensions, we will find that the equations split into two pieces. The first piece is Einstein's usual equations, which we retrieve as expected. But the second piece will not be the theory of Maxwell. We find that the remainder is precisely the Yang–Mills theory, which forms the basis of all subatomic physics! This is the key to turning the symmetries of wood into the symmetries of marble.

At first, it seems almost mystical that the symmetries of wood, which were discovered painfully by trial and error—that is, by painstakingly examining the debris from atom smashers—emerge almost automatically from higher dimensions. It is miraculous that the symmetries found by shuffling quarks and leptons among themselves should arise from hyperspace. An analogy may help us understand this. Matter may be likened to clay, which is formless and lumpy. Clay lacks any of the beautiful symmetries that are inherent in geometric figures. However, clay may be pressed into a mold, which can have symmetries. For example, the mold may preserve its shape if it is rotated by a certain angle. Then the clay will also inherit the symmetry of the mold. Clay, like matter, inherits its symmetry because the mold, like space–time, has a symmetry.

If correct, then this means that the strange symmetries we see among the quarks and leptons, which were discovered largely by accident over several decades, can now be seen as by-products of vibrations in hyperspace. For example, if the unseen dimensions have the symmetry SU(5), then we can write SU(5) GUT as a Kaluza–Klein theory.

This can also be seen from Riemann's metric tensor. We recall that it resembles Faraday's field except that it has many more components. It can be arranged like the squares of a checkerboard. By separating out the fifth column and row of the checkerboard, we can split off Maxwell's field from Einstein's field. Now perform the same trick with Kaluza–Klein theory in $(4 + N)$-dimensional space. If you split off the N columns and rows from the first four columns and rows, then you obtain a metric tensor that describes both Einstein's theory and Yang–Mills theory. In Figure 6.2, we have carved up the metric tensor of a $(4 + N)$-dimensional

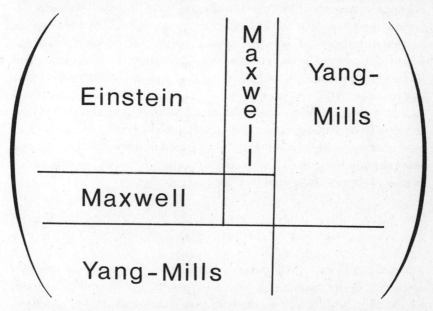

Figure 6.2. If we go to the N*th dimension, then the metric tensor is a series of* N^2 *numbers that can be arranged in an* N × N *block. By slicing off the fifth and higher columns and rows, we can extract the Maxwell electromagnetic field and the Yang–Mills field. Thus, in one stroke, the hyperspace theory allows us to unify the Einstein field (describing gravity), the Maxwell field (describing the electromagnetic force), and the Yang–Mills field (describing the weak and strong force). The fundamental forces fit together exactly like a jigsaw puzzle.*

Kaluza–Klein theory, splitting off Einstein's field from the Yang–Mills field.

Apparently, one of the first physicists to perform this reduction was University of Texas physicist Bryce DeWitt, who has spent many years studying quantum gravity. Once this trick of splitting up the metric tensor was discovered, the calculation for extracting the Yang–Mills field is straightforward. DeWitt felt that extracting the Yang–Mills field from N-dimensional gravity theory was such a simple mathematical exercise that he assigned it as a homework problem at the Les Houches Physics Summer School in France in 1963. [Recently, it was revealed by Peter Freund that Oskar Klein had independently discovered the Yang–Mills field in 1938, preceding the work of Yang, Mills, and others by several decades. In a conference held in Warsaw titled "New Physical Theories," Klein announced that he was able to generalize the work of Maxwell to include a higher symmetry, $O(3)$. Unfortunately, because of the chaos unleashed by World War II and because Kaluza–Klein theory was buried by the excitement generated by quantum theory, this important work was forgotten. It is ironic that Kaluza–Klein theory was killed by the emergence of quantum theory, which is now based on the Yang–Mills field, which was first discovered by analyzing Kaluza–Klein theory. In the excitement to develop quantum theory, physicists had ignored a central discovery coming from Kaluza–Klein theory.]

Extracting the Yang–Mills field out of Kaluza–Klein theory was only the first step. Although the symmetries of wood could now be seen as arising from the hidden symmetries of unseen dimensions, the next step was to create wood itself (made of quarks and leptons) entirely out of marble. This next step would be called supergravity.

Supergravity

Turning wood into marble still faced formidable problems because, according to the Standard Model, all particles are "spinning." Wood, for example, we now know is made of quarks and leptons. They, in turn, have ½ unit of quantum spin (measured in units of Planck's constant $\hbar$. Particles with half-integral spin (½, ³⁄₂, ⁵⁄₂, and so on) are called *fermions* (named after Enrico Fermi, who first investigated their strange properties). However, forces are described by quanta with integral spin. For example, the photon, the quantum of light, has one unit of spin. So does the Yang–Mills field. The graviton, the hypothetical packet of gravity, has two units of spin. They are called *bosons* (after the Indian physicist Satyendra Bose).

Traditionally, quantum theory kept fermions and bosons strictly apart. Indeed, any attempt to turn wood into marble would inevitably come to grips with the fact that fermions and bosons are worlds apart in their properties. For example, SU(N) may shuffle quarks among one another, but fermions and bosons were never supposed to mix. It came as a shock, therefore, when a new symmetry, called *supersymmetry,* was discovered, that did exactly that. Equations that are supersymmetric allow the interchange of a fermion with a boson and still keep the equations intact. In other words, one multiplet of supersymmetry consists of equal numbers of bosons and fermions. By shuffling the bosons and fermions within the same multiplet, the supersymmetric equations remain the same.

This gives us the tantalizing possibility of putting *all* the particles in the universe into one multiplet! As Nobel laureate Abdus Salam has emphasized, "Supersymmetry is the ultimate proposal for a complete unification of all particles."

Supersymmetry is based on a new kind of number system that would drive any schoolteacher insane. Most of the operations of multiplication and division that we take for granted fail for supersymmetry. For example, if a and b are two "super numbers," then $a \times b = -b \times a$. This, of course, is strictly impossible for ordinary numbers. Normally, any schoolteacher would throw these super numbers out the window, because you can show that $a \times a = -a \times a$, or, in other words, $a \times a = 0$. If these were ordinary numbers, then this means that $a = 0$, and the number system collapses. However, with super numbers, the system does not collapse; we have the rather astonishing statement that $a \times a = 0$ even when $a \neq 0$. Although these super numbers violate almost everything we have learned about numbers since childhood, they can be shown to yield a self-consistent and highly nontrivial system. Remarkably, an entirely new system of super calculus can be based on them.

Soon, three physicists (Daniel Freedman, Sergio Ferrara, and Peter van Nieuwenhuizen, at the State University of New York at Stony Brook) wrote down the theory of supergravity in 1976. Supergravity was the first realistic attempt to construct a world made entirely of marble. In a supersymmetric theory, all particles have super partners, called *sparticles.* The supergravity theory of the Stony Brook group contains just two fields: the spin-two graviton field (which is a boson) and its spin-3/2 partner, called the *gravitino* (which means "little gravity"). Since this is not enough particles to include the Standard Model, attempts were made to couple the theory to more complicated particles.

The simplest way to include matter is to write down the supergravity theory in 11-dimensional space. In order to write down the super

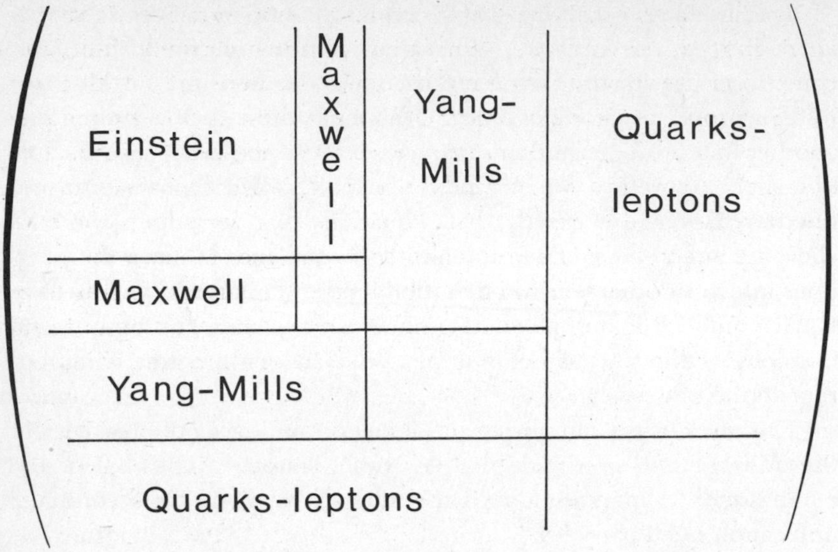

Figure 6.3. Supergravity almost fulfills Einstein's dream of giving a purely geometric derivation of all the forces and particles in the universe. To see this, notice that if we add supersymmetry to the Riemann metric tensor, the metric doubles in size, giving us the super Riemann metric. The new components of the super Riemann tensor correspond to quarks and leptons. By slicing the super Riemann tensor into its components, we find that it includes almost all the fundamental particles and forces in nature: Einstein's theory of gravity, the Yang–Mills and Maxwell fields, and the quarks and leptons. But the fact that certain particles are missing in this picture forces us to go a more powerful formalism: superstring theory.

Kaluza–Klein theory in 11 dimensions, one has to increase the components within the Riemann tensor vastly, which now becomes the super Riemann tensor. To visualize how supergravity converts wood into marble, let us write down the metric tensor and show how supergravity manages to fit the Einstein field, the Yang–Mills field, and the matter fields into one supergravity field (Figure 6.3). The essential feature of this diagram is that matter, along with the Yang–Mills and Einstein equations, is now included in the same 11-dimensional supergravity field. Supersymmetry is the symmetry that reshuffles the wood into marble and vice versa within the supergravity field. Thus they are all manifestations of the same force, the superforce. Wood no longer exists as a single, isolated entity. It is now merged with marble, to form supermarble (Figure 6.4)!

Physicist Peter van Nieuwenhuizen, one of supergravity's creators, was deeply impressed by the implication of this superunification. He wrote that supergravity "may unify grand unified theories . . . with gravity, leading to a model with almost no free parameters. It is the unique theory with a local gauge symmetry between fermions and bosons. It is the most beautiful gauge theory known, so beautiful, in fact, that Nature should be aware of it!"[6]

I fondly remember attending and giving lectures at many of these supergravity conferences. There was an intense, exhilarating feeling that we were on the verge of something important. At one meeting in Moscow, I remember well, a series of lively toasts were made to the continued success of the supergravity theory. It seemed that we were finally on the verge of carrying out Einstein's dream of a universe of marble after 60 years of neglect. Some of us jokingly called it "Einstein's revenge."

On April 29, 1980, when cosmologist Stephen Hawking assumed the Lucasian Professorship (previously held by some of the immortals of physics, including Isaac Newton and P. A. M. Dirac), he gave a lecture with the auspicious title "Is the End in Sight for Theoretical Physics?"

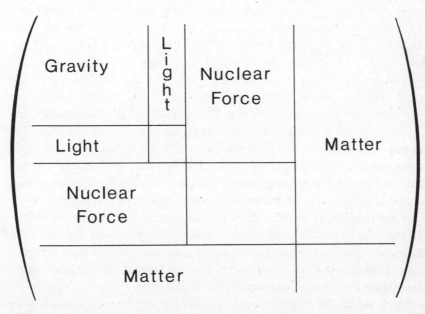

Figure 6.4. In supergravity, we almost get a unification of all the known forces (marble) with matter (wood). Like a jigsaw puzzle, they fit inside Riemann's metric tensor. This almost fulfills Einstein's dream.

A student read for him: "[W]e have made a lot of progress in recent years and, as I shall describe, there are some grounds for cautious optimism that we may see a complete theory within the lifetime of some of those present here."

Supergravity's fame gradually spread into the general public and even began to have a following among religious groups. For example, the concept of "unification" is a central belief within the transcendental meditation movement. Its followers therefore published a large poster containing the complete equations describing 11-dimensional supergravity. Each term in the equation, they claimed, represented something special, such as "harmony," "love," "brotherhood," and so on. (This poster hangs on the wall of the theoretical institute at Stony Brook. This is the first time that I am aware of that an abstract equation from theoretical physics has inspired a following among a religious group!)

Super Metric Tensors

Peter van Nieuwenhuizen cuts a rather dashing figure in physics circles. Tall, tanned, athletic looking, and well dressed, he looks more like an actor promoting suntan lotion on television than one of the original creators of supergravity. He is a Dutch physicist who is now a professor at Stony Brook; he was a student of Veltman, as was 't Hooft, and was therefore long interested in the question of unification. He is one of the few physicists I have ever met with a truly inexhaustible capacity for mathematical punishment. Working with supergravity requires an extraordinary amount of patience. We recall that the simple metric tensor introduced by Riemann in the nineteenth century had only ten components. Riemann's metric tensor has now been replaced by the super metric tensor of supergravity, which has literally hundreds of components. This is not surprising, since any theory that has higher dimensions and makes the claim of unifying all matter has to have enough components to describe it, but this vastly increases the mathematical complexity of the equations. (Sometimes I wonder what Riemann would think, knowing that after a century his metric tensor would blossom into a super metric many times larger than anything a nineteenth-century mathematician could conceive.)

The coming of supergravity and super metric tensors has meant that the amount of mathematics a graduate student must master has exploded within the past decade. As Steven Weinberg observes, "Look what's happened with supergravity. The people who've been working on it for the past ten years are enormously bright. Some of them are brighter than anyone I knew in my early years."[7]

Peter is not only a superb calculator, but also a trendsetter. Because calculations for a single supergravity equation can easily exceed a sheet of paper, he eventually started using large, oversize artist's sketch boards. I went to his house one day, and saw how he operated. He would start at the upper-left-hand corner of the pad, and start writing his equations in his microscopic handwriting. He would then proceed to work across and down the sketch pad until it was completely filled, and then turn the page and start again. This process would then go on for hours, until the calculation was completed. The only time he would ever be interrupted was when he inserted his pencil into a nearby electric pencil sharpener, and then within seconds he would resume his calculation without missing a symbol. Eventually, he would store these artist's notepads on his shelf, as though they were volumes of some scientific journal. Peter's sketch pads gradually became notorious around campus. Soon, a fad started; all the graduate students in physics began to buy these bulky artist's sketch pads and could be seen on campus hauling them awkwardly but proudly under their arms.

One time, Peter, his friend Paul Townsend (now at Cambridge University), and I were collaborating on an exceptionally difficult supergravity problem. The calculation was so difficult that it consumed several hundred pages. Since none of us totally trusted our calculations, we decided to meet in my dining room and collectively check our work. We faced a daunting challenge: Several thousand terms had to sum up to exactly zero. (Usually, we theoretical physicists can "visualize" blocks of equations in our heads and manipulate them without having to use paper. However, because of the sheer length and delicacy of this problem, we had to check every single minus sign in the calculation.)

We then divided the problem into several large chunks. Sitting around the dining-room table, each of us would busily calculate the same chunk. After an hour or so, we would then cross-check our results. Usually two out of three would get it right, and the third would be asked to find his mistake. Then we would go to the next chunk, and repeat the same process until all three of us agreed on the same answer. This repetitive cross-checking went on late into the night. We knew that even one mistake in several hundred pages would give us a totally worthless calculation. Finally, well past midnight we checked the last and final term. It was zero, as we had hoped. We then toasted our result. (The arduous calculation must have exhausted even an indefatigable workhorse like Peter. After leaving my apartment, he promptly forgot where his wife's new apartment was in Manhattan. He knocked on several doors of an apartment house, but got only angry responses; he had chosen the wrong building. After a futile search, Peter and Paul reluctantly headed back to Stony Brook. But because Peter had forgotten to replace a clutch cable, the cable snapped, and they had to push his car. They eventually straggled into Stony Brook in their broken car at 5:00 in the morning!)

The Decline of Supergravity

The critics, however, gradually began to see problems with supergravity. After an intensive search, sparticles were not seen in any experiment. For example, the spin-1/2 electron does not have any spin-0 partner. In fact, there is, at the present, not one shred of experimental evidence for sparticles in our low-energy world. However, the firm belief of physicists working in this area is that, at the enormous energies found at the instant of Creation, all particles were accompanied by their super partners. Only at this incredible energy do we see a perfectly supersymmetric world.

But after a few years of fervent interest and scores of international conferences, it became clear that this theory could not be quantized correctly, thus temporarily derailing the dream of creating a theory purely out of marble. Like every other attempt to construct a theory of matter entirely from marble, supergravity failed for a very simple reason: Whenever we tried to calculate numbers from these theories, we would arrive at meaningless infinities. The theory, although it had fewer infinities than the original Kaluza–Klein theory, was still nonrenormalizable.

There were other problems. The highest symmetry that supergravity could include was called $O(8)$, which was too small to accommodate the symmetry of the Standard Model. Supergravity, it appeared, was just another step in the long journey toward a unified theory of the universe. It cured one problem (turning wood into marble), only to fall victim to several other diseases. However, just as interest in supergravity began to wane, a new theory came along that was perhaps the strangest but most powerful physical theory ever proposed: the ten-dimensional superstring theory.

7

Superstrings

String theory is twenty-first century physics that fell accidentally into the twentieth century.

Edward Witten

EDWARD Witten, of the Institute for Advanced Study in Princeton,
New Jersey, dominates the world of theoretical physics. Witten is
currently the "leader of the pack," the most brilliant high-energy physicist, who sets trends in the physics community the way Picasso would set
trends in the art world. Hundreds of physicists follow his work religiously
to get a glimmer of his path-breaking ideas. A colleague at Princeton,
Samuel Treiman, says, "He's head and shoulders above the rest. He's
started whole groups of people on new paths. He produces elegant,
breathtaking proofs which people gasp at, which leave them in awe."
Treiman then concludes, "We shouldn't toss comparisons with Einstein
around too freely, but when it comes to Witten . . ."[1]

Witten comes from a family of physicists. His father is Louis
Witten, professor of physics at the University of Cincinnati and a
leading authority on Einstein's theory of general relativity. (His
father, in fact, sometimes states that his greatest contribution to
physics was producing his son.) His wife is Chiara Nappi, also a
theoretical physicist at the institute.

Witten is not like other physicists. Most of them begin their romance
with physics at an early age (such as in junior high school or even elementary school). Witten has defied most conventions, starting out as a

151

history major at Brandeis University with an intense interest in linguistics. After graduating in 1971, he worked on George McGovern's presidential campaign. McGovern even wrote him a letter of recommendation for graduate school. Witten has published articles in *The Nation* and the *New Republic*. (*Scientific American,* in an interview with Witten, commented, "yes, a man who is arguably the smartest person in the world is a liberal Democrat."[2])

But once Witten decided that physics was his chosen profession, he learned physics with a vengeance. He became a graduate student at Princeton, taught at Harvard, and then rocketed a full professorship at Princeton at the age of 28. He also received the prestigious MacArthur Fellowship (sometimes dubbed the "genius" award by the press). Spinoffs from his work have also deeply affected the world of mathematics. In 1990, he was awarded the Fields Medal, which is as prestigious as the Nobel Prize in the world of mathematics.

Most of the time, however, Witten sits and stares out the window, manipulating and rearranging vast arrays of equations in his head. His wife notes, "He never does calculations except in his mind. I will fill pages with calculations before I understand what I'm doing. But Edward will sit down only to calculate a minus sign, or a factor of two."[3] Witten says, "Most people who haven't been trained in physics probably think of what physicists do as a question of incredibly complicated calculations, but that's not really the essence of it. The essence of it is that physics is about concepts, wanting to understand the concepts, the principles by which the world works."[4]

Witten's next project is the most ambitious and daring of his career. A new theory called superstring theory has created a sensation in the world of physics, claiming to be the theory that can unite Einstein's theory of gravity with the quantum theory. Witten is not content, however, with the way superstring theory is currently formulated. He has set for himself the problem of finding the *origin* of superstring theory, which may prove to be a decisive development toward explaining the very instant of Creation. The key aspect of this theory, the factor that gives it its power as well as uniqueness, is its unusual geometry: Strings can vibrate self-consistently only in ten and 26 dimensions.

What Is a Particle?

The essence of string theory is that it can explain the nature of both matter and space–time—that is, the nature of wood and marble. String theory answers a series of puzzling questions about particles, such as why there are so many of them in nature. The deeper we probe into the nature of subatomic particles, the more particles we find. The current "zoo" of subatomic particles numbers several hundred, and their properties fill entire volumes. Even with the Standard Model, we are left with a bewildering number of "elementary particles." String theory answers this question because the string, about 100 billion billion times smaller than a proton, is vibrating; each mode of vibration represents a distinct resonance or particle. The string is so incredibly tiny that, from a distance, a resonance of a string and a particle are indistinguishable. Only when we somehow magnify the particle can we see that it is not a point at all, but a mode of a vibrating string.

In this picture, each subatomic particle corresponds to a distinct resonance that vibrates only at a distinct frequency. The idea of a resonance is a familiar one from daily life. Think of the example of singing in the shower. Although our natural voice may be frail, tinny, or shaky, we know that we suddenly blossom into opera stars in the privacy of our showers. This is because our sound waves bounce rapidly back and forth between the walls of the shower. Vibrations that can fit easily within the shower walls are magnified many times, producing that resonant sound. The specific vibrations are called resonances, while other vibrations (whose waves are of an incorrect size) are canceled out.

Or think of a violin string, which can vibrate at different frequencies, creating musical notes like A, B, and C. The only modes that can survive on the string are those that vanish at the endpoint of the violin string (because it is bolted down at the ends) and undulate an integral number of times between the endpoints. In principle, the string can vibrate at any of an infinite number of different frequencies. We know that the notes themselves are not fundamental. The note A is no more fundamental than the note B. However, what is fundamental is the string itself. There is no need to study each note in isolation of the others. By understanding how a violin string vibrates, we immediately understand the properties of an infinite number of musical notes.

Likewise, the particles of the universe are not, by themselves, fundamental. An electron is no more fundamental than a neutrino. They appear to be fundamental only because our microscopes are not powerful enough to reveal their structure. According to string theory, if we could somehow magnify a point particle, we would actually see a small vibrating string. In fact, according to this theory, matter is nothing but the harmonies created by this vibrating string. Since there are an infinite number of harmonies that can be composed for the violin, there are an infinite number of forms of matter that can be constructed out of vibrating strings. This explains the richness of the particles in nature. Likewise, the laws of physics can be compared to the laws of harmony allowed on the string. The universe itself, composed of countless vibrating strings, would then be comparable to a symphony.

String theory can explain not only the nature of particles, but that of space–time as well. As a string moves in space–time, it executes a complicated set of motions. The string can, in turn, break into smaller strings or collide with other strings to form longer strings. The key point is that all these quantum corrections or loop diagrams are finite and calculable. This is the first quantum theory of gravity in the history of physics to have finite quantum corrections. (*All* known previous theories, we recall—including Einstein's original theory, Kaluza–Klein theory, and supergravity—failed this key criterion.)

In order to execute these complicated motions, a string must obey a large set of self-consistency conditions. These self-consistency conditions are so stringent that they place extraordinarily restrictive conditions on space–time. In other words, the string cannot self-consistently travel in any arbitrary space–time, like a point particle.

When the constraints that the string places on space–time were first calculated, physicists were shocked to find Einstein's equations emerging from the string. This was remarkable; without assuming any of Einstein's equations, physicists found that they emerged out of the string theory, as if by magic. Einstein's equations were no longer found to be fundamental; they could be derived from string theory.

If correct, then string theory solves the long-standing mystery about the nature of wood and marble. Einstein conjectured that marble alone would one day explain all the properties of wood. To Einstein, wood was just a kink or vibration of space–time, nothing more or less. Quantum physicists, however, thought the opposite. They thought that marble could be turned into wood—that is, that Einstein's metric tensor could be turned into a graviton, the discrete packet of energy that carries the gravitational force. These are two diametrically opposite points of view, and it was long thought that a compromise between them was impossible. The string, however, is precisely the "missing link" between wood and marble.

String theory can derive the particles of matter as resonances vibrating on the string. And string theory can also derive Einstein's equations by demanding that the string move self-consistently in space–time. In this way, we have a comprehensive theory of both matter–energy and space–time.

These self-consistency constraints are surprisingly rigid. For example, they forbid the string to move in three or four dimensions. We will see that these self-consistency conditions force the string to move in a specific number of dimensions. In fact, the only "magic numbers" allowed by string theory are ten and 26 dimensions. Fortunately, a string theory defined in these dimensions has enough "room" to unify all fundamental forces.

String theory, therefore, is rich enough to explain all the fundamental laws of nature. Starting from a simple theory of a vibrating string, one can extract the theory of Einstein, Kaluza–Klein theory, supergravity, the Standard Model, and even GUT theory. It seems nothing less than a miracle that, starting from some purely geometric arguments from a string, one is able to rederive the entire progress of physics for the past 2 millennia. All the theories so far discussed in this book are automatically included in string theory.

The current interest in string theory stems from the work of John Schwarz of the California Institute of Technology and his collaborator Michael Green of Queen Mary's College in London. Previously, it was thought that the string might possess defects that would prevent a fully self-consistent theory. Then in 1984, these two physicists proved that all self-consistency conditions on the string can be met. This, in turn, ignited the current stampede among young physicists to solve the theory and win potential recognition. By the late 1980s, a veritable "gold rush" began among physicists. (The competition among hundreds of the world's brightest theoretical physicists to solve the theory has become quite fierce. In fact, the cover of *Discover* recently featured string theorist D. V. Nanopoulous of Texas, who openly boasted that he was hot on the trail of winning the Nobel Prize in physics. Rarely has such an abstract theory aroused such passions.)

Why Strings?

I once had lunch with a Nobel Prize winner in physics at a Chinese restaurant in New York. While we were passing the sweet and sour pork, the subject of superstring theory came up. Without warning, he launched into a long personal discussion of why superstring theory was not the correct path for young theoretical physicists. It was a wild-goose chase, he claimed. There had never been anything like it in the history of physics, so he found it too bizarre for his tastes. It was too alien, too orthogonal to all the previous trends in science. After a long discussion, it boiled down to one question: Why strings? Why not vibrating solids or blobs?

The physical world, he reminded me, uses the same concepts over and over again. Nature is like a work by Bach or Beethoven, often starting with a central theme and making countless variations on it that are scattered throughout the symphony. By this criterion, it appears that strings are not fundamental concepts in nature.

The concept of orbits, for example, occurs repeatedly in nature in different variations; since the work of Copernicus, orbits have provided an essential theme that is constantly repeated throughout nature in different variations, from the largest galaxy to the atom, to the smallest subatomic particle. Similarly, Faraday's fields have proved to be one of nature's favorite themes. Fields can describe the galaxy's magnetism and gravitation, or they can describe the electromagnetic theory of Maxwell, the metric theory of Riemann and Einstein, and the Yang–Mills fields found in the Standard Model. Field theory, in fact, has emerged as the universal language of subatomic physics, and perhaps the universe as well. It is the single most powerful weapon in the arsenal of theoretical physics. All known forms of matter and energy have been expressed in terms of field theory. Patterns, then, like themes and variations in a symphony, are constantly repeated.

But strings? Strings do not seem to be a pattern favored by nature in designing the heavens. We do not see strings in outer space. In fact, my colleague explained to me, we do not see strings anywhere.

A moment's thought, however, will reveal that nature has reserved the string for a special role, as a basic building block for other forms. For example, the essential feature of life on earth is the stringlike DNA molecule, which contains the complex information and coding of life itself. When building the stuff of life, as well as subatomic matter, strings seem to be the perfect answer. In both cases, we want to pack a large amount of information into a relatively simple, reproducible structure. The distinguishing feature of a string is that it is one of the most compact ways of storing vast amounts of data in a way in which information can be replicated.

For living things, nature uses the double strands of the DNA molecule, which unwind and form duplicate copies of each other. Also, our bodies contain billions upon billions of protein strings, formed of amino acid building blocks. Our bodies, in some sense, can be viewed as a vast collection of strings—protein molecules draped around our bones.

The String Quartet

Currently, the most successful version of string theory is the one created by Princeton physicists David Gross, Emil Martinec, Jeffrey Harvey, and Ryan Rohm, who are sometimes called the Princeton string quartet. The most senior of them is David Gross. At most seminars in Princeton, Witten may ask questions in his soft voice, but Gross's voice is unmistakable: loud, booming, and demanding. Anyone who gives a seminar at Princeton lives in fear of the sharp, rapid-fire questions that Gross will shoot at them. What is remarkable is that his questions are usually on the mark. Gross and his collaborators proposed what is called the *heterotic string*. Today, it is precisely the heterotic string, of all the various Kaluza–Klein-type theories that have been proposed in the past, that has the greatest potential of unifying all the laws of nature into one theory.

Gross believes that string theory solves the problem of turning wood into marble: "To build matter itself from geometry—that in a sense is what string theory does. It can be thought of that way, especially in a theory like the heterotic string which is inherently a theory of gravity in which the particles of matter as well as the other forces of nature emerge in the same way that gravity emerges from geometry."[5]

The most remarkable feature of string theory, as we have emphasized, is that Einstein's theory of gravity is automatically contained in it. In fact, the graviton (the quantum of gravity) emerges as the smallest vibration of the closed string. While GUTs strenuously avoided any mention of Einstein's theory of gravity, the superstring theories demand that Einstein's theory be included. For example, if we simply drop Einstein's theory of gravity as one vibration of the string, then the theory becomes inconsistent and useless. This, in fact, is the reason why Witten was attracted to string theory in the first place. In 1982, he read a review article by John Schwarz and was stunned to realize that gravity emerges from superstring theory from self-consistency requirements alone. He recalls that it was "the greatest intellectual thrill of my life." Witten says, "String theory is extremely attractive because gravity is forced upon us. All known consistent string theories include gravity, so while gravity is impossible in quantum field theory as we have known it, it's obligatory in string theory."[6]

Gross takes satisfaction in believing that Einstein, if he were alive, would love superstring theory. He would love the fact that the beauty and simplicity of superstring theory ultimately come from a geometric principle, whose precise nature is still unknown. Gross claims, "Einstein would have been pleased with this, at least with the goal, if not the realization. . . . He would have liked the fact that there is an underlying geometrical principle—which, unfortunately, we don't really understand."[7]

Witten even goes so far as to say that "all the really great ideas in physics" are "spinoffs" of superstring theory. By this, he means that all the great advances in theoretical physics are included within superstring theory. He even claims that Einstein's general relativity theory being discovered before superstring theory was "a mere accident of the development on planet Earth." He claims that, somewhere in outer space, "other civilizations in the universe" might have discovered superstring theory first, and derived general relativity as a by-product.[8]

Compactification and Beauty

String theory is such a promising candidate for physics because it gives a simple origin of the symmetries found in particle physics as well as general relativity.

We saw in Chapter 6 that supergravity was both nonrenormalizable and too small to accommodate the symmetry of the Standard Model. Hence, it was not self-consistent and did not begin to realistically describe the known particles. However, string theory does both. As we shall soon see, it banishes the infinities found in quantum gravity, yielding a finite theory of quantum gravity. That alone would guarantee that string theory should be taken as a serious candidate for a theory of the universe. However, there is an added bonus. When we compactify some of the dimensions of the string, we find that there is "enough room" to accommodate the symmetries of the Standard Model and even the GUTs.

The heterotic string consists of a closed string that has two types of vibrations, clockwise and counterclockwise, which are treated differently. The clockwise vibrations live in a ten-dimensional space. The counterclockwise live in a 26-dimensional space, of which 16 dimensions have been compactified. (We recall that in Kaluza's original five-dimensional theory, the fifth dimension was compactified by being wrapped up into a circle.) The heterotic string owes its name to the fact that the clockwise and the counterclockwise vibrations live in two different dimensions but are combined to produce a single superstring theory. That is why it is named after the Greek word for *heterosis*, which means "hybrid vigor."

The 16-dimensional compactified space is by far the most interesting. In Kaluza–Klein theory, we recall that the compactified N-dimensional space can have a symmetry associated with it, much like a beach ball. Then all the vibrations (or fields) defined on the N-dimensional space automatically inherit these symmetries. If the symmetry is $SU(N)$, then all the vibrations on the space must obey $SU(N)$ symmetry (in the same way that clay inherits the symmetries of the mold). In this way, Kaluza–Klein theory could accommodate the symmetries of the Standard Model. However, in this way it could also be determined that the supergravity was "too small" to contain all the particles of the symmetries found in the Standard Model. This was sufficient to kill the supergravity theory as a realistic theory of matter and space–time.

But when the Princeton string quartet analyzed the symmetries of the 16-dimensional space, they found that it is a monstrously large symmetry, called $E(8) \times E(8)$, which is much larger than any GUT symmetry that has ever been tried.[9] This was an unexpected bonus. It meant that that all the vibrations of the string would inherit the symmetry of the 16-dimensional space, which was more than enough to accommodate the symmetry of the Standard Model.

This, then, is the mathematical expression of the central theme of the book, that the laws of physics simplify in higher dimensions. In this case, the 26-dimensional space of the counterclockwise vibrations of the heterotic string has room enough to explain all the symmetries found in both Einstein's theory and quantum theory. So, for the first time, pure geometry has given a simple explanation of why the subatomic world must necessarily exhibit certain symmetries that emerge from the curling up of higher-dimensional space: *The symmetries of the subatomic realm are but remnants of the symmetry of higher-dimensional space.*

This means that the beauty and symmetry found in nature can ultimately be traced back to higher-dimensional space. For example, snowflakes create beautiful, hexagonal patterns, none of which are precisely the same. These snowflakes and crystals, in turn, have inherited their structure from the way in which their molecules have been geometrically arranged. This arrangement is mainly determined by the electron shells of the molecule, which in turn take us back to the rotational symmetries of the quantum theory, given by $O(3)$. All the symmetries of the low-energy universe that we observe in chemical elements are due to the symmetries cataloged by the Standard Model, which in turn can be derived by compactifying the heterotic string.

In conclusion, the symmetries that we see around us, from rainbows to blossoming flowers to crystals, may ultimately be viewed as manifestations of fragments of the original ten-dimensional theory.[10] Riemann and Einstein had hoped to find a geometric understanding of why forces can determine the motion and the nature of matter. But they were missing a key ingredient in showing the relationship between wood and marble. This missing link is most likely superstring theory. With the ten-dimensional string theory, we see that the geometry of the string may ultimately be responsible for both the forces and the structure of matter.

A Piece of Twenty-First-Century Physics

Given the enormous power of its symmetries, it is not surprising that superstring theory is radically different from any other kind of physics. It was, in fact, discovered quite by accident. Many physicists have commented that if this fortuitous accident had never occurred, then the theory would not have been discovered until the twenty-first century. This is because it is such a sharp departure from all the ideas tried in this century. It is not a continuous extension of trends and theories popular in this century; it stands apart.

By contrast, the theory of general relativity had a "normal" and logical evolution. First, Einstein postulated the equivalence principle. Then he reformulated this physical principle in the mathematics of a field theory of gravitation based on Faraday's fields and Riemann's metric tensor. Later came the "classical solutions," such as the black hole and the Big Bang. Finally, the last stage is the current attempt to formulate a quantum theory of gravity. Thus general relativity went through a logical progression, from a physical principle to a quantum theory:

Geometry $\rightarrow$ field theory $\rightarrow$ classical theory $\rightarrow$ quantum theory

By contrast, superstring theory has been evolving *backward* since its accidental discovery in 1968. That's why superstring theory looks so strange and unfamiliar to most physicists. We are still searching for its underlying physical principle, the counterpart to Einstein's equivalence principle.

The theory was born quite by accident in 1968 when two young theoretical physicists, Gabriel Veneziano and Mahiko Suzuki, were independently leafing through math books, looking for mathematical functions that would describe the interactions of strongly interacting particles. While studying at CERN, the European center for theoretical physics in Geneva, Switzerland, they independently stumbled on the Euler beta function, a mathematical function written down in the nineteenth century by the mathematician Leonhard Euler. They were astonished to find that the Euler beta function fit almost all the properties required to describe the strong interactions of elementary particles.

Over lunch at the Lawrence Berkeley Laboratory in California, with a spectacular view of the sun blazing down over San Francisco harbor, Suzuki once explained to me the thrill of discovering, quite by accident, a potentially important result. Physics was not supposed to happen that way.

After finding the Euler beta function in a math book, he excitedly showed his result to a senior physicist at CERN. The senior physicist, after listening to Suzuki, was not impressed. In fact, he told Suzuki that another young physicist (Veneziano) had discovered the identical function a few weeks earlier. He discouraged Suzuki from publishing his result. Today, this beta function goes by the name of the Veneziano model, which has inspired several thousand research papers, spawned a major school of physics, and now makes the claim of unifying all physical laws. (In retrospect, Suzuki, of course, should have published his result. There is a lesson to all this, I suspect: Never take too seriously the advice of your superiors.)

In 1970, the mystery surrounding the Veneziano–Suzuki model was partly explained when Yoichiro Nambu at the University of Chicago and Tetsuo Goto at Nihon University discovered that a vibrating string lies behind its wondrous properties.

Because string theory was discovered backward and by accident, physicists still do not know the physical principle that underlies string theory. The last step in the evolution of the theory (and the first step in the evolution of general relativity) is still missing.

Witten adds that

> human beings on planet Earth never had the conceptual framework that would lead them to invent string theory on purpose. . . . No one invented it on purpose, it was invented in a lucky accident. By rights, twentieth-century physicists shouldn't have had the privilege of studying this theory. By rights, string theory shouldn't have been invented until our knowledge of some of the ideas that are prerequisite for string theory had developed to the point that it was possible for us to have the right concept of what it was all about.[11]

Loops

The formula discovered by Veneziano and Suzuki, which they hoped would describe the properties of interacting subatomic particles, was still incomplete. It violated one of the properties of physics: *unitarity,* or the conservation of probability. By itself, the Veneziano–Suzuki formula would give incorrect answers for particle interactions. So the next step in the theory's evolution was to add small quantum correction terms that would restore this property. In 1969, even before the string interpretation of Nambu and Goto, three physicists (Keiji Kikkawa, Bunji Sakita, and Miguel A. Virasoro, then all at the University of Wisconsin) proposed the correct solution: adding increasingly smaller terms to the Veneziano–Suzuki formula in order to restore unitarity.

Although these physicists had to guess at how to construct the series from scratch, today it is most easily understood in the framework of the string picture of Nambu. For example, when a bumblebee flies in space, its path can be described as a wiggly line. When a piece of string drifting in the air moves in space, its path can be likened to an imaginary two-dimensional sheet. When a closed string floats in space, its path resembles a tube.

Strings interact by breaking into smaller strings and by joining with other strings. When these interacting strings move, they trace out the configurations shown in Figure 7.1. Notice that two tubes come in from the left, with one tube fissioning in half, exchange the middle tube, and then veer off to the right. This is how tubes interact with each other. This diagram, of course, is shorthand for a very complicated mathematical expression. When we calculate the numerical expression corresponding to these diagrams, we get back the Euler beta function.

In the string picture, the essential trick proposed by Kikkawa–Sakita–Virasoro (KSV) amounted to adding *all possible* diagrams where strings can collide and break apart. There are, of course, an infinite number of these diagrams. The process of adding an infinite number of "loop" diagrams, with each diagram coming closer to the final answer, is perturbation theory and is one of most important weapons in the arsenal of any quantum physicist. (These string diagrams possess a beautiful symmetry that has never been seen in physics before, which is known as *conformal symmetry* in two dimensions. This conformal symmetry allows us to treat these tubes and sheets as though they were made of rubber: We can pull, stretch, bend, and shrink these diagrams. Then, because of conformal symmetry, we can prove that all these mathematical expressions remain the same.)

KSV claimed that the sum total of all these loop diagrams would yield the precise mathematical formula explaining how subatomic particles interact. However, the KSV program consisted of a series of unproven conjectures. Someone had to construct these loops explicitly, or else these conjectures were useless.

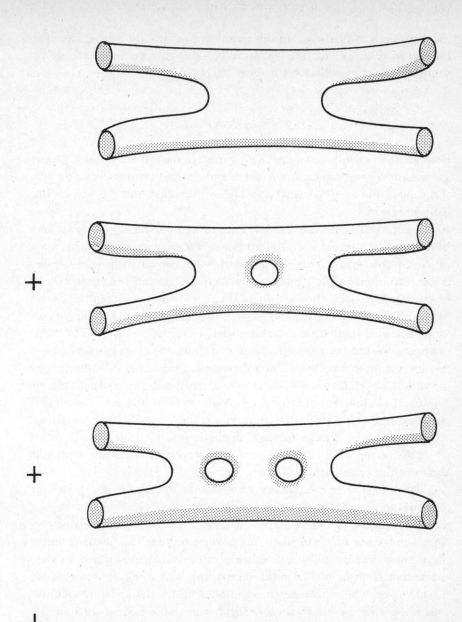

+

+

+ . . .

Figure 7.1. In string theory, the gravitational force is represented by the exchange of closed strings, which sweep out tubes in space–time. Even if we add up an infinite series of diagrams with a large number of holes, infinities never appear in the theory, giving us a finite theory of quantum gravity.

Intrigued by the program being initiated by KSV, I decided to try my luck at solving the problem. This was a bit difficult, because I was dodging machine-gun bullets at the time.

Boot Camp

I remember clearly when the KSV paper came out in 1969. KSV was proposing a program for future work, rather than giving precise details. I decided then to calculate all possible loops explicitly and complete the KSV program.

It's hard to forget those times. There was a war raging overseas, and the university campuses from Kent State to the University of Paris, were in a state of turmoil. I had graduated from Harvard the year before, when President Lyndon Johnson revoked deferments for graduate students, sending panic throughout graduate schools in the country. Chaos gripped the campuses. Suddenly, my friends were dropping out of college, teaching high school, packing their bags and heading to Canada, or trying to ruin their health in order to flunk the army physical.

Promising careers were being shattered. One of my good friends in physics from MIT vowed that he would go to jail rather than fight in Vietnam. He told us to send copies of the *Physical Review* to his jail cell so he could keep up with developments in the Veneziano model. Other friends, who quit college to teach in high schools rather than fight in the war, terminated promising scientific careers. (Many of them still teach in these high schools.)

Three days after graduation, I left Cambridge and found myself in the United States Army stationed at Fort Benning, Georgia (the largest infantry training center in the world), and later at Fort Lewis, Washington. Tens of thousands of raw recruits with no previous military training were being hammered into a fighting force and then shipped to Vietnam, replacing the 500 GIs who were dying every week.

One day, while throwing live grenades under the grueling Georgia sun and seeing the deadly shrapnel scatter in all directions, my thoughts began to wander. How many scientists throughout history had to face the punishing ravages of war? How many promising scientists were snuffed out by a bullet in the prime of their youth?

I remembered that Karl Schwarzschild had died in the kaiser's army on the Russian front during World War I just a few months after he found the basic solution to Einstein's equations used in every black hole calculation. (The Schwarzschild radius of a black hole is named in his honor. Einstein addressed the Prussian Academy in 1916 to commemorate Schwarzschild's work after his untimely death at the front lines.) And how many promising people were cut down even before they could begin their careers?

Infantry training, I discovered, is rigorous; it is designed to toughen the spirit and dull the intellect. Independence of thought is ground out of you. After all, the military does not necessarily want some wit who will question the sergeant's orders in the middle of a firefight. Understanding this, I decided to bring along some physics papers. I needed something to keep my mind active while peeling potatoes in KP or firing machine guns, so I brought along a copy of the KSV paper.

During night infantry training, I had to go past an obstacle course, which meant dodging live machine-gun bullets, froglegging under barbed wire, and crawling through thick brown mud. Because the automatic fire had tracers on them, I could see the beautiful crimson streaks made by thousands of machine-gun bullets sailing a few feet over my head. However, my thoughts kept drifting back to the KSV paper and how their program could be carried out.

Fortunately, the essential feature of the calculation was strictly topological. It was clear to me that these loops were introducing an entirely new language to physics, the language of topology. Never before in the history of physics had Möbius strips or Klein bottles been used in a fundamental way.

Because I rarely had any paper or pencils while practicing with machine guns, I forced myself to visualize in my head how strings could be twisted into loops and turned inside out. Machine-gun training was actually a blessing in disguise because it forced me to manipulate large blocks of equations in my head. By the time I finished the advanced machine-gun-training program, I was convinced that I could complete the program of calculating all loops.

Finally, I managed to squeeze time from the army to go to the University of California at Berkeley, where I furiously worked out the details that were racing in my head. I sank several hundred hours of intense thought into the question. This, in fact, became my Ph.D. dissertation.

By 1970, the final calculation took up several hundred densely filled notebook pages. Under the careful supervision of my adviser, Stanley Mandelstam, my colleague Loh-ping Yu and I successfully calculated an explicit expression for all possible loop diagrams known at that time. However, I wasn't satisfied with this work. The KSV program consisted of a hodge-podge of rules of thumb and intuition, not a rigorous set of basic principles from which these loops could be derived. String theory,

we saw, was evolving backward, since its accidental discovery by Veneziano and Suzuki. The next step in the backward evolution of the string
was to follow in the footsteps of Faraday, Riemann, Maxwell, and Einstein
and construct a *field theory of strings*.

Field Theory of Strings

Ever since the pioneering work of Faraday, every physical theory had
been written in terms of fields. Maxwell's theory of light was based on
field theory. So was Einstein's. In fact, all of particle physics was based
on field theory. The only theory *not* based on field theory was string
theory. The KSV program was more a set of convenient rules than a field
theory.

My next goal was to rectify that situation. The problem with a field
theory of strings, however, was that many of the pioneering figures in
physics argued against it. Their arguments were simple. These giants of
physics, such as Hideki Yukawa and Werner Heisenberg, had labored
for years to create a field theory that was not based on point particles.
Elementary particles, they thought, might be pulsating blobs of matter,
rather than points. However, no matter how hard they tried, field theories based on blobs always violated causality.

If we were to shake the blob at one point, the interactions would
spread faster than the speed of light throughout the blob, violating special relativity and creating all sorts of time paradoxes. Thus "nonlocal
field theories" based on blobs were known to be a monstrously difficult
problem. Many physicists, in fact, insisted that only local field theories
based on point particles could be consistent. Nonlocal field theories
must violate relativity.

The second argument was even more convincing. The Veneziano
model had many magical properties (including something called *duality*) that had never been seen before in field theory. Years earlier, Richard Feynman had given "rules" that any field theory should obey. However, these Feynman rules were in direct violation of duality. Thus many
string theorists were convinced that a field theory of strings was impossible because string theory necessarily violated the properties of the
Veneziano model. String theory, they said, was unique in all of physics
because it could not be recast as a field theory.

I collaborated with Keiji Kikkawa on this difficult but important problem. Step by step we built our field theory, in much the same way that
our predecessors had constructed field theories for other forces. Follow-

ing Faraday, we introduced a field at every point in space–time. However, for a field theory of strings, we had to generalize the concept of Faraday and postulate a field that was defined for all possible configurations of a string vibrating in space–time.

The second step was to postulate the field equations that the string obeyed. The field equation for a single string moving alone in space–time was easy. As expected, our field equations reproduced an infinite series of string resonances, each corresponding to a subatomic particle. Next, we found that the objections of Yukawa and Heisenberg were solved by string field theory. If we jiggled the string, the vibrations traveled down the string at less than the speed of light.

Soon, however, we hit a brick wall. When we tried to introduce interacting strings, we could not reproduce the Veneziano amplitude correctly. Duality and the counting of graphs given by Feynman for any field theory were in direct conflict. Just as the critics expected, the Feynman graphs were incorrect. This was disheartening. It appeared that field theory, which had formed the foundation of physics for the past century, was fundamentally incompatible with string theory.

Discouraged, I remember mulling over the problem late into the night. For hours, I began systematically to check all the possible alternatives to this problem. But the conclusion that duality had to be broken seemed inescapable. Then I remembered what Sherlock Holmes, in Arthur Conan Doyle's "The Sign of Four," said to Watson: "How often have I said to you that when you have eliminated the impossible, whatever remains, *however improbable*, must be the truth." Encouraged by this idea, I eliminated all the impossible alternatives. The only improbable alternative remaining was to violate the properties of the Veneziano–Suzuki formula. At about 3:00 A.M., the resolution finally hit me. I realized that physicists had overlooked the obvious fact that one can split the Veneziano–Suzuki formula into two pieces. Each part then corresponds to one of Feynman's diagrams, and each part violates duality, but the *sum* obeys all the correct properties of a field theory.

I quickly took out some paper and went over the calculation. I spent the next 5 hours checking and rechecking the calculation from all possible directions. The conclusion was inescapable: Field theory does violate duality, as everyone expected, but this is acceptable because the final sum reproduces the Veneziano–Suzuki formula.

I had now solved most of the problem. However, one more Feynman diagram, representing the collision of four strings, was still lacking. That year, I was teaching introductory electricity and magnetism to undergraduates at the City University of New York, and we were studying Faraday's lines of force. I would ask the students to draw the lines of force emanating from different configurations of charges, repeating the same steps pioneered by Faraday in the nineteenth century. Suddenly, it dawned on me that the squiggly lines that I was asking my students to draw had exactly the same topological structure as the collision of strings. Thus by rearranging charges in a freshman laboratory, I had found the correct configuration describing the collision of four strings.

Was it that simple?

I rushed home to check my hunch, and I was right. By employing pictorial techniques that even a freshman can use, I could show that the four-string interaction must be hidden within the Veneziano formula. By the winter of 1974, using methods dating back to Faraday, Kikkawa and I completed the field theory of strings, the first successful attempt to combine string theory with the formalism of field theory.

Our field theory, although it correctly embodied the entire information contained within string theory, still needed improvement. Because we were constructing the field theory backward, many of the symmetries were still obscure. For example, the symmetries of special relativity were present but not in an obvious way. Much more work was needed to streamline the field equations we had found. But just as we were beginning to explore the properties of our field theory, the model unexpectedly suffered a severe setback.

That year, physicist Claude Lovelace of Rutgers University discovered that the bosonic string (describing integral spins) is self-consistent only in 26 dimensions. Other physicists verified this result and showed that the superstring (describing both integral and half-integral spin) is self-consistent only in ten dimensions. It was soon realized that, in dimensions other than ten or 26 dimensions, the theory completely loses all its beautiful mathematical properties. But no one believed that a theory defined in ten or 26 dimensions had anything to do with reality. Research in string theory abruptly ground to a halt. Like Kaluza–Klein theory before it, string theory lapsed into a deep hibernation. For 10 long years, the model was banished to obscurity. (Although most string physicists, myself included, abandoned the model like a sinking ship, a few die-hards, like physicists John Schwarz and the late Joel Scherk, tried to keep the model alive by steadily making improvements. For example, string theory was originally thought to be just a theory of the strong interactions, with each mode of vibration corresponding to a resonance of the quark model. Schwarz and Scherk correctly showed that the string model was really a unified theory of all forces, not just the strong interactions.)

Research in quantum gravity went into other direction. From 1974 to 1984, when string theory was in eclipse, a large number of alternative theories of quantum gravity were successively studied. During this period, the original Kaluza–Klein theory and then the supergravity theory enjoyed great popularity, but each time the failures of these models also became apparent. For example, both Kaluza–Klein and supergravity theories were shown to be nonrenormalizable.

Then something strange happened during that decade. On the one hand, physicists became frustrated by the growing list of models that were tried and then discarded during this period. Everything failed. The realization came slowly that Kaluza–Klein theory and supergravity theory were probably on the right track, but they weren't sophisticated enough to solve the problem of nonrenormalizability. But the only theory complex enough to contain both Kaluza–Klein theory and the supergravity theory was superstring theory. On the other hand, physicists slowly became accustomed to working in hyperspace. Because of the Kaluza–Klein renaissance, the idea of hyperspace didn't seem that farfetched or forbidding anymore. Over time, even a theory defined in 26 dimensions didn't seem that outlandish. The original resistance to 26 dimensions began to slowly melt away with time.

Finally, in 1984, Green and Schwarz proved that superstring theory was the *only* self-consistent theory of quantum gravity, and the stampede began. In 1985, Edward Witten made a significant advance in the field theory of strings, which many people think is one of the most beautiful achievements of the theory. He showed that our old field theory could be derived using powerful mathematical and geometric theorems (coming from something called *cohomology theory*) with a fully relativistic form.

With Witten's new field theory, the true mathematical elegance of string field theory, which was concealed in our formalism, was revealed. Soon, almost a hundred scientific papers were written to explore the fascinating mathematical properties of Witten's field theory.[12]

No One Is Smart Enough

Assuming that string field theory is correct, in principle we should be able to calculate the mass of the proton from first principles and make contact with known data, such as the masses of the various particles. If the numerical answers are wrong, then we will have to throw the theory out the window. However, if the theory is correct, it will rank among the most significant advances in physics in 2,000 years.

After the intense, euphoric fanfare of the late 1980s (when it appeared that the theory would be completely solved within a few years and the Nobel Prizes handed out by the dozen), a certain degree of cold realism has set in. Although the theory is well defined mathematically, no one has been able to solve the theory. No one.

The problem is that *no one is smart enough to solve the field theory of strings* or any other nonperturbative approach to string theory. This is a well-defined problem, but the irony is that solving field theory requires techniques that are currently beyond the skill of any physicist. This is frustrating. Sitting before us is a perfectly well-defined theory of strings. Within it is the possibility of settling all the controversy surrounding higher-dimensional space. The dream of calculating everything from first principles is staring us in the face. The problem is how to solve it. One is reminded of Julius Caesar's famous remark in Shakespeare's play: "The fault, dear Brutus, is not in our stars, but in ourselves." For a string theorist, the fault is not in the theory, but in our primitive mathematics.

The reason for this pessimism is that our main calculational tool, perturbation theory, fails. Perturbation theory begins with a Veneziano-like formula and then calculates quantum corrections to it (which have the shape of loops). It was the hope of string theorists that they could write down a more advanced Veneziano-like formula defined in four dimensions that would uniquely describe the known spectrum of particles. In retrospect, they were too successful. The problem is that millions upon millions of Veneziano-like formulas have now been discovered. Embarrassingly, string theorists are literally drowning in these perturbative solutions.

The fundamental problem that has stalled progress in superstring theory in the past few years is that no one knows how to select the correct solution out of the millions that have been discovered. Some of these solutions come remarkably close to describing the real world. With a few modest assumptions, it is easy to extract the Standard Model as one vibration of the string. Several groups have announced, in fact, that they can find solutions that agree with the known data about subatomic particles.

The problem, we see, is that there are also millions upon millions of other solutions describing universes that do not appear anything like our universe. In some of these solutions, the universe has no quarks or too many quarks. In most of them, life as we know it cannot exist. Our universe may be lost somewhere among the millions of possible universes that have been found in string theory. To find the correct solution, we must use nonperturbative techniques, which are notoriously difficult. Since 99% of what we know about high-energy physics is based on perturbation theory, this means that we are at a total loss to find the one true solution to the theory.

There is some room for optimism, however. Nonperturbative solutions that have been found for much simpler theories show that many of the solutions are actually unstable. After a time, these incorrect, unstable solutions will make a quantum leap to the correct, stable solution. If this is true for string theory, then perhaps the millions of solutions that have been found are actually unstable and will decay over time to the correct solution.

To understand the frustration that we physicists feel, think, for a moment, of how nineteenth-century physicists might react if a portable computer were given to them. They could easily learn to turn the dials and press the buttons. They could learn to master video games or watch educational programs on the monitor. Being a century behind in technology, they would marvel at the fantastic calculational ability of the computer. Within its memory could easily be stored all known scientific knowledge of that century. In a short period of time, they could learn to perform mathematical feats that would amaze any of their colleagues. However, once they decide to open up the monitor to see what is inside, they would be horrified. The transistors and microprocessors would be totally alien to anything they could understand. There would be really nothing in their experience to compare with the electronic computer. It would be beyond their ken. They could only stare blankly at the complicated circuitry, not knowing in the slightest how it works or what it all means.

The source of their frustration would be that the computer exists and is sitting there in front of their noses, but they would have no reference frame from which to explain it. Analogously, string theory appears to be twenty-first-century physics that was discovered accidentally in our century. String field theory, too, seems to include all physical knowledge. With little effort, we are able to turn a few dials and press a few buttons with the theory, and out pops the supergravity theory, Kaluza–Klein theory, and the Standard Model. But we are at a total loss to explain why it works. String field theory exists, but it taunts us because we are not smart enough to solve it.

The problem is that while twenty-first-century physics fell accidentally into the twentieth century, twenty-first-century mathematics hasn't been invented yet. It seems that we may have to wait for twenty-first-century

mathematics before we can make any progress, or the current generation of physicists must invent twenty-first-century mathematics on their
own.

Why Ten Dimensions?

One of the deepest secrets of string theory, which is still not well understood, is why it is defined in only ten and 26 dimensions. If the theory
were three dimensional, it would not be able to unify the known laws of
physics in any sensible manner. Thus it is the geometry of higher dimensions that is the central feature of the theory.

If we calculate how strings break and re-form in N-dimensional space,
we constantly find meaningless terms cropping up that destroy the marvelous properties of the theory. Fortunately, these unwanted terms
appear multiplied by $(N - 10)$. Therefore, to make these anomalies
vanish, we have no choice but to fix N to be ten. String theory, in fact,
is the only known quantum theory that specifically demands that the
dimension of space–time be fixed at a unique number.

Unfortunately, string theorists are, at present, at a loss to explain why
ten dimensions are singled out. The answer lies deep within mathematics, in an area called *modular functions*. Whenever we manipulate the KSV
loop diagrams created by interacting strings, we encounter these strange
modular functions, where the number ten appears in the strangest
places. These modular functions are as mysterious as the man who investigated them, the mystic from the East. Perhaps if we better understood
the work of this Indian genius, we would understand why we live in our
present universe.

The Mystery of Modular Functions

Srinivasa Ramanujan was the strangest man in all of mathematics, probably in the entire history of science. He has been compared to a bursting
supernova, illuminating the darkest, most profound corners of mathematics, before being tragically struck down by tuberculosis at the age of
33, like Riemann before him. Working in total isolation from the main
currents of his field, he was able to rederive 100 years' worth of Western
mathematics on his own. The tragedy of his life is that much of his work
was wasted rediscovering known mathematics. Scattered throughout the

obscure equations in his notebooks are these modular functions, which are among the strangest ever found in mathematics. They reappear in the most distant and unrelated branches of mathematics. One function, which appears again and again in the theory of modular functions, is today called the *Ramanujan function* in his honor. This bizarre function contains a term raised to the twenty-fourth power.

In the work of Ramanujan, the number 24 appears repeatedly. This is an example of what mathematicians call magic numbers, which continually appear, where we least expect them, for reasons that no one understands. Miraculously, Ramanujan's function also appears in string theory. The number 24 appearing in Ramanujan's function is also the origin of the miraculous cancellations occurring in string theory. In string theory, each of the 24 modes in the Ramanujan function corresponds to a physical vibration of the string. Whenever the string executes its complex motions in space–time by splitting and recombining, a large number of highly sophisticated mathematical identities must be satisfied. These are precisely the mathematical identities discovered by Ramanujan. (Since physicists add two more dimensions when they count the total number of vibrations appearing in a relativistic theory, this means that space–time must have $24 + 2 = 26$ space–time dimensions.[13])

When the Ramanujan function is generalized, the number 24 is replaced by the number 8. Thus the critical number for the superstring is $8 + 2$, or 10. This is the origin of the tenth dimension. The string vibrates in ten dimensions because it requires these generalized Ramanujan functions in order to remain self-consistent. *In other words, physicists have not the slightest understanding of why ten and 26 dimensions are singled out as the dimension of the string.* It's as though there is some kind of deep numerology being manifested in these functions that no one understands. It is precisely these magic numbers appearing in the elliptic modular function that determines the dimension of space–time to be ten.

In the final analysis, the origin of the ten-dimensional theory is as mysterious as Ramanujan himself. When asked by audiences why nature might exist in ten dimensions, physicists are forced to answer, "We don't know." We know, in vague terms, why some dimension of space–time must be selected (or else the string cannot vibrate in a self-consistent quantum fashion), but we don't know why these particular numbers are selected. Perhaps the answer lies waiting to be discovered in Ramanujan's lost notebooks.

Reinventing 100 Years of Mathematics

Ramanujan was born in 1887 in Erode, India, near Madras. Although his family was Brahmin, the highest of the Hindu castes, they were destitute, living off the meager wages of Ramanujan's father's job as a clerk in a clothing merchant's office.

By the age of 10, it was clear that Ramanujan was not like the other children. Like Riemann before him, he became well known in his village for his awesome calculational powers. As a child, he had already rederived Euler's identity between trigonometric functions and exponentials.

In every young scientist's life, there is a turning point, a singular event that helps to change the course of his or her life. For Einstein, it was the fascination of observing a compass needle. For Riemann, it was reading Legendre's book on number theory. For Ramanujan, it was when he stumbled on an obscure, forgotten book on mathematics by George Carr. This book has since been immortalized by the fact that it marked Ramanujan's only known exposure to modern Western mathematics. According to his sister, "It was this book which awakened his genius. He set himself to establish the formulae given therein. As he was without the aid of other books, each solution was a piece of research so far as he was concerned. . . . Ramanujan used to say that the goddess of Namakkal inspired him with the formulae in dreams."[14]

Because of his brilliance, he was able to win a scholarship to high school. But because he was bored with the tedium of classwork and intensely preoccupied with the equations that were constantly dancing in his head, he failed to enter his senior class, and his scholarship was canceled. Frustrated, he ran away from home. He did finally return, but only to fall ill and fail his examinations again.

With the help of friends, Ramanujan managed to become a low-level clerk in the Port Trust of Madras. It was a menial job, paying a paltry £20 a year, but it freed Ramanujan, like Einstein before him at the Swiss patent office, to follow his dreams in his spare time. Ramanujan then mailed some of the results of his "dreams" to three well-known British mathematicians, hoping for contact with other mathematical minds. Two of the mathematicians, receiving this letter written by an unknown Indian clerk with no formal education, promptly threw it away. The third one was the brilliant Cambridge mathematician Godfrey H. Hardy. Because of his stature in England, Hardy was accustomed to receiving crank mail and thought dimly of the letter. Amid the dense scribbling he noticed many theorems of mathematics that were already well known.

Thinking it the obvious work of a plagiarist, he also threw it away. But something wasn't quite right. Something nagged at Hardy; he couldn't help wondering about this strange letter.

At dinner that night, January 16, 1913, Hardy and his colleague John Littlewood discussed this odd letter and decided to take a second look at its contents. It began, innocently enough, with "I beg to introduce myself to you as a clerk in the Accounts Department of the Port Trust Office of Madras on a salary of only 20 pounds per annum."[15] But the letter from the poor Madras clerk contained theorems that were totally unknown to Western mathematicians. In all, it contained 120 theorems. Hardy was stunned. He recalled that proving some of these theorems "defeated me completely." He recalled, "I had never seen anything in the least like them before. A single look at them is enough to show that they could only be written down by a mathematician of the highest class."[16]

Littlewood and Hardy reached the identical astounding conclusion: This was obviously the work of a genius engaged in rederiving 100 years of European mathematics. "He had been carrying an impossible handicap, a poor and solitary Hindu pitting his brains against the accumulated wisdom of Europe," recalled Hardy.[17]

Hardy sent for Ramanujan and, after much difficulty, arranged for his stay in Cambridge in 1914. For the first time, Ramanujan could communicate regularly with his peers, the community of European mathematicians. Then began a burst of activity: 3 short, intense years of collaboration with Hardy at Trinity College in Cambridge.

Hardy later tried to estimate the mathematical skill that Ramanujan possessed. He rated David Hilbert, universally recognized as one of the greatest Western mathematicians of the nineteenth century, an 80. To Ramanujan, he assigned a 100. (Hardy rated himself a 25.)

Unfortunately, neither Hardy nor Ramanujan seemed interested in the psychology or thinking process by which Ramanujan discovered these incredible theorems, especially when this flood of material came pouring out of his "dreams" with such frequency. Hardy noted, "It seemed ridiculous to worry him about how he had found this or that known theorem, when he was showing me half a dozen new ones almost every day."[18]

Hardy vividly recalled,

I remember going to see him once when he was lying ill in Putney. I had ridden in taxi-cab No. 1729, and remarked that the number seemed to be rather a dull one, and that I hoped that it was not an unfavorable omen. "No," he replied, "it is a very interesting number; it is the smallest number expressible as a sum of two cubes in two different ways."[19]

(It is the sum of $1 \times 1 \times 1$ and $12 \times 12 \times 12$, and also the sum of $9 \times 9 \times 9$ and $10 \times 10 \times 10$.) On the spot, he could recite complex theorems in arithmetic that would require a modern computer to prove.

Always in poor health, the austerity of the war-torn British economy prevented Ramanujan from maintaining his strict vegetarian diet, and he was constantly in and out of sanitariums. After collaborating with Hardy for 3 years, Ramanujan fell ill and never recovered. World War I interrupted travel between England and India, and in 1919 he finally managed to return home, where he died a year later.

Modular Functions

Ramanujan's legacy is his work, which consists of 4,000 formulas on 400 pages filling three volumes of notes, all densely packed with theorems of incredible power but without any commentary or, which is more frustrating, any proof. In 1976, however, a new discovery was made. One hundred and thirty pages of scrap paper, containing the output of the last year of his life, was discovered by accident in a box at Trinity College. This is now called Ramanujan's "Lost Notebook." Commenting on the Lost Notebook, mathematician Richard Askey says, "The work of that one year, while he was dying, was the equivalent of a lifetime of work for a very great mathematician. What he accomplished was unbelievable. If it were a novel, nobody would believe it." To underscore the difficulty of their arduous task of deciphering the "notebooks," mathematicians Jonathan Borwein and Peter Borwein have commented, "To our knowledge no mathematical redaction of this scope or difficulty has ever been attempted."[20]

Looking at the progression of Ramanujan's equations, it's as though we have been trained for years to listen to the Western music of Beethoven, and then suddenly we are exposed to another type of music, an eerily beautiful Eastern music blending harmonies and rhythms never heard before in Western music. Jonathan Borwein says, "He seems to have functioned in a way unlike anybody else we know of. He had such a feel for things that they just flowed out of his brain. Perhaps he didn't see them in any way that's translatable. It's like watching somebody at a feast you haven't been invited to."

As physicists know, "accidents" do not appear without a reason. When performing a long and difficult calculation, and then suddenly having thousands of unwanted terms miraculously add up to zero, physicists know that this does not happen without a deeper, underlying reason. Today, physicists know that these "accidents" are an indication that a symmetry is at work. For strings, the symmetry is called conformal symmetry, the symmetry of stretching and deforming the string's world sheet.

This is precisely where Ramanujan's work comes in. In order to protect the original conformal symmetry from being destroyed by quantum theory, a number of mathematical identities must be miraculously satisfied. These identities are precisely the identities of Ramanujan's modular function.

In summary, we have said that our fundamental premise is that the laws of nature simplify when expressed in higher dimensions. However, in light of quantum theory, we must how amend this basic theme. The correct statement should now read: The laws of nature simplify when *self-consistently* expressed in higher dimensions. The addition of the word *self-consistently* is crucial. This constraint forces us to use Ramanujan's modular functions, which fixes the dimension of space–time to be ten. This, in turn, may give us the decisive clue to explain the origin of the universe.

Einstein often asked himself whether God had any choice in creating the universe. According to superstring theorists, once we demand a unification of quantum theory and general relativity, God had no choice. Self-consistency alone, they claim, must have forced God to create the universe as he did.

Although the mathematical sophistication introduced by superstring theory has reached dizzying heights and has startled the mathematicians, the critics of the theory still pound it at its weakest point. Any theory, they claim, must be testable. Since any theory defined at the Planck energy of 10^{19} billion electron volts is not testable, superstring theory is not really a theory at all!

The main problem, as we have pointed out, is theoretical rather than experimental. If we were smart enough, we could solve the theory exactly and find the true nonperturbative solution of the theory. However, this does not excuse us from finding some means by which to verify the theory experimentally. To test the theory, we must wait for signals from the tenth dimension.

8

Signals from
the Tenth Dimension

How strange it would be if the final theory were to be discovered in our lifetimes! The discovery of the final laws of nature will mark a discontinuity in human intellectual history, the sharpest that has occurred since the beginning of modern science in the seventeenth century. Can we now imagine what that would be like?

Steven Weinberg

Is Beauty a Physical Principle?

ALTHOUGH superstring theory gives us a compelling formulation of the theory of the universe, the fundamental problem is that an experimental test of the theory seems beyond our present-day technology. In fact, the theory predicts that the unification of all forces occurs at the Planck energy, or 10^{19} billion electron volts, which is about 1 quadrillion times larger than energies currently available in our accelerators.

Physicist David Gross, commenting on the cost of generating this fantastic energy, says, "There is not enough money in the treasuries of all the countries in the world put together. It's truly astronomical."[1]

This is disappointing, because it means that experimental verification, the engine that drives progress in physics, is no longer possible with our current generation of machines or with any generation of machines

178

in the conceivable future. This, in turn, means that the ten-dimensional theory is not a theory in the usual sense, because it is untestable given the present technological state of our planet. We are then left with the question: Is beauty, by itself, a physical principle that can be substituted for the lack of experimental verification?

To some, the answer is a resounding no. They derisively call these theories "theatrical physics" or "recreational mathematics." The most caustic of the critics is Nobel Prize winner Sheldon Glashow of Harvard University. He has assumed the role of gadfly in this debate, leading the charge against the claims of other physicists that higher dimensions may exist. Glashow rails against these physicists, comparing the current epidemic to the AIDS virus; that is, it's incurable. He also compares the current bandwagon effect with former President Reagan's Star Wars program:

> Here's a riddle: Name two grand designs that are incredibly complex, require decades of research to develop, and may never work in the real world? Stars Wars and string theory. . . . Neither ambition can be accomplished with existing technology, and neither may achieve its stated objectives. Both adventures are costly in terms of scarce human resources. And, in both cases, the Russians are trying desperately to catch up.[2]

To stir up more controversy, Glashow even penned a poem, which ends:

> The Theory of Everything, if you dare to be bold,
> Might be something more than a string orbifold.
> While some of your leaders have got old and sclerotic,
> Not to be trusted alone with things heterotic,
> Please heed our advice that you are not smitten—
> The Book is not finished, the last word is not Witten.[3]

Glashow has vowed (unsuccessfully) to keep these theories out of Harvard, where he teaches. But he does admit that he is often outnumbered on this question. He regrets, "I find myself a dinosaur in a world of upstart mammals."[4] (Glashow's views are certainly not shared by other Nobel laureates, such as Murray Gell-Mann and Steven Weinberg. Physicist Weinberg, in fact, says, "String theory provides our only present source of candidates for a final theory—how could anyone expect that many of the brightest young theorists would *not* work on it?"[5])

To understand the implications of this debate concerning the uni-

fication of all forces, and also the problems with its experimental verification, it is instructive to consider the following analogy, the "parable
of the gemstone."

In the beginning, let us say, was a gemstone of great beauty, which
was perfectly symmetrical in three dimensions. However, this gemstone
was unstable. One day, it burst apart and sent fragments in all directions;
they eventually rained down on the two-dimensional world of Flatland.
Curious, the residents of Flatland embarked on a quest to reassemble
the pieces. They called the original explosion the Big Bang, but did not
understand why these fragments were scattered throughout their world.
Eventually, two kinds of fragments were identified. Some fragments were
polished and smooth on one side, and Flatlanders compared them to
"marble." Other fragments were entirely jagged and ugly, with no reg-
ularity whatsoever, and Flatlanders compared these pieces to "wood."

Over the years, the Flatlanders divided into two camps. The first
camp began to piece together the polished fragments. Slowly, some of
the polished pieces begin to fit together. Marveling at how these polished fragments were being assembled, these Flatlanders were convinced
that somehow a powerful new geometry must be operating. These Flatlanders called their partially assembled piece "relativity."

The second group devoted their efforts to assembling the jagged,
irregular fragments. They, too, had limited success in finding patterns
among these fragments. However, the jagged pieces produced only a
larger but even more irregular clump, which they called the Standard
Model. No one was inspired by the ugly mass called the Standard Model.

After years of painstaking work trying to fit these various pieces
together, however, it appeared as though there was no way to put the
polished pieces together with the jagged pieces.

Then one day an ingenious Flatlander hit upon a marvelous idea.
He declared that the two sets of pieces could be reassembled into one
piece if they were moved "up"—that is, in something he called the third
dimension. Most Flatlanders were bewildered by this new approach,
because no one could understand what "up" meant. However, he was
able to show by computer that the "marble" fragments could be viewed
as outer fragments of some object, and were hence polished, while the
"wood" fragments were the inner fragments. When both sets of fragments were assembled in the third dimension, the Flatlanders gasped at
what was revealed in the computer: a dazzling gemstone with perfect
three-dimensional symmetry. In one stroke, the artificial distinction
between the two sets of fragments was resolved by pure geometry.

This solution, however, left several questions unanswered. Some Flatlanders still wanted experimental proof, not just theoretical calculations, that the pieces could really be assembled into this gemstone. This theory gave a concrete number for the energy it would take to build powerful machines that could hoist these fragments "up" off Flatland and assemble the pieces in three-dimensional space. But the energy required was about a quadrillion times the largest energy source available to the Flatlanders.

For some, the theoretical calculation was sufficient. Even lacking experimental verification, they felt that "beauty" was more than sufficient to settle the question of unification. History had always shown, they pointed out, that the solution to the most difficult problems in nature had been the ones with the most beauty. They also correctly pointed out that the three-dimensional theory had no rival.

Other Flatlanders, however, raised a howl. A theory that cannot be tested is not a theory, they fumed. Testing this theory would drain the best minds and waste valuable resources on a wild-goose chase, they claimed.

The debate in Flatland, as well as in the real world, will persist for some time, which is a good thing. As the eighteenth-century philosopher Joseph Joubert once said, "It is better to debate a question without settling it than to settle a question without debating it."

The Superconducting Supercollider: Window on Creation

The eighteenth-century English philosopher David Hume, who was famous for advancing the thesis that every theory must be grounded on the foundation of experiment, was at a loss to explain how one can experimentally verify a theory of Creation. The essence of experiment, he claimed, is reproducibility. Unless an experiment can be duplicated over and over, in different locations and at different times with the same results, the theory is unreliable. But how can one perform an experiment with Creation itself? Since Creation, by definition, is not a reproducible event, Hume had to conclude that it is impossible to verify any theory of Creation. Science, he claimed, can answer almost all questions concerning the universe except for one, Creation, the only experiment that cannot be reproduced.

In some sense, we are encountering a modern version of the problem identified by Hume in the eighteenth century. The problem remains the same: The energy necessary to re-create Creation exceeds anything available on the planet earth. However, although direct experimental verification of the ten-dimensional theory in our laboratories is not possible, there are several ways to approach this question indirectly. The most logical approach was to hope that the superconducting supercollider (SSC) would find subatomic particles that show the distinctive signature of the superstring, such as supersymmetry. Although the SSC could not have probed the Planck energy, it might have given us strong, indirect evidence of the correctness of superstring theory.

The SSC (killed off by formidable political opposition) would have been a truly monstrous machine, the last of its type. When completed outside Dallas, Texas, around the year 2000, it would have consisted of a gigantic tube 50 miles in circumference surrounded by huge magnets. (If it were centered in Manhattan, it would have extended well into Connecticut and New Jersey.) Over 3,000 full-time and visiting scientists and staff would have conducted experiments and analyzed the data from the machine.

The purpose of the SSC was to whip two beams of protons around inside this tube until they reached a velocity very close to the speed of light. Because these beams would be traveling clockwise and counterclockwise, it would have been a simple matter to make them collide within the tube when they reached their maximum energy. The protons would have smashed into one another at an energy of 40 trillion electron volts (TeV), thereby generating an intense burst of subatomic debris analyzed by detectors. This kind of collision has not occurred since the Big Bang itself (hence the nickname for the SSC: "window on creation"). Among the debris, physicists hoped to find exotic subatomic particles that would have shed light on the ultimate form of matter.

Not surprisingly, the SSC was an extraordinary engineering and physics project, stretching the limits of known technology. Because the magnetic fields necessary to bend the protons and antiprotons within the tube are so exceptionally large (on the order of 100,000 times the earth's magnetic field), extraordinary procedures would have been necessary to generate and maintain them. For example, to reduce the heating and electrical resistance within the wires, the magnets would have been cooled down nearly to absolute zero. Then they would have been specially reinforced because the magnetic fields are so intense that otherwise they would have warped the metal of the magnet itself.

Projected to cost $11 billion, the SSC became a prized plum and a matter of intense political jockeying. In the past, the sites for atom smashers were decided by unabashed political horse trading. For example, the state of Illinois was able to land the Fermilab accelerator in Batavia, just outside Chicago, because (according to *Physics Today*) President Lyndon Johnson needed Illinois senator Everett Dirkson's crucial vote on the Vietnam War. The SSC was probably no different. Although many states vigorously competed for the project, it probably came as no surprise that in 1988 the great state of Texas landed the SSC, especially when both the president-elect of the United States and the Democratic vice-presidential candidate came from Texas.

Although billions of dollars have been spent on the SSC, it will never be completed. To the horror of the physics community, the House of Representatives voted in 1993 to cancel the project completely. Intense lobbying failed to restore funding for the project. To Congress, an expensive atom smasher can be seen in two ways. It can be a juicy plum, generating thousands of jobs and billions of dollars in federal subsidies for the state that has it. Or it can be viewed as an incredible boondoggle, a waste of money that generates no direct consumer benefits. In lean times, they argue, an expensive toy for high-energy physicists is a luxury the country cannot afford. (In all fairness, though, funding for the SSC project must be put into proper perspective. Star Wars funding for just 1 year costs $4 billion. It costs about $1 billion to refurbish an aircraft carrier. A single space-shuttle mission costs $1 billion. And a single B-2 stealth bomber costs almost $1 billion.)

Although the SSC is dead, what might we have discovered with it? At the very least, scientists hoped to find exotic particles, such as the mysterious Higgs particle predicted by the Standard Model. It is the Higgs particle that generates symmetry breaking and is therefore the origin of the mass of the quarks. Thus we hoped that the SSC would have found the "origin of mass." All objects surrounding us that have weight owe their mass to the Higgs particle.

The betting among physicists, however, was that there was an even chance that the SSC would find exotic particles beyond the Standard Model. (Possibilities included "Technicolor" particles, which lie just beyond the Standard Model, or "axions," which may help to explain the dark matter problem.) But perhaps the most exciting possibility was the sparticles, which are the supersymmetric partners of ordinary particles. The gravitino, for example, is the supersymmetric partner of the graviton. The supersymmetric partners of the quark and lepton, respectively, are the squark and the slepton.

If supersymmetric particles are eventually discovered, then there is a fighting chance that we will be seeing the remnants of the superstring itself. (Supersymmetry, as a symmetry of a field theory, was first discovered in superstring theory in 1971, even before the discovery of supergravity. In fact, the superstring is probably the only theory in which supersymmetry and gravity can be combined in a totally self-consistent way.) And even though the potential discovery of sparticles will not prove the correctness of superstring theory, it will help to quiet the skeptics who have said that there is not one shred of physical evidence for superstring theory.

Signals from Outer Space

Since the SSC will never be built, and hence will never detect particles that are low-energy resonances of the superstring, then another possibility is to measure the energy of cosmic rays, which are highly energetic subatomic particles whose origin is still unknown, but must lie deep in outer space beyond our galaxy. For example, although no one knows where they come from, cosmic rays have energies much larger than anything found in our laboratories.

Cosmic rays, unlike the controlled rays produced in atom smashers, have unpredictable energies and cannot produce precise energies on demand. In some sense, it's like trying to put out a fire by either using hose water or waiting for a rainstorm. The hose water is much more convenient: We can turn it on any time we please, we can adjust the intensity of the water at will, and all the water travels at the same uniform velocity. Water from a fire hydrant therefore corresponds to producing controlled beams in atom smashers. However, water from a rainstorm may be much more intense and effective than water from a fire hydrant. The problem, of course, is that rainstorms, like cosmic rays, are unpredictable. You cannot regulate the rainwater, nor can you predict its velocity, which may fluctuate wildly.

Cosmic rays were first discovered 80 years ago in experiments performed by the Jesuit priest Theodor Wulf atop the Eiffel Tower in Paris. From the 1900s to the 1930s, courageous physicists sailed in balloons or scaled mountains to obtain the best measurements of cosmic rays. But cosmic-ray research began to fade during the 1930s, when Ernest Lawrence invented the cyclotron and produced controlled beams in the laboratory more energetic than most cosmic rays. For example, cosmic rays, which are as energetic as 100 million electron volts, are as common as rain drops; they hit the atmosphere of the earth at the rate of a few per square inch per second. However, Lawrence's invention spawned giant machines that could exceed that energy by a factor of 10 to 100.

Cosmic-ray experiments, fortunately, have changed dramatically since Father Wulf first placed electrified jars on the Eiffel Tower. Rockets and even satellites can now send radiation counters high above the earth's surface, so that atmospheric effects are minimized. When a highly energetic cosmic ray strikes the atmosphere, it shatters the atoms in its wake. These fragments, in turn, create a shower of broken atoms, or ions, which can then be detected on the ground by this series of detectors. A collaboration between the University of Chicago and the University of Michigan has inaugurated the most ambitious cosmic-ray project yet, a vast array of 1,089 detectors scattered over about a square mile of desert, waiting for the cosmic-ray showers to trigger them. These detectors are located in an ideal, isolated area: the Dugway Proving Grounds, 80 miles southwest of Salt Lake City, Utah.

The Utah detector is sensitive enough to identify the point of origin of some of the most energetic cosmic rays. So far, Cygnus X-3 and Hercules X-1 have been identified as powerful cosmic-ray emitters. They are probably large, spinning neutron stars, or even black holes, that are slowly eating up a companion star, creating a large vortex of energy and spewing gigantic quantities of radiation (for example, protons) into outer space.

So far, the most energetic cosmic ray ever detected had an energy of 10^{20} electron volts. This figure is an incredible 10 million times the energy that would have been produced in the SSC. We do not expect to generate energies approaching this cosmic energy with our machines within the century. Although this fantastic energy is still 100 million times smaller than the energy necessary to probe the tenth dimension, we hope that energies produced deep within black holes in our galaxy will approach the Planck energy. With large, orbiting spacecraft, we should be able to probe deeper into the structure of these energy sources and detect energies even larger than this.

According to one favored theory, the largest energy source within our Milky Way galaxy—far beyond anything produced by Cygnus X-3 or Hercules X-1—lies at the center, which may consist of millions of black holes. So, because the SSC was canceled by Congress, we may find that the ultimate probe for exploring the tenth dimension may lie in outer space.

Testing the Untestable

Historically speaking, there have been many times when physicists have solemnly declared certain phenomena to be "untestable" or "unprovable." But there is another attitude that scientists can take concerning the inaccessibility of the Planck energy—unforeseen breakthroughs will make indirect experiments possible near the Planck energy.

In the nineteenth century, some scientists declared that the composition of the stars would forever be beyond the reach of experiment. In 1825, the French philosopher and social critic Auguste Comte, writing in *Cours de philosophie*, declared that we would never know the stars other than as unreachable points of light in the sky because of their enormous distance from us. The machines of the nineteenth century, or any century, he argued, were not powerful enough to escape from the earth and reach the stars.

Although determining what the stars were made of seemed beyond the capabilities of any science, ironically at almost the same time, the German physicist Joseph von Fraunhofer was doing just that. Using a prism and spectroscope, he could separate the white light emitted from the distant stars and determine the chemical composition of those stars. Since each chemical within the stars emits a characteristic "fingerprint," or spectrum of light, it was easy for Fraunhofer to perform the "impossible" and to determine that hydrogen is the most abundant element in the stars.

This, in turn, inspired poet Ian D. Bush to write:

> Twinkle, twinkle little star
> I don't wonder what you are,
> For by spectroscopic ken,
> I know that you are hydrogen.[6]

Thus although the energy necessary to reach the stars via rockets was far beyond anything available to Comte (or, for that matter, anything available to modern science), the crucial step did not involve energy. The key observation was that signals from the stars, rather than direct measurement, were sufficient to solve the problem. Similarly, we can hope that signals from the Planck energy (perhaps from cosmic rays or perhaps an as yet unknown source), rather than a direct measurement from large atom smashers, may be sufficient to probe the tenth dimension.

Another example of an "untestable" idea was the existence of atoms. In the nineteenth century, the atomic hypothesis proved to be the decisive step in understanding the laws of chemistry and thermodynamics. However, many physicists refused to believe that atoms actually exist. Perhaps they were just a mathematical device that, by accident, gave the correct description of the world. For example, the philosopher Ernst

Mach did not believe in the existence of atoms, other than as a calculational tool. (Even today, we are still unable to take direct pictures of the atom because of the Heisenberg Uncertainty Principle, although indirect methods now exist.) In 1905, however, Einstein gave the most convincing, although indirect, evidence of the existence of atoms when he showed that Brownian motion (that is, the random motion of dust particles suspended in a liquid) can be explained as random collisions between the particles and atoms in the liquid.

By analogy, we might hope for experimental confirmation of the physics of the tenth dimension using indirect methods that have not yet been discovered. Instead of photographing the object we desire, perhaps we should be satisfied with a photograph of its "shadow." The indirect approach would be to examine carefully low-energy data from an atom smasher, and try to see if ten-dimensional physics affects the data in some way.

The third "untestable" idea in physics was the existence of the elusive neutrino.

In 1930, physicist Wolfgang Pauli hypothesized a new, unseen particle called the *neutrino* in order to account for the missing component of energy in certain experiments on radioactivity that seemed to violate the conservation of matter and energy. Pauli realized, though, that neutrinos would be almost impossible to observe experimentally, because they would interact so weakly, and hence so rarely, with matter. For example, if we could construct a solid block of lead that stretched several light-years from our solar system to Alpha Centauri and placed it in the path of a beam of neutrinos, some would still come out the other end. They can penetrate the earth as though it doesn't even exist, and, in fact, trillions of neutrinos emitted from the sun are always penetrating your body, even at night. Pauli admitted, "I have committed the ultimate sin, I have predicted the existence of a particle that can never be observed."[7]

So elusive and undetectable was the neutrino that it even inspired a poem by John Updike, called "Cosmic Gall":

> Neutrinos, they are very small.
>   They have no charge and have no mass
> And do not interact at all.
>   The earth is just a silly ball
> To them, through which they simply pass,
>   Like dustmaids down a drafty hall
> Or photons though a sheet of glass.
>   They snub the most exquisite gas,

Ignore the most substantial wall,
   Cold-shoulder steel and sounding brass,
Insult the stallion in his stall,
   And scorning barriers of class,
Infiltrate you and me! Like tall
   And painless guillotines, they fall
Down through our heads into the grass.
   At night, they enter at Nepal
And pierce the lover and his lass
   From underneath the bed—you call
It wonderful; I call it crass.[8]

Although the neutrino, because it barely interacts with other materials, was once considered the ultimate "untestable" idea, today we regularly produce beams of neutrinos in atom smashers, perform experiments with the neutrinos emitted from a nuclear reactor, and detect their presence within mines far below the earth's surface. (In fact, when a spectacular supernova lit up the sky in the southern hemisphere in 1987, physicists noticed a burst of neutrinos streaming through their detectors deep in these mines. This was the first time that neutrino detectors were used to make crucial astronomical measurements.) Neutrinos, in 3 short decades, have been transformed from an "untestable" idea into one of the workhorses of modern physics.

The Problem Is Theoretical, Not Experimental

Taking the long view on the history of science, perhaps there is some cause for optimism. Witten is convinced that science will some day be able to probe down to Planck energies. He says,

It's not always so easy to tell which are the easy questions and which are the hard ones. In the 19th century, the question of why water boils at 100 degrees was hopelessly inaccessible. If you told a 19th-century physicist that by the 20th century you would be able to calculate this, it would have seemed like a fairy tale. . . . Quantum field theory is so difficult that nobody fully believed it for 25 years.

In his view, "good ideas always get tested."[9]
The astronomer Arthur Eddington even questioned whether scientists were not overstating the case when they insisted that everything should be tested. He wrote: "A scientist commonly professes to base his

beliefs on observations, not theories. . . . I have never come across any-one who carries this profession into practice. . . . Observation is not sufficient . . . theory has an important share in determining belief."[10] Nobel laureate Paul Dirac said it even more bluntly, "It is more important to have beauty in one's equations than to have them fit experiment."[11] Or, in the words of CERN physicist John Ellis, "in the words of a candy wrapper I opened a few years ago: 'It is only the optimists who achieve anything in this world.' " Nonetheless, despite arguments that uphold a certain degree of optimism, the experimental situation looks bleak. I share, along with the skeptics, the idea that the best we can hope for is indirect tests of ten-dimensional theory into the twenty-first century. This is because, in the final analysis, this theory is a theory of Creation, and hence testing it necessarily involves re-creating a piece of the Big Bang in our laboratories.

Personally, I don't think that we have to wait a century until our accelerators, space probes, and cosmic-ray counters will be powerful enough to probe the tenth dimension indirectly. Within a span of years, and certainly within the lifetime of today's physicists, someone will be clever enough to either verify or disprove the ten-dimensional theory by solving the field theory of strings or some other nonperturbative formulation. The problem is thus theoretical, not experimental.

Assuming that some bright physicist solves the field theory of strings and derives the known properties of our universe, there is still the practical problem of when we might be able to harness the power of the hyperspace theory. There are two possibilities:

1. Wait until our civilization attains the ability to master energies trillions of times larger than anything we can produce today
2. Encounter extraterrestrial civilizations that have mastered the art of manipulating hyperspace

We recall that it took about 70 years, between the work of Faraday and Maxwell to the work of Edison and his co-workers, to exploit the electromagnetic force for practical purposes. Yet modern civilization depends crucially on the harnessing of this force. The nuclear force was discovered near the turn of the century, and 80 years later we still do not have the means to harness it successfully with fusion reactors. The next leap, to harness the power of the unified field theory, requires a much greater jump in our technology, but one that will probably have vastly more important implications.

The fundamental problem is that we are forcing superstring theory to answer questions about everyday energies, when its "natural home" lies at the Planck energy. This fabulous energy was released only at the instant of Creation itself. In other words, superstring theory is naturally a theory of Creation. Like the caged cheetah, we are demanding that this superb animal dance and sing for our entertainment. The real home of the cheetah is the vast plains of Africa. The real "home" of superstring theory is the instant of Creation. Nevertheless, given the sophistication of our artificial satellites, there is perhaps one last "laboratory" in which we may experimentally probe the natural home of superstring theory, and this is the echo of Creation!

9
Before Creation

In the beginning, was the great cosmic egg. Inside the egg was chaos, and floating in chaos was P'an Ku, the divine Embryo.

P'an Ku myth (China, third century)

If God created the world, where was He before Creation? . . . Know that the world is uncreated, as time itself is, without beginning and end.

Mahapurana (India, ninth century)

"**D**ID God have a mother?"

Children, when told that God made the heavens and the earth, innocently ask whether God had a mother. This deceptively simple question has stumped the elders of the church and embarrassed the finest theologians, precipitating some of the thorniest theological debates over the centuries. All the great religions have elaborate mythologies surrounding the divine act of Creation, but none of them adequately confronts the logical paradoxes inherent in the questions that even children ask.

God may have created the heavens and the earth in 7 days, but what happened before the first day? If one concedes that God had a mother, then one naturally asks whether she, too, had a mother, and so on, forever. However, if God did not have a mother, then this answer raises even more questions: Where did God come from? Was God always in existence since eternity, or is God beyond time itself?

Over the centuries, even great painters commissioned by the church grappled with these ticklish theological debates in their works of art: When depicting God or Adam and Eve, do you give them belly buttons? Since the navel marks the point of attachment of the umbilical cord, then neither God nor Adam and Eve could be painted with belly buttons. For example, Michelangelo faced this dilemma in his celebrated depiction of Creation and the expulsion of Adam and Eve from the Garden of Eden when he painted the ceiling of the Sistine Chapel. The answer to this theological question is to be found hanging in any large museum: God and Adam and Eve simply have no belly buttons, because they were the first.

Proofs of the Existence of God

Troubled by the inconsistencies in church ideology, St. Thomas Aquinas, writing in the thirteenth century, decided to raise the level of theological debate from the vagueness of mythology to the rigor of logic. He proposed to solve these ancient questions in his celebrated "proofs of the existence of God."

Aquinas summarized his proofs in the following poem:

> Things are in motion, hence there is a first mover
> Things are caused, hence there is a first cause
> Things exist, hence there is a creator
> Perfect goodness exists, hence it has a source
> Things are designed, hence they serve a purpose.[1]

(The first three lines are variations of what is called the *cosmological proof;* the fourth argues on moral grounds; and the fifth is called the *teleological proof.* The moral proof is by far the weakest, because morality can be viewed in terms of evolving social customs.)

Aquinas's "cosmological" and "teleological" proofs of the existence of God have been used by the church for the past 700 years to answer this sticky theological question. Although these proofs have since been shown to be flawed in light of the scientific discoveries made over the past 7 centuries, they were quite ingenious for their time and show the influence of the Greeks, who were the first to introduce rigor into their speculations about nature.

Aquinas began the cosmological proof by postulating that God was the First Mover and First Maker. He artfully dodged the question of "who made God" by simply asserting that the question made no sense. God had no maker because he was the First. Period. The cosmological proof states that everything that moves must have had something push it, which in turn must have had something push it, and so on. But what started the first push?

Imagine, for the moment, idly sitting in the park and seeing a wagon moving in front of you. Obviously, you think, there is a young child pushing the wagon. You wait a moment, only to find another wagon pushing the first wagon. Curious, you wait a bit longer for the child, but there is a third wagon pushing the first two wagons. As time goes by, you witness hundreds of wagons, each one pushing the others, with no child in sight. Puzzled, you look out into the distance. You are surprised to see an infinite sequence of wagons stretching into the horizon, each wagon pushing the others, with no child at all. If it takes a child to push a wagon, then can an infinite sequence of wagons be pushed without the First Pusher? Can an infinite sequence of wagons push itself? No. Therefore, God must exist.

The teleological proof is even more persuasive. It states that there has to be a First Designer. For example, imagine walking on the sands of Mars, where the winds and dust storms have worn even the mountains and giant craters. Over tens of millions of years, nothing has escaped the corrosive, grinding effect of the sand storms. Then, to your surprise, you find a beautiful camera lying in the sand dunes. The lens is smoothly polished and the shutter mechanism delicately crafted. Surely, you think, the sands of Mars could not have created such a beautiful piece of craftsmanship. You conclude that someone intelligent obviously made this camera. Then, after wandering on the surface of Mars some more, you come across a rabbit. Obviously, the eye of the rabbit is infinitely more intricate than the eye of the camera. The muscles of the rabbit's eye are infinitely more elaborate than the shutter of the camera. Therefore, the maker of this rabbit must be infinitely more advanced than the maker of the camera. This maker must therefore be God.

Now imagine the machines on the earth. There is no question that these machines were made by something even greater, such as humans. There is no question that a human is infinitely more complicated than a machine. Therefore, the person who created us must be infinitely more complicated than we are. So therefore God must exist.

In 1078, St. Anselm, the archbishop of Canterbury, cooked up perhaps the most sophisticated proof of the existence of God, the *ontological proof*, which does not depend on First Movers or First Designers at all.

St. Anselm claimed that he could prove the existence of God from pure logic alone. He defined God as the most perfect, most powerful being imaginable. It is, however, possible to conceive of two types of God. The first God, we imagine, does not exist. The second God, we imagine, actually does exist and can perform miracles, such as parting the rivers and raising the dead. Obviously, the second God (who exists) is more powerful and more perfect than the first God (who does not exist).

However, we defined God to be the most perfect and powerful being imaginable. By the definition of God, the second God (who exists) is the more powerful and more perfect one. Therefore, the second God is the one who fits the definition. The first God (who does not exist) is weaker and less perfect than the second God, and therefore does not fit the definition of God. Hence God must exist. In other words, if we define God as "that being nothing greater than which can be conceived," then God must exist because if he didn't, it's possible to conceive of a much greater God who does exist. This rather ingenious proof, unlike those of St. Thomas Aquinas, is totally independent of the act of Creation and rests solely on the definition of the perfect being.

Remarkably, these "proofs" of the existence of God lasted for over 700 years, defying the repeated challenges of scientists and logicians. The reason for this is that not enough was known about the fundamental laws of physics and biology. In fact, only within the past century have new laws of nature been discovered that can isolate the potential flaws in these proofs.

The flaw in the cosmological proof, for example, is that the conservation of mass and energy is sufficient to explain motion without appealing to a First Mover. For example, gas molecules may bounce against the walls of a container without requiring anyone or anything to get them moving. In principle, these molecules can move forever, requiring no beginning or end. Thus there is no necessity for a First or a Last Mover as long as mass and energy are conserved.

For the teleological proof, the theory of evolution shows that it is possible to create higher and more complex life forms from more primitive ones through natural selection and chance. Ultimately, we can trace the origin of life itself back to the spontaneous formation of protein molecules in the early earth's oceans without appealing to a higher intelligence. Studies performed by Stanley L. Miller in 1955 have shown that sparks sent through a flask containing methane, ammonia, and other gases found in the early earth's atmosphere can spontaneously create complex hydrocarbon molecules and eventually amino acids (precursors to protein molecules) and other complex organic molecules. Thus a First Designer is not necessary to create the essentials for life, which can apparently emerge naturally out of inorganic chemicals if they are given enough time.

And, finally, Immanuel Kant was the first to isolate the error in the ontological proof after centuries of confusion. Kant pointed out that stating that an object exists does not make it more perfect. For example, this proof can be used to prove the existence of the unicorn. If we define the unicorn to be the most perfect horse imaginable, and if unicorns don't exist, then it's possible to imagine a unicorn that does exist. But saying that it exists does not mean that it is more perfect than a unicorn that does not exist. Therefore, unicorns do not necessarily have to exist. And neither does God.

Have we made any progress since the time of St. Thomas Aquinas and St. Anselm?

Yes and no. We can say that present-day theories of Creation are built on two pillars: quantum theory and Einstein's theory of gravity. We can say that, for the first time in a thousand years, religious "proofs" of the existence of God are being replaced by our understanding of thermodynamics and particle physics. However, by replacing God's act of Creation with the Big Bang, we have supplanted one problem with another. Aquinas thought he solved the problem of what came before God by defining him as the First Mover. Today, we are still struggling with the question of what happened before the Big Bang.

Unfortunately, Einstein's equations break down at the enormously small distances and large energies found at the origin of the universe. At distances on the order of 10^{-33} centimeter, quantum effects take over from Einstein's theory. Thus to resolve the philosophical questions involving the beginning of time, we must necessarily invoke the ten-dimensional theory.

Throughout this book, we have emphasized the fact that the laws of physics unify when we add higher dimensions. When studying the Big Bang, we see the precise reverse of this statement. The Big Bang, as we shall see, perhaps originated in the breakdown of the original ten-dimensional universe into a four- and a six-dimensional universe. Thus we can view the history of the Big Bang as the history of the breakup of ten-dimensional space and hence the breakup of previously unified symmetries. This, in turn, is the theme of this book in reverse.

It is no wonder, therefore, that piecing together the dynamics of the Big Bang has been so difficult. In effect, by going backward in time, we are reassembling the pieces of the ten-dimensional universe.

Experimental Evidence for the Big Bang

Every year, we find more experimental evidence that the Big Bang occurred roughly 15 to 20 billion years ago. Let us review some of these experimental results.

First, the fact that the stars are receding from us at fantastic velocities has been repeatedly verified by measuring the distortion of their starlight (called the red shift). (The starlight of a receding star is shifted to longer wavelengths—that is, toward the red end of the spectrum—in the same way that the whistle of a receding train sounds higher than normal when approaching and lower when receding. This is called the Doppler effect. Also, Hubble's Law states that the farther from us the star or galaxy, the faster it is receding from us. This fact, first announced by the astronomer Edwin Hubble in 1929, has been experimentally verified over the past 50 years.) We do not see any blue shift of the distant galaxies, which would mean a collapsing universe.

Second, we know that the distribution of the chemical elements in our galaxy are in almost exact agreement with the prediction of heavy-element production in the Big Bang and in the stars. In the original Big Bang, because of the enormous heat, elemental hydrogen nuclei banged into one another at large enough velocities to fuse them, forming a new element: helium. The Big Bang theory predicts that the ratio of helium to hydrogen in the universe should be approximately 25% helium to 75% hydrogen. This agrees with the observational result for the abundance of helium in the universe.

Third, the earliest objects in the universe date back 10 to 15 billion years, in agreement with the rough estimate for the Big Bang. We do not see any evidence for objects older than the Big Bang. Since radioactive materials decay (for example, via the weak interactions) at a precisely known rate, it is possible to tell the age of an object by calculating the relative abundance of certain radioactive materials. For example, half of a radioactive substance called carbon-14 decays every 5,730 years, which allows us to determine the age of archeological artifacts that contain carbon. Other radioactive elements (like uranium-238, with a half-life of over 4 billion years) allow us to determine the age of moon rocks (from the *Apollo* mission). The oldest rocks and meteors found on earth date to about 4 to 5 billion years, which is the approximate age of the solar system. By calculating the mass of certain stars whose evolution is known, we can show that the oldest stars in our galaxy date back about 10 billion years.

Fourth, and most important, the Big Bang produced a cosmic "echo" reverberating throughout the universe that should be measurable by our instruments. In fact, Arno Penzias and Robert Wilson of the Bell Telephone Laboratories won the Nobel Prize in 1978 for detecting this echo of the Big Bang, a microwave radiation that permeates the known universe. The fact that the echo of the Big Bang should be circulating around the universe billions of years after the event was first predicted by George Gamow and his students Ralph Alpher and Robert Herman, but no one took them seriously. The very idea of measuring the echo of Creation seemed outlandish when they first proposed this idea soon after World War II.

Their logic, however, was very compelling. Any object, when heated, gradually emits radiation. This is the reason why iron gets red hot when placed in a furnace. The hotter the iron, the higher the frequency of radiation it emits. A precise mathematical formula, the Stefan–Boltzmann law, relates the frequency of light (or the color, in this case) to the temperature. (In fact, this is how scientists determine the surface temperature of a distant star, by examining its color.) This radiation is called *blackbody radiation*.

When the iron cools, the frequency of the emitted radiation also decreases, until the iron no longer emits in the visible range. The iron returns to its normal color, but it continues to emit invisible infrared radiation. This is how the army's night glasses operate in the dark. At night, relatively warm objects such as enemy soldiers and tank engines may be concealed in the darkness, but they continue to emit invisible blackbody radiation in the form of infrared radiation, which can be picked up by special infrared goggles. This is also why your sealed car gets hot during the summer. Sunlight penetrates the glass of your car and heats the interior. As it gets hot, it begins to emit blackbody radiation in the form of infrared radiation. However, infrared radiation does not penetrate glass very well, and hence is trapped inside your car, dramatically raising its temperature. (Similarly, blackbody radiation drives the greenhouse effect. Like glass, rising levels of carbon dioxide in the atmosphere, caused by the burning of fossil fuels, can trap the infrared blackbody radiation of the earth and thereby gradually heat the planet.)

Gamow reasoned that the Big Bang was initially quite hot, and hence would be an ideal blackbody emitter of radiation. Although the technology of the 1940s was too primitive to pick up this faint signal from Creation, he could calculate the temperature of this radiation and confidently predict that one day our instruments would be sensitive enough to detect this "fossil" radiation. The logic behind his thinking was as follows: About 300,000 years after the Big Bang, the universe cooled to the point where atoms could begin to condense; electrons could begin to circle protons and form stable atoms that would no longer be broken up by the intense radiation permeating the universe. Before this time, the universe was so hot that atoms were continually ripped apart by radiation as soon as they were formed. This meant that the universe was opaque, like a thick, absorbing, and impenetrable fog. After 300,000 years, however, the radiation was no longer sufficiently strong to break up the atoms, and hence light could travel long distances without being scattered. In other words, the universe suddenly became black and transparent after 300,000 years. (We are so used to hearing about the "blackness of outer space" that we forget that the early universe was not transparent at all, but filled with turbulent, opaque radiation.)

After 300,000 years, electromagnetic radiation no longer interacted so strongly with matter, and hence became blackbody radiation. Gradually, as the universe cooled, the frequency of this radiation decreased. Gamow and his students calculated that the radiation would be far below the infrared range, into the microwave region. Gamow reasoned that by scanning the heavens for a uniform, isotropic source of microwave radiation, one should be able to detect this microwave radiation and discover the echo of the Big Bang.

Gamow's prediction was forgotten for many decades, until the microwave background radiation was discovered quite by accident in 1965. Penzias and Wilson found a mysterious background radiation permeating all space when they turned on their new horn reflector antenna in Holmdel, New Jersey. At first, they thought this unwanted radiation was due to electrical static caused by contaminants, such as bird droppings on their antenna. But when they disassembled and cleaned large portions of the antenna, they found that the "static" persisted. At the same time, physicists Robert Dicke and James Peebles at Princeton University were rethinking Gamow's old calculation. When Penzias and Wilson were finally informed of the Princeton physicists' work, it was clear that there was a direct relationship between their results. When they realized that this background radiation might be the echo of the original Big Bang, they are said to have exclaimed, "Either we've seen a pile of bird s——t, or the creation of the universe!" They discovered that this uniform background radiation was almost exactly what had been predicted years earlier by George Gamow and his collaborators if the Big Bang had left a residual blanket of radiation that had cooled down to 3°K.

COBE and the Big Bang

Perhaps the most spectacular scientific confirmation of the Big Bang theory came in 1992 with the results of the *COBE* (*Cosmic Background Explorer*) satellite. On April 23, newspaper headlines across the country heralded the findings of a team of scientists at the University of California at Berkeley, led by George Smoot, who announced the most dramatic, convincing argument for the Big Bang theory. Journalists and columnists, with no background in physics or theology, were suddenly waxing eloquent about the "face of God" in their dispatches.

The *COBE* satellite was able to improve vastly the earlier work of Penzias, Wilson, Peebles, and Dicke by many orders of magnitude, sufficient to rule out all doubt that the fossil radiation emitted by the Big Bang had been conclusively found. Princeton cosmologist Jeremiah P. Ostriker declared, "When fossils were found in the rocks, it made the origin of species absolutely clear-cut. Well, *COBE* found its fossils."[2] Launched in late 1989, the *COBE* satellite was specifically designed to analyze the microscopic details in the structure of the microwave background radiation first postulated by George Gamow and his colleagues. The mission of *COBE* also had a new task: to resolve an earlier puzzle arising from the background radiation.

The original work of Penzias and Wilson was crude; they could show only that the background radiation was smooth to 10%. When scientists analyzed the background radiation in more detail, they found that it was exceptionally smooth, with no apparent ripples, kinks, or blotches. In fact, it was *too* smooth. The background radiation was like a smooth, invisible fog filling up the universe, so uniform that scientists had difficulty reconciling it with known astronomical data.

In the 1970s, astronomers turned their great telescopes to systematically map enormous collections of galaxies across large portions of the sky. To their surprise, they found that, 1 billion years after the Big Bang, the universe had already exhibited a pattern of condensing into galaxies and even large clusters of galaxies and huge, empty spaces called voids. The clusters were enormous, containing billions of galaxies at a time, and the voids stretched across millions of light-years.

But here lay a cosmic mystery: If the Big Bang was exceptionally smooth and uniform, then 1 billion years was not enough time to develop the clumpiness that we see among the galactic clusters. The gross mismatch between the original smooth Big Bang and the lumpiness of the universe 1 billion years later was a nagging problem that gnawed at every cosmologist. The Big Bang theory itself was never in any doubt; what was in trouble was our understanding of the post–Big Bang evolution 1 billion years after Creation. But without sensitive satellites that could measure the cosmic background radiation, the problem festered over the years. In fact, by 1990, journalists without a rigorous science background began to write sensational articles saying erroneously that scientists had found a fatal flaw in the Big Bang theory itself. Many journalists wrote that the Big Bang theory was about to be overthrown. Long-discredited alternatives to the Big Bang theory began to resurface in the press. Even the *New York Times* published a major article saying that the Big Bang theory was in serious trouble (which was scientifically incorrect).

This pseudocontroversy surrounding the Big Bang theory made the announcement of the *COBE* data all the more interesting. With unprecedented accuracy, capable of detecting variations as small as one part in 100,000, the *COBE* satellite was able to scan the heavens and radio back the most accurate map of the cosmic background radiation ever constructed. The *COBE* results reconfirmed the Big Bang theory, and more.

COBE's data, however, were not easy to analyze. The team led by Smoot had to face enormous problems. For example, they had to subtract carefully the effect of the earth's motion in the background radiation. The solar system drifts at a velocity of 370 kilometers per second relative to the background radiation. There is also the relative motion of the solar system with respect to the galaxy, and the galaxy's complex motions with respect to galactic clusters. Nevertheless, after painstaking computer enhancement, several stunning results came out of the analysis. First, the microwave background fit the earlier prediction of George Gamow (adjusted with more accurate experimental numbers) to within 0.1% (Figure 9.1). The solid line represents the prediction; the x's mark the data points measured by the *COBE* satellite. When this graph was flashed on the screen for the first time to a meeting of about a thousand astronomers, everyone in the room erupted in a standing ovation. This was perhaps the first time in the history of science that a simple graph received such a thunderous applause from so many distinguished scientists.

Second, Smoot's team was able to show that tiny, almost microscopic blotches did, in fact, appear in the microwave background. These tiny blotches were precisely what was needed to explain the clumpiness and voids found 1 billion years after the Big Bang. (If these blotches had not been found by *COBE*, then a major revision in the post–Big Bang analysis would have had to be made.)

Third, the results were consistent with, but did not prove, the so-

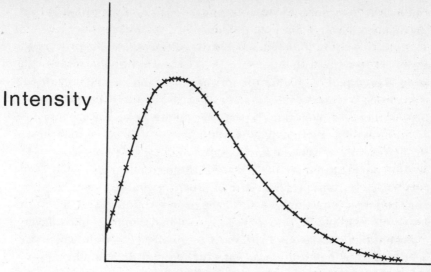

Figure 9.1. The solid line represents the prediction made by the Big Bang theory, which predicts that the background cosmic radiation should resemble blackbody radiation in the microwave region. The x's represent the actual data collected by the COBE *satellite, giving us one of the most convincing proofs of the Big Bang theory.*

called *inflation theory*. (This theory, proposed by Alan Guth of MIT, states that there was a much more explosive expansion of the universe at the initial instant of Creation than the usual Big Bang scenario; it holds that the visible universe we see with our telescopes is only the tiniest part of a much bigger universe whose boundaries lie beyond our visible horizon.)

Before Creation: Orbifolds?

The *COBE* results have given physicists confidence that we understand the origin of the universe to within a fraction of a second after the Big Bang. However, we are still left with the embarrassing questions of what preceded the Big Bang and why it occurred. General relativity, if taken to its limits, ultimately yields nonsensical answers. Einstein, realizing that general relatively simply breaks down at those enormously small distances, tried to extend general relativity into a more comprehensive theory that could explain these phenomena.

At the instant of the Big Bang, we expect quantum effects to be the dominant force, overwhelming gravity. The key to the origin of the Big Bang, therefore, is a quantum theory of gravity. So far, the only theory that can claim to solve the mystery of what happened before the Big Bang is the ten-dimensional superstring theory. Scientists are just now conjecturing how the ten-dimensional universe split into a four- and a six-dimensional universe. What does our twin universe look like?

One physicist who is struggling with these cosmic questions is Cumrum Vafa, a Harvard professor who has spent several years studying how our ten-dimensional universe may have been torn into two smaller universes. He is, ironically, also a physicist torn between two worlds. Living in Cambridge, Massachusetts, Vafa is originally from Iran, which has been racked by political convulsions for the past decade. On the one hand, he wishes eventually to return to his native Iran, perhaps when the social tumult has calmed down. On the other hand, his research takes him far from that troubled region of the world, all the way to the far reaches of six-dimensional space, long before the tumult in the early universe had a chance to stabilize.

"Imagine a simple video game," he says. A rocket ship can travel in the video screen, he points out, until it veers too far to the right. Any video-game player knows that the rocket ship then suddenly appears from the left side of the screen, at exactly the same height. Similarly, if the rocket ship wanders too far and falls off the bottom of the screen, it rematerializes at the top of the screen. Thus, Vafa explains, there is an entirely self-contained universe in that video screen. You can never leave the universe defined by that screen. Even so, most teenagers have never asked themselves what that universe is actually shaped like. Vafa points out, surprisingly enough, that the topology of the video screen is that of an inner tube!

Think of the video screen as a sheet of paper. Since points at the top of the screen are identical to the points at the bottom, we can seal the top and bottom sides together with glue. We now have rolled the sheet of paper into a tube. But the points on the left side of the tube are identical to the points on the right side of the tube. One way to glue these two ends is to bend the tube carefully into a circle, and seal the two open ends together with glue (Figure 9.2).

What we have done is to turn a sheet of paper into a doughnut. A rocket ship wandering on the video screen can be described as moving on the surface of an inner tube. Every time the rocket vanishes off the

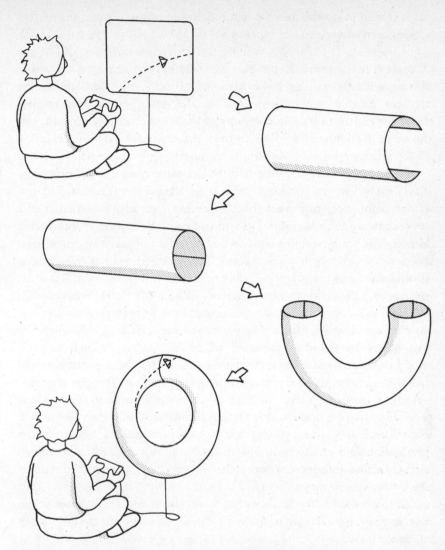

Figure 9.2. If a rocket disappears off the right side of a video-game screen, it re-emerges on the left. If it disappears at the top, it re-emerges at the bottom. Let us now wrap the screen so that identical points match. We first match the top and bottom points by wrapping up the screen. Then we match the points on the left- and right-hand sides by rolling up the screen like a tube. In this way, we can show that a video-game screen has the topology of a doughnut.

video screen and reappears on the other side of the screen, this corresponds to the rocket ship moving across the glued joint of the inner tube.

Vafa conjectures that our sister universe has the shape of some sort of twisted six-dimensional torus. Vafa and his colleagues have pioneered the concept that our sister universe can be described by what mathematicians call an *orbifold*. In fact, his proposal that our sister universe has the topology of an orbifold seems to fit the observed data rather well.[3]

To visualize an orbifold, think of moving 360 degrees in a circle. Everyone knows that we come back to the same point. In other words, if I dance 360 degrees around a May pole, I know that I will come back to the same spot. In an orbifold, however, if we move less than 360 degrees around the May pole, we will still come back to the same point. Although this may sound preposterous, it is easy to construct orbifolds. Think of Flatlanders living on a cone. If they move less than 360 degrees around the apex of the cone, they arrive at the same spot. Thus an orbifold is a higher-dimensional generalization of a cone (Figure 9.3).

To get a feel for orbifolds, imagine that some Flatlanders live on what is called a Z-orbifold, which is equivalent to the surface of a square bean bag (like those found at carnivals and country fairs). At first, nothing seems different from living in Flatland itself. As they explore the surface, however, they begin to find strange happenings. For example, if a Flatlander walks in any direction long enough, he returns to his original position as though he walked in a circle. However, Flatlanders also notice that there is something strange about certain points in their universe (the four points of the bean bag). When walking around any of these four points by 180 degrees (not 360 degrees), they return to the same place from which they started.

The remarkable thing about Vafa's orbifolds is that, with just a few assumptions, we can derive many of the features of quarks and other subatomic particles. (This is because, as we saw earlier, the geometry of space in Kaluza–Klein theory forces the quarks to assume the symmetry of that space.) This gives us confidence that we are on the right track. If these orbifolds gave us totally meaningless results, then our intuition would tell us that there is something fundamentally wrong with this construction.

If none of the solutions of string theory contains the Standard Model, then we must throw away superstring theory as another promising but ultimately incorrect theory. However, physicists are excited by the fact that it is possible to obtain solutions that are tantalizingly close to the Standard Model.

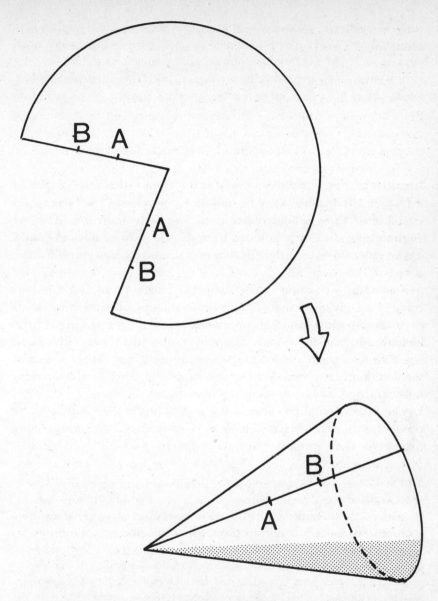

Figure 9.3. If we join points A and B, then we form a cone, which is the simplest example of an orbifold. In string theory, our four-dimensional universe may have a six-dimensional twin, which has the topology of an orbifold. However, the six-dimensional universe is so small that it is unobservable.

Mathematicians for the past 80 years have been working out the properties of these weird surfaces in higher dimensions, ever since the French mathematician Henri Poincaré pioneered the subject of topology in the early twentieth century. Thus the ten-dimensional theory is able to incorporate a large body of modern mathematics that previously seemed quite useless.

Why Are There Three Generations?

In particular, the rich storehouse of mathematical theorems compiled by mathematicians over the past century are now being used to explain why there are three families of particles. As we saw earlier, one disastrous feature of the GUTs is that there are three identical families of quarks and leptons. However, orbifolds may explain this disconcerting feature of the GUTs.[4]

Vafa and his co-workers have discovered many promising solutions to the string equations that appear to resemble the physical world. With a remarkably small set of assumptions, in fact, they can rederive the Standard Model, which is an important step for the theory. This is, in fact, both the strength and the weakness of superstring theory. Vafa and his co-workers have been, in a way, too successful: They have found millions of other possible solutions to the string equations.

The fundamental problem facing superstring theory is this: *Of the millions of possible universes that can be mathematically generated by superstring theory, which is the correct one?* As David Gross has said,

[T]here are millions and millions of solutions that have three spatial dimensions. There is an enormous abundance of possible classical solutions. . . . This abundance of riches was originally very pleasing because it provided evidence that a theory like the heterotic string could look very much like the real world. These solutions, in addition to having four space–time dimensions, had many other properties that resemble our world—the right kinds of particles such as quarks and leptons, and the right kinds of interactions. . . . That was a source of excitement two years ago.[5]

Gross cautions that although some of these solutions are very close to the Standard Model, other solutions produce undesirable physical properties: "It is, however, slightly embarrassing that we have so many solutions but no good way of choosing among them. It seems even more embarrassing that these solutions have, in addition to many desired properties, a few potentially disastrous properties."[6] A layperson, hearing this for the first time, may be puzzled and ask: Why don't you just calculate which solution the string prefers? Since string theory is a well-defined theory, it seems puzzling that physicists cannot calculate the answer.

The problem is that the perturbation theory, one of the main tools in physics, is of no use. Perturbation theory (which adds up increasingly small quantum corrections) fails to break the ten-dimensional theory down to four and six dimensions. Thus we are forced to use nonperturbative methods, which are notoriously difficult to use. This, then, is the reason why we cannot solve string theory. As we said earlier, string field theory, developed by Kikkawa and me and further improved by Witten, cannot at present be solved nonperturbatively. No one is smart enough.

I once had a roommate who was a graduate student in history. I remember one day he warned me about the computer revolution, which eventually might put physicists out of a job. "After all," he said, "computers can calculate everything, can't they?" To him, it was only a matter of time before mathematicians put all physics questions in the computer and physicists got on the unemployment line.

I was taken aback by the comment, because to a physicist a computer is nothing more than a sophisticated adding machine, an impeccable idiot. It makes up in speed what it lacks in intelligence. You have to input the theory into the computer before it can make a calculation. The computer cannot generate new theories by itself.

Furthermore, even if a theory is known, the computer may take an infinite amount of time to solve a problem. In fact, computing all the really interesting questions in physics would take an infinite amount of computer time. This is the problem with string theory. Although Vafa and his colleagues have produced millions of possible solutions, it would take an infinite amount of time to decide which of the millions of possibilities was the correct one, or to calculate solutions to quantum problems involving the bizarre process of tunneling, one of the most difficult of quantum phenomena to solve.

Tunneling Through Space and Time

In the final analysis, we are asking the same question posed by Kaluza in 1919—Where did the fifth dimension go?—except on a much higher level. As Klein pointed out in 1926, the answer to this question has to do with quantum theory. Perhaps the most startling (and complex) phenomenon in quantum theory is tunneling.

For example, I am now sitting in a chair. The thought of my body suddenly zapping through the molecules of the wall next to me and reassembling, uninvited, in someone else's living room is an unpleasant one. Also an unlikely one. However, quantum mechanics postulates that there is a finite (although small) probability that even the most unlikely, bizarre events—such as waking up one morning and finding our bed in the middle of the Amazon jungle—will actually happen. All events, no matter how strange, are reduced by quantum theory to probabilities.

This tunneling process sounds more like science fiction than real science. However, tunneling can be measured in the laboratory and, in fact, solves the riddle of radioactive decay. Normally, the nucleus of an atom is stable. The protons and neutrons within the nucleus are bound together by the nuclear force. However, there is a small probability that the nucleus might fall apart, that the protons and neutrons might escape by tunneling past the large energy barrier, the nuclear force, that binds the nucleus together. Ordinarily, we would say that all nuclei must therefore be stable. But it is an undeniable fact that uranium nuclei do, in fact, decay when they shouldn't; in fact, the conservation of energy law is briefly violated as the neutrons in the nucleus tunnel their way through the barrier.

The catch, however, is that these probabilities are vanishingly small for large objects, such as humans. The probability of our tunneling through a wall within the lifetime of the known universe is infinitesimally small. Thus I can safely assume that I will not be ungraciously transported through the wall, at least within my own lifetime. Similarly, our universe, which originally might have begun as a ten-dimensional universe, was not stable; it tunneled and exploded into a four- and a six-dimensional universe.

To understand this form of tunneling, think of an imaginary Charlie Chaplin film, in which Chaplin is trying to stretch a bed sheet around an oversize bed. The sheet is the kind with elastic bands on the corners. But it is too small, so he has to strain to wrap the elastic bands around each corner of the mattress, one at a time. He grins with satisfaction once he has stretched the bed sheet smoothly around all four corners of the bed. But the strain is too great; one elastic band pops off one corner, and the bed sheet curls up. Frustrated, he pulls this elastic around the corner, only to have another elastic pop off another corner. Every time he yanks an elastic band around one corner, another elastic pops off another corner.

This process is called *symmetry breaking*. The smoothly stretched bed sheet possesses a high degree of symmetry. You can rotate the bed 180 degrees along any axis, and the bed sheet remains the same. This highly symmetrical state is called the *false vacuum*. Although the false vacuum appears quite symmetrical, it is not stable. The sheet does not want to be in this stretched condition. There is too much tension. The energy is too high. Thus one elastic pops off, and the bed sheet curls up. The symmetry is broken, and the bed sheet has gone to a lower-energy state with less symmetry. By rotating the curled-up bed sheet 180 degrees around an axis, we no longer return to the same sheet.

Now replace the bed sheet with ten-dimensional space–time, the space–time of ultimate symmetry. At the beginning of time, the universe was perfectly symmetrical. If anyone was around at that time, he could freely pass through any of the ten dimensions without problem. At that time, gravity and the weak, the strong, and the electromagnetic forces were all unified by the superstring. All matter and forces were part of the same string multiplet. However, this symmetry couldn't last. The ten-dimensional universe, although perfectly symmetrical, was unstable, just like the bed sheet, and in a false vacuum. Thus tunneling to a lower-energy state was inevitable. When tunneling finally occurred, a phase transition took place, and symmetry was lost.

Because the universe began to split up into a four- and a six-dimensional universe, the universe was no longer symmetrical. Six dimensions have curled up, in the same way that the bed sheet curls up when one elastic pops off a corner of a mattress. But notice that there are four ways in which the bed sheet can curl up, depending on which corner pops off first. For the ten-dimensional universe, however, there are apparently millions of ways in which to curl up. To calculate which state the ten-dimensional universe prefers, we need to solve the field theory of strings using the theory of phase transitions, the most difficult problem in quantum theory.

Symmetry Breaking

Phase transitions are nothing new. Think of our own lives. In her book *Passages,* Gail Sheehy stresses that life is not a continuous stream of experiences, as it often appears, but actually passes through several stages, characterized by specific conflicts that must be resolved and goals that must be achieved.

The psychologist Erik Erikson even proposed a theory of the psychological stages of development. A fundamental conflict characterizes each phase. When this conflict is correctly resolved, we move on to the next phase. If this conflict is not resolved, it may fester and even cause regression to an earlier period. Similarly, the psychologist Jean Piaget showed that early childhood mental development is also not a smooth process of learning, but is actually typified by abrupt stages in a child's ability to conceptualize. One month, a child may give up looking for a ball once it has rolled out of view, not understanding that an object exists even if you can no longer see it. The next month, this is obvious to the child.

This is the essence of dialectics. According to this philosophy, all objects (people, gases, the universe itself) go through a series of stages. Each stage is characterized by a conflict between two opposing forces. The nature of this conflict, in fact, determines the nature of the stage. When the conflict is resolved, the object goes to a higher stage, called the synthesis, where a new contradiction begins, and the process starts over again at a higher level.

Philosophers call this the transition from "quantity" to "quality." Small quantitative changes eventually build up until there is a qualitative rupture with the past. This theory applies to societies as well. Tensions in a society can rise dramatically, as they did in France in the late eighteenth century. The peasants faced starvation, spontaneous food riots took place, and the aristocracy retreated behind its fortresses. When the tensions reached the breaking point, a phase transition occurred from the quantitative to the qualitative: The peasants took up arms, seized Paris, and stormed the Bastille.

Phase transitions can also be quite explosive affairs. For example, think of a river that has been dammed up. A reservoir quickly fills up behind the dam with water under enormous pressure. Because it is unstable, the reservoir is in the false vacuum. The water would prefer to be in its true vacuum, meaning it would prefer to burst the dam and wash downstream, to a state of lower energy. Thus a phase transition would involve a dam burst, which could have disastrous consequences.

An even more explosive example is an atomic bomb. The false vacuum corresponds to stable uranium nuclei. Although the uranium nucleus appears stable, there are enormous, explosive energies trapped within the uranium nucleus that are a million times more powerful, pound for pound, than a chemical explosive. Once in a while, the nucleus tunnels to a lower state, which means that the nucleus spontaneously splits apart all by itself. This is called radioactive decay. However, it is possible, by shooting neutrons at the uranium nucleus, to release this pent-up energy all at once. This, of course, is an atomic explosion.

The new feature discovered by scientists about phase transitions is that they are usually accompanied by a symmetry breaking. Nobel laureate Abdus Salam likes the following illustration: Consider a circular banquet table, where all the guests are seated with a champagne glass on either side. There is a symmetry here. Looking at the banquet table through a mirror, we see the same thing: each guest seated around the table, with champagne glasses on either side. Similarly, we can rotate the circular banquet table, and the arrangement is still the same.

Now break the symmetry. Assume that the first diner picks up the glass on his or her right. By custom, all the other guests pick up the champagne glass to their right. Notice that the image of the banquet table as seen in the mirror produces the opposite situation. Every diner has picked up the glass to his or her left. Thus left–right symmetry has been broken.

Another example of symmetry breaking comes from an ancient fairy tale. This fable concerns a princess who is trapped on top of a polished crystal sphere. Although there are no iron bars confining her to the sphere, she is a prisoner because if she makes the slightest move, she will slip off the sphere and kill herself. Numerous princes have tried to rescue the princess, but each has failed to scale the sphere because it is too smooth and slippery. This is an example of symmetry breaking. While the princess is atop the sphere, she is in a perfectly symmetrical state. There is no preferred direction for the sphere. We can rotate the sphere at any angle, and the situation remains the same. Any false move off the center, however, will cause the princess to fall, thereby breaking the symmetry. If she falls to the west, for example, the symmetry of rotation is broken. The westerly direction is now singled out.

Thus the state of maximum symmetry is often also an unstable state, and hence corresponds to a false vacuum. The true vacuum state corresponds to the princess falling off the sphere. So a phase transition (falling off the sphere) corresponds to symmetry breaking (selecting the westerly direction).

Regarding superstring theory, physicists assume (but cannot yet prove) that the original ten-dimensional universe was unstable and tunneled its way to a four- and a six-dimensional universe. Thus the original universe was in the state of the false vacuum, the state of maximum symmetry, while today we are in the broken state of the true vacuum.

This raises a disturbing question: What would happen if our universe were actually not the true vacuum? What would happen if the superstring only temporarily chose our universe, but the true vacuum lay among the millions of possible orbifolds? This would have disastrous consequences. In many other orbifolds, we find that the Standard Model is not present. Thus if the true vacuum were actually a state where the

Standard Model was not present, then all the laws of chemistry and physics, as we know them, would come tumbling down.

If this occurred, a tiny bubble might suddenly appear in our universe. Within this bubble, the Standard Model would no longer hold, so a different set of chemical and physical laws would apply. Matter inside the bubble would disintegrate and perhaps re-form in different ways. This bubble would then expand at the speed of light, swallowing up entire star systems, galaxies, and galactic clusters, until it gobbled up the entire universe.

We would never see it coming. Traveling at the speed of light, it could never be observed beforehand. We would never know what hit us.

From Ice Cubes to Superstrings

Consider an ordinary ice cube sitting in a pressure cooker in our kitchen. We all know what happens if we turn on the stove. But what happens to an ice cube if we heat it up to *trillions upon trillions* of degrees?

If we heat the ice cube on the stove, first it melts and turns into water; that is, it undergoes a phase transition. Now let us heat the water until it boils. It then undergoes another phase transition and turns into steam. Now continue to heat the steam to enormous temperatures. Eventually, the water molecules break up. The energy of the molecules exceeds the binding energy of the molecules, which are ripped apart into elemental hydrogen and oxygen gas.

Now we continue to heat it past 3,000°K, until the atoms of hydrogen and oxygen are ripped apart. The electrons are pulled from the nucleus, and we now have a plasma (an ionized gas), often called the fourth state of matter (after gases, liquids, and solids). Although a plasma is not part of common experience, we can see it every time we look at the sun. In fact, plasma is the most common state of matter in the universe.

Now continue to heat the plasma on the stove to 1 billion°K, until the nuclei of hydrogen and oxygen are ripped apart, and we have a "gas" of individual neutrons and protons, similar to the interior of a neutron star.

If we heat the "gas" of nucleons even further to 10 trillion°K, these subatomic particles will turn into disassociated quarks. We will now have a gas of quarks and leptons (the electrons and neutrinos).

If we heat this gas to 1 quadrillion°K, the electromagnetic force and the weak force will become united. The symmetry $SU(2) \times U(1)$ will emerge at this temperature. At 10^{28} °K, the electroweak and strong forces become united, and the GUT symmetries [SU(5), O(10), or E(6)] appear.

Finally, at a fabulous 10^{32} °K, gravity unites with the GUT force, and all the symmetries of the ten-dimensional superstring appear. We now have a gas of superstrings. At that point, so much energy will have gone into the pressure cooker that the geometry of space–time may very well begin to distort, and the dimensionality of space–time may change. The space around our kitchen may very well become unstable, a rip may form in the fabric of space, and a wormhole may appear in the kitchen. At this point, it may be advisable to leave the kitchen.

Cooling the Big Bang

Thus by heating an ordinary ice cube to fantastic temperatures, we can retrieve the superstring. The lesson here is that matter goes through definite stages of development as we heat it up. Eventually, more and more symmetry becomes restored as we increase the energy.

By reversing this process, we can appreciate how the Big Bang occurred as a sequence of different stages. Instead of heating an ice cube, we now cool the superhot matter in the universe through different stages. Beginning with the instant of Creation, we have the following stages in the evolution of our universe.

10^{-43} seconds The ten-dimensional universe breaks down to a four- and a six-dimensional universe. The six-dimensional universe collapses down to 10^{-32} centimeter in size. The four-dimensional universe inflates rapidly. The temperature is 10^{32} °K.

10^{-35} seconds The GUT force breaks; the strong force is no longer united with the electroweak interactions. SU(3) breaks off from the GUT symmetry. A small speck in the larger universe becomes inflated by a factor of 10^{50}, eventually becoming our visible universe.

10^{-9} seconds The temperature is now 10^{15} °K, and the electroweak symmetry breaks into SU(2) and U(1).

10^{-3} seconds Quarks begin to condense into neutrons and protons. The temperature is roughly 10^{14} °K.

3 minutes The protons and neutrons are now condensing into stable nuclei. The energy of random collisions is no longer powerful enough to break up the nucleus of the emerging nuclei. Space is still opaque to light because ions do not transmit light well.

300,000 years Electrons begin to condense around nuclei. Atoms begin to form. Because light is no longer scattered or absorbed as much, the universe becomes transparent to light. Outer space becomes black.

3 billion years   The first quasars appear.

5 billion years   The first galaxies appear.

10 to 15 billion years   The solar system is born. A few billion years after that, the first forms of life appear on earth.

It seems almost incomprehensible that we, as intelligent apes on the third planet of a minor star in a minor galaxy, would be able to reconstruct the history of our universe going back almost to the instant of its birth, where temperatures and pressures exceeded anything ever found in our solar system. Yet the quantum theory of the weak, electromagnetic, and strong interactions reveals this picture to us.

As startling as this picture of Creation is, perhaps stranger still is the possibility that wormholes can act as gateways to another universe and perhaps even as time machines into the past and future. Armed with a quantum theory of gravity, physicists may be able to answer the intriguing questions: Are there parallel universes? Can the past be changed?

PART III

Wormholes: Gateways to Another Universe?

10
Black Holes and Parallel Universes

Listen, there's a hell of a universe next door: let's go!
e. e. cummings

Black Holes: Tunnels Through Space and Time

BLACK holes have recently seized the public's imagination. Books and documentaries have been devoted to exploring this strange prediction of Einstein's equations, the final stage in the death of a collapsed star. Ironically, the public remains largely unaware of perhaps the most peculiar feature of black holes, that they may be *gateways to an alternative universe*. Furthermore, there is also intense speculation in the scientific community that a black hole may open up a tunnel in time.

To understand black holes and how difficult they are to find, we must first understand what makes the stars shine, how they grow, and how they eventually die. A star is born when a massive cloud of hydrogen gas many times the size of our solar system is slowly compressed by the force of gravity. The gravitational force compressing the gas gradually heats up the gas, as gravitational energy is converted into the kinetic energy of the hydrogen atoms. Normally, the repulsive charge of the protons within the hydrogen gas is sufficient to keep them apart. But at a certain point, when the temperature rises to 10 to 100 million°K, the kinetic

217

energy of the protons (which are hydrogen nuclei) overcomes their electrostatic repulsion, and they slam into one another. The nuclear force then takes over from the electromagnetic force, and the two hydrogen nuclei "fuse" into helium, releasing vast quantities of energy.

In other words, a star is a nuclear furnace, burning hydrogen fuel and creating nuclear "ash" in the form of waste helium. A star is also a delicate balancing act between the force of gravity, which tends to crush the star into oblivion, and the nuclear force, which tends to blow the star apart with the force of trillions of hydrogen bombs. A star then matures and ages as it exhausts its nuclear fuel.

To see how energy is extracted from the fusion process and to understand the stages in the life of a star leading to a black hole, we must analyze Figure 10.1, which shows one of the most important curves in modern science, sometimes called the *binding energy curve*. On the horizontal scale is the atomic weight of the various elements, from hydrogen to uranium. On the vertical scale, crudely speaking, is the approximate average "weight" of each proton in the nucleus. Notice that hydrogen and uranium have protons that weigh, on average, more than the protons of other elements in the center of the diagram.

Our sun is an ordinary yellow star, consisting mainly of hydrogen. Like the original Big Bang, it fuses hydrogen and forms helium. However, because the protons in hydrogen weigh more than the protons in helium, there is an excess of mass, which is converted into energy via Einstein's $E = mc^2$ formula. This energy is what binds the nuclei together. This is also the energy released when hydrogen is fused into helium. This is why the sun shines.

However, as the hydrogen is slowly used up over several billion years, a yellow star eventually builds up too much waste helium, and its nuclear furnace shuts off. When that happens, gravity eventually takes over and crushes the star. As temperatures soar, the star soon becomes hot enough to burn waste helium and convert it into the other elements, like lithium and carbon. Notice that energy can still be released as we descend down the curve to the higher elements. In other words, it is still possible to burn waste helium (in the same way that ordinary ash can still be burned under certain conditions). Although the star has decreased enormously in size, its temperature is quite high, and its atmosphere expands greatly in size. In fact, when our own sun exhausts its hydrogen supply and starts to burn helium, its atmosphere may extend out to the orbit of Mars. This is what is called a *red giant*. This means, of course, that the earth will be vaporized in the process. Thus the curve also predicts the ultimate fate of the earth. Since our sun is a middle-

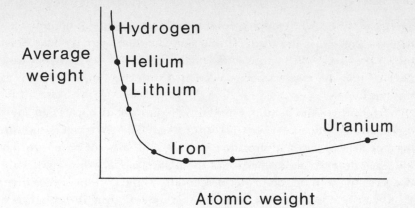

Figure 10.1. The average "weight" of each proton of lighter elements, such as hydrogen and helium, is relatively large. Thus if we fuse hydrogen to form helium inside a star, we have excess mass, which is converted to energy via Einstein's equation $E = mc^2$. This is the energy that lights up the stars. But as stars fuse heavier and heavier elements, eventually we reach iron, and we cannot extract any more energy. The star then collapses, and the tremendous heat of collapse creates a supernova. This colossal explosion rips the star apart and seeds the interstellar space, in which new stars are formed. The process then starts all over again, like a pinball machine.

aged star about 5 billion years old, it still has another 5 billion years before it consumes the earth. (Ironically, the earth was originally born out of the same swirling gas cloud that created our sun. Physics now predicts that the earth, which was created with the sun, will return to the sun.)

Finally, when the helium is used up, the nuclear furnace again shuts down, and gravity takes over to crush the star. The red giant shrinks to become a *white dwarf*, a miniature star with the mass of an entire star squeezed down to about the size of the planet earth.[1] White dwarfs are not very luminous because, after descending to the bottom of the curve, there is only a little excess energy one can squeeze from it through $E = mc^2$. The white dwarf burns what little there is left at the bottom of the curve.

Our sun will eventually turn into a white dwarf and, over billions of years, slowly die as it exhausts its nuclear fuel. It will eventually become a dark, burned-out dwarf star. However, it is believed that if a star is sufficiently massive (several times the mass of our sun), then most of the elements in the white dwarf will continue to be fused into increasingly heavier elements, eventually reaching iron. Once we reach iron, we are near the very bottom of the curve. We can no longer extract any more energy from the excess mass, so the nuclear furnace shuts off. Gravity once again takes over, crushing the star until temperatures rise explosively a thousandfold, reaching trillions of degrees. At this point, the iron core collapses and the outer layer of the white dwarf blows off, releasing the largest burst of energy known in the galaxy, an exploding star called a *supernova*. Just one supernova can temporarily outshine an entire galaxy of 100 billion stars.

In the aftermath of the supernova, we find a totally dead star, a *neutron star* about the size of Manhattan. The densities in a neutron star are so great that, crudely speaking, all the neutrons are "touching" one another. Although neutron stars are almost invisible, we can still detect them with our instruments. Because they emit some radiation while they are rotating, they act like a cosmic lighthouse in outer space. We see them as a blinking star, or *pulsar*. (Although this scenario sounds like science fiction, well over 400 pulsars have been observed since their initial discovery in 1967.)

Computer calculations have shown that most of the heavier elements beyond iron can be synthesized in the heat and pressure of a supernova. When the star explodes, it releases vast amounts of stellar debris, consisting of the higher elements, into the vacuum of space. This debris eventually mixes with other gases, until enough hydrogen gas is accumulated to begin the gravitational contraction process once again. Second-generation stars that are born out of this stellar gas and dust contain an abundance of heavy elements. Some of these stars (like our sun) will have planets surrounding them that also contain these heavy elements.

This solves a long-standing mystery in cosmology. Our bodies are made of heavy elements beyond iron, but our sun is not hot enough to forge them. If the earth and the atoms of our bodies were originally from the same gas cloud, then where did the heavy elements of our bodies come from? The conclusion is inescapable: The heavy elements in our bodies were synthesized in a supernova that blew up *before* our sun was created. In other words, a nameless supernova exploded billions of years ago, seeding the original gas cloud that created our solar system.

The evolution of a star can be roughly pictured as a pinball machine, as in Figure 10.1, with the shape of the binding energy curve. The ball starts at the top and bounces from hydrogen, to helium, from the lighter elements to the heavier elements. Each time it bounces along the curve, it becomes a different type of star. Finally, the ball bounces to the bottom of the curve, where it lands on iron, and is ejected explosively in a supernova. Then as this stellar material is collected again into a new hydrogen-rich star, the process starts all over again on the pinball.

Notice, however, that there are two ways for the pinball to bounce down the curve. It can also start at the other side of the curve, at uranium, and go down the curve in a single bounce by fissioning the uranium nucleus into fragments. Since the average weight of the protons in fission products, like cesium and krypton, is smaller than the average weight of the protons in uranium, the excess mass has been converted into energy via $E = mc^2$. This is the source of energy behind the atomic bomb.

Thus the curve of binding energy not only explains the birth and death of stars and the creation of the elements, but also makes possible the existence of hydrogen and atomic bombs! (Scientists are often asked whether it would be possible to develop nuclear bombs other than atomic and hydrogen bombs. From the curve of binding energy, we can see that the answer is no. Notice that the curve excludes the possibility of bombs made of oxygen or iron. These elements are near the bottom of the curve, so there is not enough excess mass to create a bomb. The various bombs mentioned in the press, such as neutron bombs, are only variations on uranium and hydrogen bombs.)

When one first hears the life history of stars, one may be a bit skeptical. After all, no one has ever lived 10 billion years to witness their evolution. However, since there are uncountable stars in the heavens, it is a simple matter to see stars at practically every stage in their evolution. (For example, the 1987 supernova, which was visible to the naked eye in the southern hemisphere, yielded a treasure trove of astronomical data that matched the theoretical predictions of a collapsing dwarf with an iron core. Also, the spectacular supernova observed by ancient Chinese astronomers on July 4, 1054, left behind a remnant, which has now been identified as a neutron star.)

In addition, our computer programs have become so accurate that we can essentially predict the sequence of stellar evolution numerically. I once had a roommate in graduate school who was an astronomy major. He would invariably disappear in the early morning and return late at night. Just before he would leave, he would say that he was putting a star in the oven to watch it grow. At first, I thought he said this in jest. However, when I pressed him on this point, he said with all seriousness that he was putting a star into the computer and watching it evolve during the day. Since the thermodynamic equations and the fusion equations were well known, it was just a matter of telling the computer to start with a certain mass of hydrogen gas and then letting it numerically solve for the evolution of this gas. In this way, we can check that our theory of stellar evolution can reproduce the known stages of star life that we see in the heavens with our telescopes.

Black Holes

If a star was ten to 50 times the size of our sun, then gravity will continue to squeeze it even after it becomes a neutron star. Without the force of fusion to repel the gravitational pull, there is nothing to oppose the final collapse of the star. At this point, it becomes the famous black hole.

In some sense, black holes must exist. A star, we recall, is the byproduct of two cosmic forces: gravity, which tries to crush the star, and fusion, which tries to blow the star apart like in a hydrogen bomb. All the various phases in the life history of a star are a consequence of this delicate balancing act between gravity and fusion. Sooner or later, when all the nuclear fuel in a massive star is finally exhausted and the star is a mass of pure neutrons, there is nothing known that can then resist the powerful force of gravity. Eventually, the gravitational force will take over and crush the neutron star into nothingness. The star has come full circle: It was born when gravity first began to compress hydrogen gas in the heavens into a star, and it will die when the nuclear fuel is exhausted and gravity collapses it.

The density of a black hole is so large that light, like a rocket launched from the earth, will be forced to orbit around it. Since no light can escape from the enormous gravitational field, the collapsed star becomes black in color. In fact, that is the usual definition of a black hole, a collapsed star from which no light can escape.

To understand this, we note that all heavenly bodies have what is called an *escape velocity*. This is the velocity necessary to escape permanently the gravitational pull of that body. For example, a space probe must reach an escape velocity of 25,000 miles per hour in order to leave the gravitational pull of the earth and go into deep space. Our space probes like the *Voyager* that have ventured into deep space and have completely left the solar system (carrying good-will messages to any aliens who might pick them up) have reached the escape velocity of our sun. (The fact that we breathe oxygen is because the oxygen atoms do not have enough velocity to escape the earth's gravitational field. The fact that Jupiter and the other gas giants are made mainly of hydrogen is because their escape velocity is large enough to capture the primordial hydrogen of the early solar system. Thus escape velocity helps to explain the planetary evolution of the planets of our solar system over the past 5 billion years.)

Newton's theory of gravity, in fact, gives the precise relationship between the escape velocity and the mass of the star. The heavier the planet or star and the smaller its radius, the larger the escape velocity necessary to escape its gravitational pull. As early as 1783, the English astronomer John Michell used this calculation to propose that a super massive star might have an escape velocity equal to the speed of light. The light emitted by such a massive star could never escape, but would orbit around it. Thus, to an outside observer, the star would appear totally black. Using the best knowledge available in the eighteenth century, he actually calculated the mass of such a black hole.* Unfortunately, his theory was considered to be crazy and was soon forgotten. Nevertheless, today we tend to believe that black holes exist because our telescopes and instruments have seen white dwarfs and neutron stars in the heavens.

There are two ways to explain why black holes are black. From the

---

*In the *Philosophical Transactions of the Royal Society,* he wrote, "If the semi-diameter of a sphere of the same density with the Sun were to exceed that of the Sun in the proportion of 500 to 1, a body falling from an infinite height towards it, would have acquired at its surface greater velocity than that of light, and consequently supposing light to be attracted by the same force in proportion to its *vis inertiae,* with other bodies, all light emitted from such a body would be made to return to it by its own proper gravity."[2]

pedestrian point of view, the "force" between the star and a light beam is so great that its path is bent into a circle. Or one can take the Einsteinian point of view, in which case the "shortest distance between two points is a curved line." Bending a light beam into a full circle means that space itself has been bent full circle. This can happen only if the black hole has completely pinched a piece of space–time along with it, so the light beam is circulating in a hypersphere. This piece of space–time has now disconnected itself from the space–time around it. Space itself has now "ripped."

The Einstein–Rosen Bridge

The relativistic description of the black hole comes from the work of Karl Schwarzschild. In 1916, barely a few months after Einstein wrote down his celebrated equations, Schwarzschild was able to solve Einstein's equations exactly and calculate the gravitational field of a massive, stationary star.

Schwarzschild's solution has several interesting features. First, a "point of no return" surrounds the black hole. Any object that comes closer than this radius will inevitably be sucked into the black hole, with no possibility of escape. Inexorably, any person unfortunate enough to come within the Schwarzschild radius would be captured by the black hole and crushed to death. Today, this distance from the black hole is called the *Schwarzschild radius,* or the *horizon* (the farthest visible point).

Second, anyone who fell within the Schwarzschild radius would be aware of a "mirror universe" on the "other side" of space–time (Figure 10.2). Einstein was not worried about the existence of this bizarre mirror universe because communication with it was impossible. Any space probe sent into the center of a black hole would encounter infinite curvature; that is, the gravitational field would be infinite, and any material object would be crushed. The electrons would be ripped off atoms, and even the protons and neutrons within the nuclei themselves would be torn apart. Also, to penetrate through to the alternative universe, the probe would have to go faster than the speed of light, which is not possible. Thus although this mirror universe is mathematically necessary to make sense of the Schwarzschild solution, it could never be observed physically.

Consequently, the celebrated *Einstein–Rosen bridge* connecting these two universes (named after Einstein and his collaborator, Nathan Rosen) was considered a mathematical quirk. The bridge was necessary

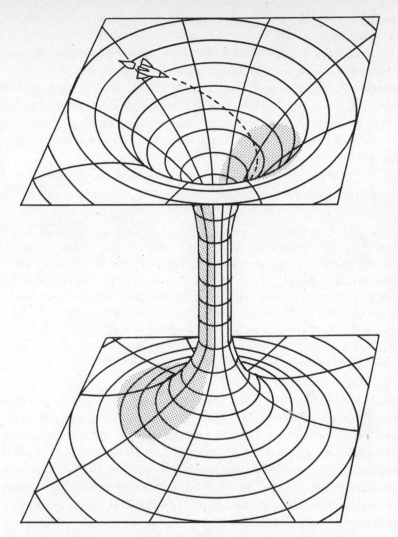

Figure 10.2. The Einstein–Rosen bridge connects two different universes. Einstein believed that any rocket that entered the bridge would be crushed, thereby making communication between these two universes impossible. However, more recent calculations show that travel through the bridge might be very difficult, but perhaps possible.

to have a mathematically consistent theory of the black hole, but it was impossible to reach the mirror universe by traveling through the Einstein–Rosen bridge. Einstein–Rosen bridges were soon found in other solutions of the gravitational equations, such as the Reissner–Nordstrom solution describing an electrically charged black hole. However, the Einstein–Rosen bridge remained a curious but forgotten footnote in the lore of relativity.

Things began to change with the work of New Zealand mathematician Roy Kerr, who in 1963 found another exact solution to Einstein's equations. Kerr assumed that any collapsing star would be rotating. Like a spinning skater who speeds up when bringing in his or her hands, a rotating star would necessarily accelerate as it began to collapse. Thus the stationary Schwarzschild solution for a black hole was not the most physically relevant solution of Einstein's equations.

Kerr's solution created a sensation in the field of relativity when it was proposed. Astrophysicist Subrahmanyan Chandrasekhar once said,

> In my entire scientific life, extending over forty-five years, the most shattering experience has been the realization that an exact solution of Einstein's equations of general relativity, discovered by the New Zealand mathematician Roy Kerr, provides the *absolutely exact representation* of untold numbers of massive black holes that populate the universe. This "shuddering before the beautiful," this incredible fact that a discovery motivated by a search after the beautiful in mathematics should find its exact replica in Nature, persuades me to say that beauty is that to which the human mind responds at its deepest and most profound level.[3]

Kerr found, however, that a massive rotating star does not collapse into a point. Instead, the spinning star flattens until it eventually is compressed into a ring, which has interesting properties. If a probe were shot into the black hole from the side, it would hit the ring and be totally demolished. The curvature of space–time is still infinite when approaching the ring from the side. There is still a "ring of death," so to speak, surrounding the center. However, if a space probe were shot into the ring from the top or bottom, it would experience a large but finite curvature; that is, the gravitational force would not be infinite.

This rather surprising conclusion from Kerr's solution means that any space probe shot through a spinning black hole along its axis of rotation might, in principle, survive the enormous but finite gravitational fields at the center, and go right on through to the mirror universe without being destroyed by infinite curvature. The Einstein–Rosen bridge acts like a tunnel connecting two regions of space–time; it is a wormhole. Thus the Kerr black hole is a gateway to another universe.

Now imagine that your rocket has entered the Einstein–Rosen bridge. As your rocket approaches the spinning black hole, it sees a ring-shaped spinning star. At first, it appears that the rocket is headed for a disastrous crash landing as it descends toward the black hole from the north pole. However, as we get closer to the ring, light from the mirror universe reaches our sensors. Since all electromagnetic radiation, including radar, orbits the black hole, our radar screens are detecting signals that have been circulating around the black hole a number of times. This effect resembles a hall of mirrors, in which we are fooled by the multiple images that surround us. Light goes ricocheting across numerous mirrors, creating the illusion that there are numerous copies of ourselves in the hall.

The same effect occurs as we pass through the Kerr black hole. Because the same light beam orbits the black hole numerous times, our rocket's radar detects images that have gone spinning around the black hole, creating the illusion of objects that aren't really there.

Warp Factor 5

Does this mean that black holes can be used for travel throughout the galaxy, as in *Star Trek* and other science-fiction movies?

As we saw earlier, the curvature in a certain space is determined by the amount of matter–energy contained in that space (Mach's principle). Einstein's famous equation gives us the precise degree of space–time bending caused by the presence of matter–energy.

When Captain Kirk takes us soaring through hyperspace at "warp factor 5," the "dilithium crystals" that power the *Enterprise* must perform miraculous feats of warping space and time. This means that the dilithium crystals have the magical power of bending the space–time continuum into pretzels; that is, they are tremendous storehouses of matter and energy.

If the *Enterprise* travels from the earth to the nearest star, it does not physically move to Alpha Centauri—rather, Alpha Centauri comes to the *Enterprise*. Imagine sitting on a rug and lassoing a table several feet away. If we are strong enough and the floor is slick enough, we can pull the lasso until the carpet begins to fold underneath us. If we pull hard enough, the table comes to us, and the "distance" between the table and us disappears into a mass of crumpled carpeting. Then we simply hop across this "carpet warp." In other words, we have hardly moved; the space between us and the table has contracted, and we just step across this contracted distance. Similarly, the *Enterprise* does not really cross the entire space to Alpha Centauri; it simply moves across the crumpled space–time—through a wormhole. To better understand what happens when one falls down the Einstein–Rosen bridge, let us now discuss the topology of wormholes.

To visualize these multiply connected spaces, imagine that we are strolling down New York's Fifth Avenue one bright afternoon, minding our own business, when a strange floating window opens up in front of us, much like Alice's looking glass. (Never mind for the moment that the energy necessary to open this window might be enough to shatter the earth. This is a purely hypothetical example.)

We step up to the hovering window to take a closer look, and are horrified to find ourselves staring at the head of a nasty-looking *Tyrannosaurus rex*. We are about to run for our lives, when we notice that the tyrannosaur has no body. He can't hurt us because his entire body is clearly on the other side of the window. When we look below the window to find the dinosaur's body, we can see all the way down the street, as though the dinosaur and the window weren't there at all. Puzzled, we slowly circle the window and are relieved to find that the tyrannosaur is nowhere to be found. However, when we peer into the window from the back side, we see the head of a brontosaur staring us in the face (Figure 10.3)!

Frightened, we walk around the window once more, staring at the window sideways. Much to our surprise, all traces of the window, the tyrannosaur, and the brontosaur are gone. We now take a few more turns around the floating window. From one direction, we see the head of the tyrannosaur. From the other direction, we see the head of the brontosaur. And when we look from the side, we find that both the mirror and the dinosaurs have disappeared.

What's happening?

In some faraway universe, the tyrannosaur and the brontosaur have squared off in a life-and-death confrontation. As they face each other, a floating window suddenly appears between them. When the tyrannosaur peers into the floating mirror, he is startled to see the head of a puny, skinny-looking mammal, with frizzy hair and a tiny face: a human. The head is clearly visible, but it has no body. However, when the brontosaur stares into the same window from the other direction, he sees Fifth Avenue, with its shops and traffic. Then the tyrannosaur finds that this human creature in the window has disappeared, only to appear on the side of the window facing the brontosaur.

Now let us say that suddenly the wind blows our hat into the window. We see the hat sailing into the sky of the other universe, but it is nowhere to be seen along Fifth Avenue. We take one long gulp, and then, in

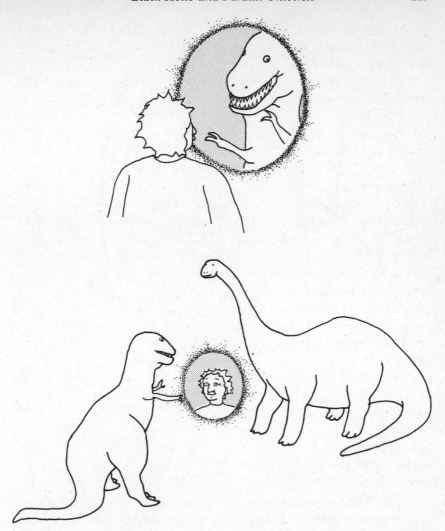

Figure 10.3. In this purely hypothetical example, a "window" or wormhole has opened up in our universe. If we look into the window from one direction, we see one dinosaur. If we look into the other side of the window, we see another dinosaur. As seen from the other universe, a window has opened up between the two dinosaurs. Inside the window, the dinosaurs see a strange small animal (us).

desperation, we stick our hand into the window to retrieve the hat. As seen by the tyrannosaur, a hat blows out the window, appearing from nowhere. Then he sees a disembodied hand reaching out the window, desperately groping for the hat.

The wind now changes direction, and the hat is carried in the other

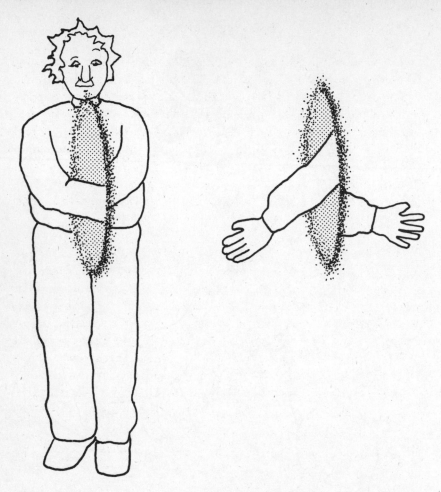

Figure 10.4. If we insert our hands into the window from two different directions, then it appears as though our hands have disappeared. We have a body, but no hands. In the alternative universe, two hands have emerged from either side of the window but they are not attached to a body.

direction. We stick our other hand into the window, but from the other side. We are now in an awkward position. Both our hands are sticking into the window, but from different sides. But we can't see our fingers. Instead, it appears to us that both hands have disappeared.

How does this appear to the dinosaurs? They see two wiggling, tiny hands dangling from the window, from either side. But there is no body (Figure 10.4).

This example illustrates some of the delicious distortions of space and time that one can invent with multiply connected spaces.

Closing the Wormhole

It seems remarkable that such a simple idea—that higher dimensions can unify space with time, and that a "force" can be explained by the warping of that space–time—would lead to such a rich diversity of physical consequences. However, with the wormhole and multiply connected spaces, we are probing the very limits of Einstein's theory of general relativity. In fact, the amount of matter–energy necessary to create a wormhole or dimensional gateway is so large that we expect quantum effects to dominate. Quantum corrections, in turn, may actually close the opening of the wormhole, making travel through the gateway impossible.

Since neither quantum theory nor relativity is powerful enough to settle this question, we will have to wait until the ten-dimensional theory is completed to decide whether these wormholes are physically relevant or just another crazy idea. However, before we discuss the question of quantum corrections and the ten-dimensional theory, let us now pause and consider perhaps the most bizarre consequence of wormholes. Just as physicists can show that wormholes allow for multiply connected spaces, we can also show that they allow for time travel as well.

Let us now consider perhaps the most fascinating, and speculative, consequence of multiply connected universes: building a time machine.

11

To Build a Time Machine

People like us, who believe in physics, know that the distinction between past, present, and future is only a stubbornly persistent illusion.

Albert Einstein

Time Travel

CAN we go backward in time?

Like the protagonist in H. G. Wells's *The Time Machine,* can we spin the dial of a machine and leap hundreds of thousands of years to the year 802,701? Or, like Michael J. Fox, can we hop into our plutonium-fired cars and go back to the future?

The possibility of time travel opens up a vast world of interesting possibilities. Like Kathleen Turner in *Peggy Sue Got Married,* everyone harbors a secret wish somehow to relive the past and correct some small but vital mistake in one's life. In Robert Frost's poem "The Road Not Taken," we wonder what might have happened, at key junctures in our lives, if we had made different choices and taken another path. With time travel, we could go back to our youth and erase embarrassing events from our past, choose a different mate, or enter different careers; or we could even change the outcome of key historical events and alter the fate of humanity.

For example, in the climax of *Superman,* our hero is emotionally devastated when an earthquake ravages most of California and crushes his lover under hundreds of tons of rock and debris. Mourning her horrible death, he is so overcome by anguish that he rockets into space and violates his oath not to tamper with the course of human history. He increases his velocity until he shatters the light barrier, disrupting the fabric of space and time. By traveling at the speed of light, he forces time to slow down, then to stop, and finally to go backward, to a time before Lois Lane was crushed to death.

This trick, however, is clearly not possible. Although time does slow down when you increase your velocity, you cannot go faster than the speed of light (and hence make time go backward) because special relativity states that your mass would become infinite in the process. Thus the faster-than-light travel method preferred by most science-fiction writers contradicts the special theory of relativity.

Einstein himself was well aware of this impossibility, as was A. H. R. Buller when he published the following limerick in *Punch*[1]:

> There was a young lady girl named Bright,
> Whose speed was far faster than light,
> She traveled one day,
> In a relative way,
> And returned on the previous night.

Most scientists, who have not seriously studied Einstein's equations, dismiss time travel as poppycock, with as much validity as lurid accounts of kidnappings by space aliens. However, the situation is actually quite complex.

To resolve the question, we must leave the simpler theory of special relativity, which forbids time travel, and embrace the full power of the *general* theory of relativity, which may permit it. General relativity has much wider validity than special relativity. While special relativity describes only objects moving at constant velocity far away from any stars, the general theory of relativity is much more powerful, capable of describing rockets accelerating near supermassive stars and black holes. The general theory therefore supplants some of the simpler conclusions of the special theory. For any physicist who has seriously analyzed the mathematics of time travel within Einstein's general theory of relativity, the final conclusion is, surprisingly enough, far from clear.

Proponents of time travel point out that Einstein's equations for general relativity do allow some forms of time travel. They acknowledge, however, that the energies necessary to twist time into a circle are so great that Einstein's equations break down. In the physically interesting region where time travel becomes a serious possibility, quantum theory takes over from general relativity.

Einstein's equations, we recall, state that the curvature or bending of space and time is determined by the matter–energy content of the universe. It is, in fact, possible to find configurations of matter–energy powerful enough to force the bending of time and allow for time travel. However, the concentrations of matter–energy necessary to bend time backward are so vast that general relativity breaks down and quantum corrections begin to dominate over relativity. Thus the final verdict on time travel cannot be answered within the framework of Einstein's equations, which break down in extremely large gravitational fields, where we expect quantum theory to become dominant.

This is where the hyperspace theory can settle the question. Because both quantum theory and Einstein's theory of gravity are united in ten-dimensional space, we expect that the question of time travel will be settled decisively by the hyperspace theory. As in the case of wormholes and dimensional windows, the final chapter will be written when we incorporate the full power of the hyperspace theory.

Let us now describe the controversy surrounding time travel and the delicious paradoxes that inevitably arise.

Collapse of Causality

Science-fiction writers have often wondered what might happen if a single individual went back in time. Many of these stories, on the surface, appear plausible. But imagine the chaos that would arise if time machines were as common as automobiles, with tens of millions of them commercially available. Havoc would soon break loose, tearing at the fabric of our universe. Millions of people would go back in time to meddle with their own past and the past of others, rewriting history in the process. A few might even go back in time armed with guns to shoot down the parents of their enemies before they were born. It would thus be impossible to take a simple census to see how many people there were at any given time.

If time travel is possible, then the laws of causality crumble. In fact, all of history as we know it might collapse as well. Imagine the chaos caused by thousands of people going back in time to alter key events that changed the course of history. All of a sudden, the audience at Ford's Theater would be crammed with people from the future bickering among themselves to see who would have the honor of preventing

Lincoln's assassination. The landing at Normandy would be botched as thousands of thrill seekers with cameras arrived to take pictures.

The key battlefields of history would be changed beyond recognition. Consider Alexander the Great's decisive victory over the Persians, led by Darius III, in 331 B.C. at the Battle of Gaugamela. This battle led to the collapse of the Persian forces and ended their rivalry with the West, which helped allow the flourishing of Western civilization and culture over the world for the next 1,000 years. But consider what would happen if a small band of armed mercenaries equipped with small rockets and modern artillery were to enter the battle. The slightest display of modern firepower would rout Alexander's terrified soldiers. This meddling in the past would cripple the expansion of Western influence in the world.

Time travel would mean that any historical event could never be completely resolved. History books could never be written. Some diehard would always be trying to assassinate General Ulysses S. Grant or give the secret of the atomic bomb to the Germans in the 1930s.

What would happen if history could be rewritten as casually as erasing a blackboard? Our past would be like the shifting sands at the seashore, constantly blown this way or that by the slightest breeze. History would be constantly changing every time someone spun the dial of a time machine and blundered his or her way into the past. History, as we know it, would be impossible. It would cease to exist.

Most scientists obviously do not relish this unpleasant possibility. Not only would it be impossible for historians to make any sense out of "history," but genuine paradoxes immediately arise whenever we enter the past or future. Cosmologist Stephen Hawking, in fact, has used this situation to provide "experimental" evidence that time travel is not possible. He believes that time travel is not possible by "the fact that we have not been invaded by hoardes of tourists from the future."

Time Paradoxes

To understand the problems with time travel, it is first necessary to classify the various paradoxes. In general, most can be broken down into one of two principal types:

1. Meeting your parents before you are born
2. The man with no past

The first type of time travel does the most damage to the fabric of space–time because it alters previously recorded events. For example, remember that in *Back to the Future,* our young hero goes back in time and meets his mother as a young girl, just before she falls in love with his father. To his shock and dismay, he finds that he has inadvertently prevented the fateful encounter between his parents. To make matters worse, his young mother has now become amorously attracted to him! If he unwittingly prevents his mother and father from falling in love and is unable to divert his mother's misplaced affections, he will disappear because his birth will never happen.

The second paradox involves events without any beginning. For example, let's say that an impoverished, struggling inventor is trying to construct the world's first time machine in his cluttered basement. Out of nowhere, a wealthy, elderly gentleman appears and offers him ample funds and the complex equations and circuitry to make a time machine. The inventor subsequently enriches himself with the knowledge of time travel, knowing beforehand exactly when stock-market booms and busts will occur before they happen. He makes a fortune betting on the stock market, horse races, and other events. Decades later, as a wealthy, aging man, he goes back in time to fulfill his destiny. He meets himself as a young man working in his basement, and gives his younger self the secret of time travel and the money to exploit it. The question is: Where did the idea of time travel come from?

Perhaps the craziest of these time travel paradoxes of the second type was cooked up by Robert Heinlein in his classic short story "All You Zombies—."

A baby girl is mysteriously dropped off at an orphanage in Cleveland in 1945. "Jane" grows up lonely and dejected, not knowing who her parents are, until one day in 1963 she is strangely attracted to a drifter. She falls in love with him. But just when things are finally looking up for Jane, a series of disasters strike. First, she becomes pregnant by the drifter, who then disappears. Second, during the complicated delivery, doctors find that Jane has both sets of sex organs, and to save her life, they are forced to surgically convert "her" to a "him." Finally, a mysterious stranger kidnaps her baby from the delivery room.

Reeling from these disasters, rejected by society, scorned by fate, "he" becomes a drunkard and drifter. Not only has Jane lost her parents and her lover, but he has lost his only child as well. Years later, in 1970, he stumbles into a lonely bar, called Pop's Place, and spills out his pathetic story to an elderly bartender. The sympathetic bartender offers the drifter the chance to avenge the stranger who left her pregnant and abandoned, on the condition that he join the "time travelers corps." Both of them enter a time machine, and the bartender drops off the drifter in 1963. The drifter is strangely attracted to a young orphan woman, who subsequently becomes pregnant.

The bartender then goes forward 9 months, kidnaps the baby girl from the hospital, and drops off the baby in an orphanage back in 1945. Then the bartender drops off the thoroughly confused drifter in 1985, to enlist in the time travelers corps. The drifter eventually gets his life together, becomes a respected and elderly member of the time travelers corps, and then disguises himself as a bartender and has his most diffi-cult mission: a date with destiny, meeting a certain drifter at Pop's Place in 1970.

The question is: Who is Jane's mother, father, grandfather, grand-mother, son, daughter, granddaughter, and grandson? The girl, the drifter, and the bartender, of course, are all the same person. These paradoxes can made your head spin, especially if you try to untangle Jane's twisted parentage. If we draw Jane's family tree, we find that all the branches are curled inward back on themselves, as in a circle. We come to the astonishing conclusion that she is her own mother and father! She is an entire family tree unto herself.

World Lines

Relativity gives us a simple method to sort through the thorniest of these paradoxes. We will make use of the "world line" method, pioneered by Einstein.

For example, say our alarm clock wakes us up one day at 8:00 A.M., and we decide to spend the morning in bed instead of going to work. Although it appears that we are doing nothing by loafing in bed, we are actually tracing out a "world line."

Take a sheet of graph paper, and on the horizontal scale put "dis-tance" and on the vertical scale put "time." If we simply lie in bed from 8:00 to 12:00, our world line is a straight vertical line. We went 4 hours into the future, but traveled no distance. Even engaging in our favorite pastime, doing nothing, creates a world line. (If someone ever criticizes us for loafing, we can truthfully claim that, according to Einstein's theory of relativity, we are tracing out a world line in four-dimensional space–time.)

Now let's say that we finally get out of bed at noon and arrive at work at 1:00 P.M. Our world line becomes slanted because we are moving in space as well as time. In the lower left corner is our home, and on the upper right is our office (Figure 11.1).If we take the car to work, though, we arrive at the office earlier, at 12:30. This means that the faster we travel, the more our world line deviates from the vertical. (Notice that there is also a "forbidden region" in the diagram that our world line can't enter because we would have to be going faster than the speed of light.)

One conclusion is immediate. Our world line never really begins or ends. Even when we die, the world lines of the molecules in our bodies keep going. These molecules may disperse into the air or soil, but they will trace out their own never-ending world lines. Similarly, when we are born, the world lines of the molecules coming from our mother coalesce into a baby. At no point do world lines break off or appear from nothing.

To see how this all fits together, take the simple example of our own personal world line. In 1950, say, our mother and father met, fell in love, and produced a baby (us). Thus the world lines of our mother and father collided and produced a third world line (ours). Eventually, when someone dies, the world lines forming the person disperse into billions of world lines of our molecules. From this point of view, a human being can be defined as a temporary collection of world lines of molecules. These world lines were scattered before we were born, came together to form our bodies, and will rescatter after we die. The Bible says, "from dust to dust." In this relativistic picture, we might say, "from world lines to world lines."

Our world line thus contains the entire body of information concerning our history. Everything that ever happened to us—from our first bicycle, to our first date, to our first job—is recorded in our world line. In fact, the great Russian cosmologist George Gamow, who was famous for approaching Einstein's work with wit and whimsy, aptly titled his autobiography *My World Line.*

With the aid of the world line, we can now picture what happens when we go back in time. Let's say we enter a time machine and meet our mother before we are born. Unfortunately, she falls in love with us and jilts our father. Do we really disappear, as depicted in *Back to the Future?* On a world line, we now see why this is impossible. When we disappear, our world line disappears. However, according to Einstein, world lines cannot be cut. Thus altering the past is not possible in relativity.

The second paradox, involving re-creating the past, poses interesting problems, however. For example, by going back in time, we are fulfilling the past, not destroying it. Thus the world line of the inventor of time

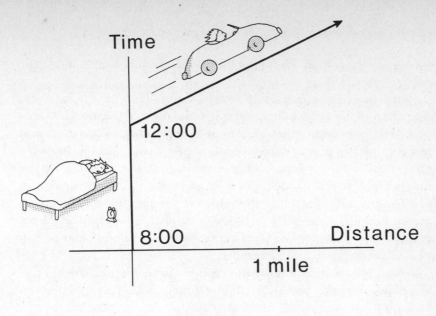

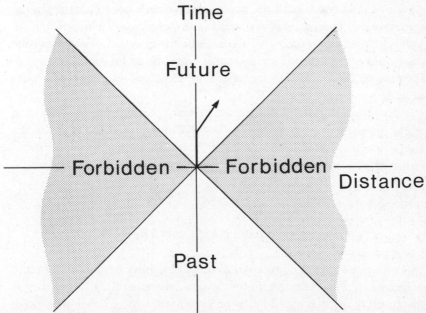

Figure 11.1. Our world line summarizes our entire history, from birth to death. For example, if we lie in bed from 8:00 A.M. to 12:00, our world line is a vertical line. If we travel by car to work, then our world becomes a slanted line. The faster we move, the more slanted our world line becomes. The fastest we can travel, however, is the speed of light. Thus part of this space–time diagram is "forbidden"; that is, we would have to go faster than the speed of light to enter into this forbidden zone.

travel is a closed loop. His world line *fulfills,* rather than changes, the past.

Much more complicated is the world line of "Jane," the woman who is her own mother and father and son and daughter (Figure 11.2).

Notice, once again, that we cannot alter the past. When our world line goes back in time, it simply fulfills what is already known. In such a universe, therefore, it is possible to meet yourself in the past. If we live through one cycle, then sooner or later we meet a young man or woman who happens to be ourselves when we were younger. We tell this young person that he or she looks suspiciously familiar. Then, thinking a bit, we remember that when we were young, we met a curious, older person who claimed that we looked familiar.

Thus perhaps we can fulfill the past, but never alter it. World lines, as we have stressed, cannot be cut and cannot end. They can perhaps perform loops in time, but never alter it.

These light cone diagrams, however, have been presented only in the framework of special relativity, which can describe what happens if we enter the past, but is too primitive to settle the question of whether time travel makes any sense. To answer this larger question, we must turn to the general theory of relativity, where the situation becomes much more delicate.

With the full power of general relativity, we see that these twisted world lines might be physically allowed. These closed loops go by the scientific name *closed timelike curves* (CTCs). The debate in scientific circles is whether CTCs are allowed by general relativity and quantum theory.

Spoiler of Arithmetic and General Relativity

In 1949, Einstein was concerned about a discovery by one of his close colleagues and friends, the Viennese mathematician Kurt Gödel, also at the Institute for Advanced Study at Princeton, where Einstein worked. Gödel found a disturbing solution to Einstein's equations that allowed for violations of the basic tenets of common sense: His solution allowed for certain forms of time travel. For the first time in history, time travel was given a mathematical foundation.

In some quarters, Gödel was known as a spoiler. In 1931, he became famous (or, actually, infamous) when he proved, contrary to every expectation, that you cannot prove the self-consistency of arithmetic. In the process, he ruined a 2,000-year-old dream, dating back to Euclid and

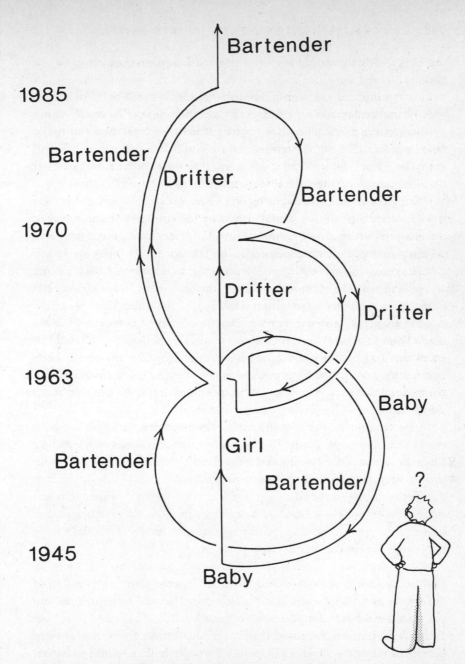

1985

1970

1963

1945

Bartender

Bartender
Drifter
Bartender

Drifter
Drifter

Baby

Girl
Bartender

Bartender

?

Baby

Figure 11.2. If time travel is possible, then our world line becomes a closed loop. In 1945, the girl is born. In 1963, she has a baby. In 1970, he is a drifter, who goes back to 1945 to meet himself. In 1985, he is a time traveler, who picks himself up in a bar in 1970, takes himself back to 1945, kidnaps the baby and takes her back to 1945, to start all over again. The girl is her own mother, father, grandfather, grandmother, son, daughter, and so on.

the Greeks, which was to have been the crowning achievement of mathematics: to reduce all of mathematics to a small, self-consistent set of axioms from which everything could be derived.

In a mathematical tour de force, Gödel showed that there will always be theorems in arithmetic whose correctness or incorrectness can never be demonstrated from the axioms of arithmetic; that is, arithmetic will always be incomplete. Gödel's result was the most startling, unexpected development in mathematical logic in perhaps a thousand years.

Mathematics, once thought to be the purest of all sciences because it was precise and certain, untarnished by the unpleasant crudeness of our material world, now became uncertain. After Gödel, the fundamental basis for mathematics seemed to be left adrift. (Crudely speaking, Gödel's remarkable proof began by showing that there are curious paradoxes in logic. For example, consider the statement "This sentence is false." If the sentence is true, then it follows that it is false. If the sentence is false, then the sentence is true. Or consider the statement "I am a liar." Then I am a liar only if I tell the truth. Gödel then formulated the statement "This sentence cannot be proved true." If the sentence is correct, then it cannot be proved to be correct. By carefully building a complex web of such paradoxes, Gödel showed that there are true statements that cannot be proved using arithmetic.)

After demolishing one of the most cherished dreams of all of mathematics, Gödel next shattered the conventional wisdom surrounding Einstein's equations. He showed that Einstein's theory contains some surprising pathologies, including time travel.

He first assumed that the universe was filled with gas or dust that was slowly rotating. This seemed reasonable, since the far reaches of the universe do seem to be filled with gas and dust. However, Gödel's solution caused great concern for two reasons.

First, his solution violated Mach's principle. He showed that *two* solutions of Einstein's equations were possible with the same distribution of dust and gas. (This meant that Mach's principle was somehow incomplete, that hidden assumptions were present.)

More important, he showed that certain forms of time travel were permitted. If one followed the path of a particle in a Gödel universe, eventually it would come back and meet itself in the past. He wrote, "By making a round trip on a rocket ship in a sufficiently wide curve, it is possible in these worlds to travel into any region of the past, present, and future, and back again."[2] Thus Gödel found the first CTC in general relativity.

Previously, Newton considered time to be moving like a straight arrow, which unerringly flies forward toward its target. Nothing could deflect or change the course of this arrow once it was shot. Einstein, however, showed that time was more like a mighty river, moving forward but often meandering through twisting valleys and plains. The presence of matter or energy might momentarily shift the direction of the river, but overall the river's course was smooth: It never abruptly ended or jerked backward. However, Gödel showed that the river of time could be smoothly bent backward into a circle. Rivers, after all, have eddy currents and whirlpools. In the main, a river may flow forward, but at the edges there are always side pools where water flows in a circular motion.

Gödel's solution could not be dismissed as the work of a crackpot because Gödel had used Einstein's own field equations to find strange solutions in which time bent into a circle. Because Gödel had played by the rules and discovered a legitimate solution to his equations, Einstein was forced to take the evasive route and dismiss it because it did not fit the experimental data.

The weak spot in Gödel's universe was the assumption that the gas and dust in the universe were slowly rotating. Experimentally, we do not see any rotation of the cosmic dust and gas in space. Our instruments have verified that the universe is expanding, but it does not appear to be rotating. Thus the Gödel universe can be safely ruled out. (This leaves us with the rather disturbing, although plausible, possibility that if our universe did rotate, as Gödel speculated, then CTCs and time travel would be physically possible.)

Einstein died in 1955, content that disturbing solutions to his equations could be swept under the rug for experimental reasons and that people could not meet their parents before they were born.

Living in the Twilight Zone

Then, in 1963, Ezra Newman, Theodore Unti, and Louis Tamburino discovered a new solution to Einstein's equations that was even crazier than Gödel's. Unlike the Gödel universe, their solution was not based on a rotating dust-filled universe. On the surface, it resembled a typical black hole.

As in the Gödel solution, their universe allowed for CTCs and time travel. Moreover, when going 360 degrees around the black hole, you would not wind up where you originally started. Instead, like living on a universe with a Riemann cut, you would wind up on another sheet of the universe. The topology of a Newman–Unti–Tamburino universe might be compared to living on a spiral staircase. If we move 360 degrees around the staircase, we do not arrive at the same point at which we started, but on another landing of the staircase. Living in such a universe would surpass our worst nightmare, with common sense being completely thrown out the window. In fact, this bizarre universe was so pathological that it was quickly coined the NUT universe, after the initials of its creators.

At first, relativists dismissed the NUT solution in the same way they had dismissed the Gödel solution; that is, our universe didn't seem to evolve in the way predicted by these solutions, so they were arbitrarily discarded for experimental reasons. However, as the decades went by, there was a flood of such bizarre solutions to Einstein's equations that allowed for time travel. In the early 1970s, Frank J. Tipler at Tulane University in New Orleans reanalyzed an old solution to Einstein's equations found by W. J. van Stockum in 1936, even before Gödel's solution. This solution assumed the existence of an infinitely long, rotating cylinder. Surprisingly enough, Tipler was able to show that this solution also violated causality.

Even the Kerr solution (which represents the most physically realistic description of black holes in outer space) was shown to allow for time travel. Rocket ships that pass through the center of the Kerr black hole (assuming they are not crushed in the process) could violate causality.

Soon, physicists found that NUT-type singularities could be inserted into any black hole or expanding universe. In fact, it now became possible to cook up an infinite number of pathological solutions to Einstein's equations. For example, every wormhole solution to Einstein's equations could be shown to allow some form of time travel.

According to relativist Frank Tipler, "solutions to the field equations can be found which exhibit virtually any type of bizarre behavior."[3] Thus an explosion of pathological solutions to Einstein's equations was discovered that certainly would have horrified Einstein had he still been alive.

Einstein's equations, in some sense, were like a Trojan horse. On the surface, the horse looks like a perfectly acceptable gift, giving us the observed bending of starlight under gravity and a compelling explanation of the origin of the universe. However, inside lurk all sorts of strange demons and goblins, which allow for the possibility of interstellar travel through wormholes and time travel. The price we had to pay for peering into the darkest secrets of the universe was the potential downfall of some of our most commonly held beliefs about our world—that its space is simply connected and its history is unalterable.

But the question still remained: Could these CTCs be dismissed on purely experimental grounds, as Einstein did, or could someone show that they were theoretically possible and then actually build a time machine?

To Build a Time Machine

In June 1988, three physicists (Kip Thorne and Michael Morris at the California Institute of Technology and Ulvi Yurtsever at the University of Michigan) made the first serious proposal for a time machine. They convinced the editors of *Physical Review Letters,* one of the most distinguished publications in the world, that their work merited serious consideration. (Over the decades, scores of crackpot proposals for time travel have been submitted to mainstream physics journals, but all have been rejected because they were not based on sound physical principles or Einstein's equations.) Like experienced scientists, they presented their arguments in accepted field theoretical language and then carefully explained where their weakest assumptions were.

To overcome the skepticism of the scientific community, Thorne and his colleagues realized that they would have to overcome the standard objections to using wormholes as time machines. First, as mentioned earlier, Einstein himself realized that the gravitational forces at the center of a black hole would be so enormous that any spacecraft would be torn apart. Although wormholes were mathematically possible, they were, in practice, useless.

Second, wormholes might be unstable. One could show that small disturbances in wormholes would cause the Einstein–Rosen bridge to collapse. Thus a spaceship's presence inside a black hole would be sufficient to cause a disturbance that would close the entrance to the wormhole.

Third, one would have to go faster than the speed of light actually to penetrate the wormhole to the other side.

Fourth, quantum effects would be so large that the wormhole might close by itself. For example, the intense radiation emitted by the entrance to the black hole not only would kill anyone who tried to enter the black hole, but also might close the entrance.

Fifth, time slows down in a wormhole and comes to a complete stop at the center. Thus wormholes have the undesirable feature that as seen by someone on the earth, a space traveler appears to slow down and come to a total halt at the center of the black hole. The space traveler looks like he or she is frozen in time. In other words, it takes an infinite amount of time for a space traveler to go through a wormhole. Assuming, for the moment, that one could somehow go through the center of the wormhole and return to earth, the distortion of time would still be so great that millions or even billions of years may have passed on the earth.

For all these reasons, the wormhole solutions were never taken seriously.

Thorne is a serious cosmologist, one who might normally view time machines with extreme skepticism or even derision. However, Thorne was gradually drawn into this quest in the most curious way. In the summer of 1985, Carl Sagan sent to Thorne the prepublication draft of his next book, a novel called *Contact,* which seriously explores the scientific and political questions surrounding an epoch-making event: making contact with the first extraterrestrial life in outer space. Every scientist pondering the question of life in outer space must confront the question of how to break the light barrier. Since Einstein's special theory of relativity explicitly forbids travel faster than the speed of light, traveling to the distant stars in a conventional spaceship may take thousands of years, thereby making interstellar travel impractical. Since Sagan wanted to make his book as scientifically accurate as possible, he wrote to Thorne asking whether there was any scientifically acceptable way of evading the light barrier.

Sagan's request piqued Thorne's intellectual curiosity. Here was an honest, scientifically relevant request made by one scientist to another that demanded a serious reply. Fortunately, because of the unorthodox nature of the request, Thorne and his colleagues approached the question in a most unusual way: They worked *backward.* Normally, physicists start with a certain known astronomical object (a neutron star, a black hole, the Big Bang) and then solve Einstein's equations to find the curvature of the surrounding space. The essence of Einstein's equations, we recall, is that the matter and energy content of an object determines the amount of curvature in the surrounding space and time. Proceeding in this way, we are guaranteed to find solutions to Einstein's equations for astronomically relevant objects that we expect to find in outer space.

However, because of Sagan's strange request, Thorne and his colleagues approached the question backward. They started with a rough idea of what they wanted to find. They wanted a solution to Einstein's equations in which a space traveler would not be torn apart by the tidal effects of the intense gravitational field. They wanted a wormhole that would be stable and not suddenly close up in the middle of the trip. They wanted a wormhole in which the time it takes for a round trip would be measured in days, not millions or billions of earth years, and so on. In fact, their guiding principle was that they wanted a time traveler to have a reasonably comfortable ride back through time after entering the wormhole. Once they decided what their wormhole would look like, then, and only then, did they begin to calculate the amount of energy necessary to create such a wormhole.

From their unorthodox point of view, they did not particularly care if the energy requirements were well beyond twentieth-century science. To them, it was an engineering problem for some future civilization actually to construct the time machine. They wanted to prove that it was scientifically feasible, not that it was economical or within the bounds of present-day earth science:

> Normally, theoretical physicists ask, "What are the laws of physics?" and/ or "What do those laws predict about the Universe?" In this Letter, we ask, instead, "What constraints do the laws of physics place on the activities of an arbitrarily advanced civilization?" This will lead to some intriguing queries about the laws themselves. We begin by asking whether the laws of physics permit an arbitrarily advanced civilization to construct and maintain wormholes for interstellar travel.[4]

The key phrase, of course, is "arbitrarily advanced civilization." The laws of physics tell us what is possible, not what is practical. The laws of physics are independent of what it might cost to test them. Thus what is theoretically possible may exceed the gross national product of the planet earth. Thorne and his colleagues were careful to state that this mythical civilization that can harness the power of wormholes must be "arbitrarily advanced"—that is, capable of performing all experiments that are possible (even if they are not practical for earthlings).

Much to their delight, with remarkable ease they soon found a surprisingly simple solution that satisfied all their rigid constraints. It was *not* a typical black hole solution at all, so they didn't have to worry about all the problems of being ripped apart by a collapsed star. They christened their solution the "transversible wormhole," to distinguish it from the other wormhole solutions that are not transversible by spaceship. They were so excited by their solution that they wrote back to Sagan, who then incorporated some of their ideas in his novel. In fact, they were so surprised by the simplicity of their solution that they were convinced that a beginning graduate student in physics would be able to understand their solution. In the autumn of 1985, on the final exam in a course on general relativity given at Caltech, Thorne gave the wormhole solution to the students without telling them what it was, and they were asked to deduce its physical properties. (Most students gave detailed mathematical analyses of the solution, but they failed to grasp that they were looking at a solution that permitted time travel.)

If the students had been a bit more observant on that final exam, they would have been able to deduce some rather astonishing properties of the wormhole. In fact, they would have found that a trip through this transversible wormhole would be as comfortable as a trip on an airplane. The maximum gravitational forces experienced by the travelers would not exceed 1 g. In other words, their apparent weight would not exceed their weight on the earth. Furthermore, the travelers would never have to worry about the entrance of the wormhole closing up during the journey. Thorne's wormhole is, in fact, permanently open. Instead of taking a million or a billion years, a trip through the transversible wormhole would be manageable. Morris and Thorne write that "the trip will be fully comfortable and will require a total of about 200 days," or less.[5]

So far, Thorne notes that the time paradoxes that one usually encounters in the movies are not to be found: "From exposure to science fiction scenarios (for example, those in which one goes back in time and kills oneself) one might expect CTCs to give rise to initial trajectories with zero multiplicities" (that is, trajectories that are impossible).[6] However, he has shown that the CTCs that appear in his wormhole seem to *fulfill* the past, rather than change it or initiate time paradoxes.

Finally, in presenting these surprising results to the scientific community, Thorne wrote, "A new class of solutions of the Einstein field equations is presented, which describe wormholes that, in principle, could be traversed by human beings."

There is, of course, a catch to all this, which is one reason why we do not have time machines today. The last step in Thorne's calculation was to deduce the precise nature of the matter and energy necessary to create this marvelous transversible wormhole. Thorne and his colleagues found that at the center of the wormhole, there must be an "exotic" form of matter that has unusual properties. Thorne is quick to point out that this "exotic" form of matter, although unusual, does not seem to violate any of the known laws of physics. He cautions that, at some future point, scientists may prove that exotic matter does not exist. However, at present, exotic matter seems to be a perfectly acceptable form of matter *if* one has access to sufficiently advanced technology. Thorne writes confidently that "from a single wormhole an arbitrarily advanced civilization can construct a machine for backward time travel."

Blueprint for a Time Machine

Anyone who has read H. G. Wells's *The Time Machine,* however, may be disappointed with Thorne's blueprint for a time machine. You do not sit in a chair in your living room, turn a few dials, see blinking lights, and witness the vast panorama of history, including destructive world wars, the rise and fall of great civilizations, or the fruits of futuristic scientific marvels.

One version of Thorne's time machine consists of two chambers, each containing two parallel metal plates. The intense electric fields created between each pair of plates (larger than anything possible with today's technology) rips the fabric of space–time, creating a hole in space that links the two chambers. One chamber is then placed in a rocket ship and is accelerated to near-light velocities, while the other chamber stays on the earth. Since a wormhole can connect two regions of space with different times, a clock in the first chamber ticks slower than a clock in the second chamber. Because time would pass at different rates at the two ends of the wormhole, anyone falling into one end of the wormhole would be instantly hurled into the past or the future.

Another time machine might look like the following. If exotic matter can be found and shaped like metal, then presumably the ideal shape would be a cylinder. A human stands in the center of the cylinder. The exotic matter then warps the space and time surrounding it, creating a wormhole that connects to a distant part of the universe in a different time. At the center of the vortex is the human, who then experiences no more than 1 *g* of gravitational stress as he or she is then sucked into the wormhole and finds himself or herself on the other end of the universe.

On the surface, Thorne's mathematical reasoning is impeccable. Einstein's equations indeed show that wormhole solutions allow for time to pass at different rates on either side of the wormhole, so that time travel, in principle, is possible. The trick, of course, is to create the wormhole in the first place. As Thorne and his collaborators are quick to point out, the main problem is how to harness enough energy to create and maintain a wormhole with exotic matter.

Normally, one of the basic tenets of elementary physics is that all objects have positive energy. Vibrating molecules, moving cars, flying birds, and soaring rockets all have positive energy. (By definition, the empty vacuum of space has zero energy.) However, if we can produce objects with "negative energies" (that is, something that has an energy content less than the vacuum), then we might be able to generate exotic configurations of space and time in which time is bent into a circle.

This rather simple concept goes by a complicated-sounding title: the *averaged weak energy condition* (AWEC). As Thorne is careful to point out, the AWEC must be violated; energy must become temporarily *negative* for time travel to be successful. However, negative energy has historically been anathema to relativists, who realize that negative energy would make possible antigravity and a host of other phenomena that have never been seen experimentally.

But Thorne is quick to point out that there is a way to obtain negative energy, and this is through quantum theory. In 1948, the Dutch physicist Henrik Casimir demonstrated that quantum theory can create negative energy: Just take two large, uncharged parallel metal plates. Ordinarily, common sense tells us that these two plates, because they are electrically neutral, have no force between them. But Casimir proved that the vacuum separating these two plates, because of the Heisenberg Uncertainty Principle, is actually teeming with activity, with trillions of particles and antiparticles constantly appearing and disappearing. They appear out of nowhere and disappear back into the vacuum. Because they are so fleeting, they are, for the most part, unobservable, and they do not violate any of the laws of physics. These "virtual particles" create a net attractive force between these two plates that Casimir predicted was measurable.

When Casimir first published his paper, it met with extreme skepticism. After all, how can two electrically neutral objects attract each other, thereby violating the usual laws of classical electricity? This was unheard of. However, in 1958 physicist M. J. Sparnaay observed this effect in the laboratory, exactly as Casimir had predicted. Since then, it has been christened the *Casimir effect*.

One way of harnessing the Casimir effect is to place two large conducting parallel plates at the entrance of each wormhole, thereby creating negative energy at each end. As Thorne and his colleagues conclude, "It may turn out that the average weak energy condition can never be violated, in which case there could be no such things as transversible wormholes, time travel, or a failure of causality. It's premature to try to cross a bridge before you come to it."[7]

At present, the jury is still out on Thorne's time machine. The decisive factor, all agree, is to have a fully quantized theory of gravity settle the matter once and for all. For example, Stephen Hawking has pointed out that the radiation emitted at the wormhole entrance will be quite large and will contribute back into the matter–energy content of Einstein's equations. This feedback into Einstein's equations will distort the entrance to the wormhole, perhaps even closing it forever. Thorne, however, disagrees that the radiation will be sufficient to close the entrance.

This is where superstring theory comes in. Because superstring theory is a fully quantum-mechanical theory that includes Einstein's theory of general relativity as a subset, it can be used to calculate corrections to the original wormhole theory. In principle, it will allow us to determine whether the AWEC condition is physically realizable, and whether the wormhole entrance stays open for time travelers to enjoy a trip to the past.

Hawking has expressed reservations about Thorne's wormholes. However, this is ironic because Hawking himself has proposed a new theory of wormholes that is even more fantastic. Instead of connecting the present with the past, Hawking proposes to use wormholes to connect our universe with an infinite number of parallel universes!

12
Colliding Universes

[Nature is] not only queerer than we suppose, it is queerer
than we can suppose.

                                                    J. B. S. Haldane

COSMOLOGIST Stephen Hawking is one of the most tragic figures
in science. Dying of an incurable, degenerative disease, he has
relentlessly pursued his research activities in the face of almost insurmountable obstacles. Although he has lost control of his hands, legs,
tongue, and finally his vocal cords, he has spearheaded new avenues of
research while confined to a wheelchair. Any lesser physicist would have
long ago given up the struggle to tackle the great problems of science.

Unable to grasp a pencil or pen, he performs all his calculations in
his head, occasionally aided by an assistant. Bereft of vocal cords, he uses
mechanical devices to communicate with the outside world. But he not
only maintains a vigorous research program, but still took time to write
a best-selling book, *A Brief History of Time*, and to lecture around the
world.

I once visited Hawking in his home just outside Cambridge University
when I was invited to speak at a physics conference he was organizing.
Walking through his living room, I was surprised by the impressive array
of ingenious gadgets that he uses to continue his research. For example,
I saw on his desk a device much like those used by musicians to hold
music sheets. This one, however, was much more elaborate and had the
ability to grab each page and carefully turn it for reading a book. (I
shivered to ponder, as I think many physicists have, whether I would

252

have the stamina and sheer willpower to continue research without arms, legs, or a voice even if I had the finest mechanical aids available.)

Hawking is the Lucasian Professor of Physics at Cambridge University, the same chair held by Isaac Newton. And like his illustrious predecessor, Hawking has embarked on the greatest quest of the century, the final unification of Einstein's theory of gravity and quantum theory. As a result, he, too, has marveled at the elegant, self-consistency of the ten-dimensional theory, and in fact closes his best-selling book with a discussion of it.

Hawking no longer spends the bulk of his creative energy on the field that made him world-famous—black holes—which are by now passé. He is hunting bigger game—the unified field theory. String theory, we recall, began as a quantum theory and then later absorbed Einstein's theory of gravity. Hawking, starting as a pure classical relativist rather than a quantum theorist, approaches the problem from the other point of view. He and his colleague James Hartle start with Einstein's classical universe, and then quantize the entire universe!

Wave Function of the Universe

Hawking is one of the founders of a new scientific discipline, called *quantum cosmology*. At first, this seems like a contradiction in terms. The word *quantum* applies to the infinitesimally small world of quarks and neutrinos, while *cosmology* signifies the almost limitless expanse of outer space. However, Hawking and others now believe that the ultimate questions of cosmology can be answered only by quantum theory. Hawking takes quantum cosmology to its ultimate quantum conclusion, allowing the existence of infinite numbers of parallel universes.

The starting point of quantum theory, we recall, is a wave function that describes all the various possible states of a particle. For example, imagine a large, irregular thundercloud that fills up the sky. The darker the thundercloud, the greater the concentration of water vapor and dust at that point. Thus by simply looking at a thundercloud, we can rapidly estimate the probability of finding large concentrations of water and dust in certain parts of the sky.

The thundercloud may be compared to a single electron's wave function. Like a thundercloud, it fills up all space. Likewise, the greater its value at a point, the greater the probability of finding the electron there. Similarly, wave functions can be associated with large objects, like people. As I sit in my chair in Princeton, I know that I have a Schödinger probability wave function. If I could somehow see my own wave function, it would resemble a cloud very much in the shape of my body. However, some of the cloud would spread out over all space, out to Mars and even beyond the solar system, although it would be vanishingly small there. This means that there is very large likelihood that I am, in fact, sitting in my chair and not on the planet Mars. Although part of my wave function has spread even beyond the Milky Way galaxy, there is only an infinitesimal chance that I am sitting in another galaxy.

Hawking's new idea was to treat the entire universe as though it were a quantum particle. By repeating some simple steps, we are led to some eye-opening conclusions.

We begin with a wave function describing the *set of all possible universes*. This means that the starting point of Hawking's theory must be an infinite set of parallel universes, the *wave function of the universe*. Hawking's rather simple analysis, replacing the word *particle* with *universe*, has led to a conceptual revolution in our thinking about cosmology.

According to this picture, the wave function of the universe spreads out over all possible universes. The wave function is assumed to be quite large near our own universe, so there is a good chance that our universe is the correct one, as we expect. However, the wave function spreads out over all other universes, even those that are lifeless and incompatible with the familiar laws of physics. Since the wave function is supposedly vanishingly small for these other universes, we do not expect that our universe will make a quantum leap to them in the near future.

The goal facing quantum cosmologists is to verify this conjecture mathematically, to show that the wave function of the universe is large for our present universe and vanishingly small for other universes. This would then prove that our familiar universe is in some sense unique and also stable. (At present, quantum cosmologists are unable to solve this important problem.)

If we take Hawking seriously, it means that we must begin our analysis with an infinite number of all possible universes, coexisting with one another. To put it bluntly, the definition of the word *universe* is no longer "all that exists." It now means "all that can exist." For example, in Figure 12.1 we see how the wave function of the universe can spread out over several possible universes, with our universe being the most likely one but certainly not the only one. Hawking's quantum cosmology also assumes that the wave function of the universe allows these universes to collide. Wormholes can develop and link these universes. However, these wormholes are not like the ones we encountered in the previous

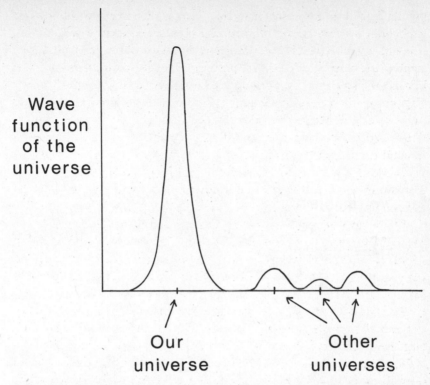

Wave
function
of the
universe

Our
universe

Other
universes

Figure 12.1. In Hawking's wave function of the universe, the wave function is most likely concentrated around own universe. We live in our universe because it is the most likely, with the largest probability. However, there is a small but non-vanishing probability that the wave function prefers neighboring, parallel universes. Thus transitions between universes may be possible (although with very low probability).

chapters, which connect different parts of three-dimensional space with itself—these wormholes connect different universes with one another.

Think, for example, of a large collection of soap bubbles, suspended in air. Normally, each soap bubble is like a universe unto itself, except that periodically it bumps into another bubble, forming a larger one, or splits into two smaller bubbles. The difference is that each soap bubble is now an entire ten-dimensional universe. Since space and time can exist only on each bubble, there is no such thing as space and time between the bubbles. Each universe has its own self-contained "time." It is meaningless to say that time passes at the same rate in all these universes. (We should, however, stress that travel between these universes is not open to us because of our primitive technological level. Furthermore,

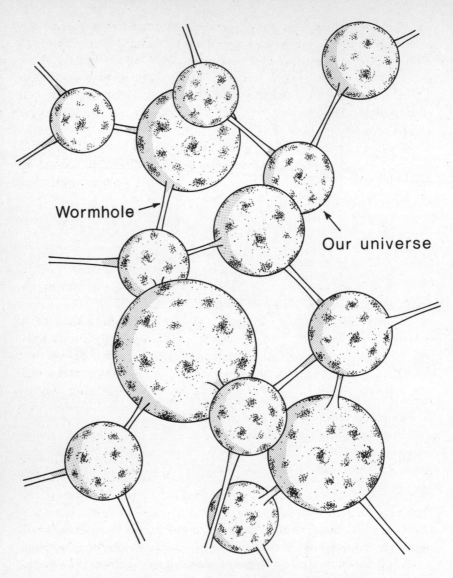

Wormhole

Our universe

Figure 12.2. - Our universe may be one of an infinite number of parallel universes, each connected to the others by an infinite series of wormholes. Travel between these wormholes is possible but extremely unlikely.

we should also stress that large quantum transitions on this scale are extremely rare, probably much larger than the lifetime of our universe.) Most of these universes are dead universes, devoid of any life. On these universes, the laws of physics were different, and hence the physical conditions that made life possible were not satisfied. Perhaps, among the billions of parallel universes, only one (ours) had the right set of physical laws to allow life (Figure 12.2).

Hawking's "baby universe" theory, although not a practical method of transportation, certainly raises philosophical and perhaps even religions questions. Already, it has stimulated two long-simmering debates among cosmologists.

Putting God Back in the Universe?

The first debate concerns the *anthropic principle*. Over the centuries, scientists have learned to view the universe largely independent of human bias. We no longer project our human prejudices and whims onto every scientific discovery. Historically, however, early scientists often committed the fallacy of anthropomorphism, which assumes that objects and animals have humanlike qualities. This error is committed by anyone who sees human emotions and feelings being exhibited by their pets. (It is also committed by Hollywood scriptwriters who regularly assume that beings similar to us must populate planets orbiting the stars in the heavens.)

Anthropomorphism is an age-old problem. The Ionian philosopher Xenophanes once lamented, "Men imagine gods to be born, and to have clothes and voices and shapes like theirs. . . . Yea, the gods of the Ethiopians are black and flat-nosed, and the gods of the Thracians are red-haired and blue-eyed." Within the past few decades, some cosmologists have been horrified to find anthropomorphism creeping back into science, under the guise of the anthropic principle, some of whose advocates openly declare that they would like to put God back into science.

Actually, there is some scientific merit to this strange debate over the anthropic principle, which revolves around the indisputable fact that if the physical constants of the universe were altered by the smallest amount, life in the universe would be impossible. Is this remarkable fact just a fortunate coincidence, or does it show the work of some Supreme Being?

There are two versions of the anthropic principle. The "weak" version states that the fact that intelligent life (us) exists in the universe should be taken as an experimental fact that helps us understand the constants of the universe. As Nobel laureate Steven Weinberg explains it, "the world is the way it is, at least in part, because otherwise there would be no one to ask why it is the way it is."[1] Stated in this way, the weak version of the anthropic principle is hard to argue with.

To have life in the universe, you need a rare conjunction of many coincidences. Life, which depends on a variety of complex biochemical reactions, can easily be rendered impossible if we change some of the constants of chemistry and physics by a small amount. For example, if the constants that govern nuclear physics were changed even slightly, then nucleosynthesis and the creation of the heavy elements in the stars and supernovae might become impossible. Then atoms might become unstable or impossible to create in supernovae. Life depends on the heavy elements (elements beyond iron) for the creation of DNA and protein molecules. Thus the smallest change in nuclear physics would make the heavy elements of the universe impossible to manufacture in the stars. We are children of the stars; however, if the laws of nuclear physics change in the slightest, then our "parents" are incapable of having "children" (us). As another example, it is safe to say that the creation of life in the early oceans probably took 1 to 2 billion years. However, if we could somehow shrink the lifetime of the proton to several million years, then life would be impossible. There would not be enough time to create life out of random collisions of molecules.

In other words, the very fact that we exist in the universe to ask these questions about it means that a complex sequence of events must necessarily have happened. It means that the physical constants of nature must have a certain range of values, so that the stars lived long enough to create the heavy elements in our bodies, so that protons don't decay too rapidly before life has a chance to germinate, and so on. In other words, the existence of humans who can ask questions about the universe places a huge number of rigid constraints on the physics of the universe—for example, its age, its chemical composition, its temperature, its size, and its physical processes.

Remarking on these cosmic coincidences, physicist Freeman Dyson once wrote, "As we look out into the Universe and identify the many accidents of physics and astronomy that have worked together to our benefit, it almost seems as if the Universe must in some sense have known that we were coming." This takes us to the "strong" version of the anthropic principle, which states that all the physical constants of the universe have been precisely chosen (by God or some Supreme Being) so that life is possible in our universe. The strong version, because it raises questions about a deity, is much more controversial among scientists.

Conceivably, it might have been blind luck if only a few constants of nature were required to assume certain values to make life possible. However, it appears that a large set of physical constants must assume a narrow band of values in order for life to form in our universe. Since accidents of this type are highly improbable, perhaps a divine intelligence (God) precisely chose those values in order to create life.

When scientists first hear of some version of the anthropic principle, they are immediately taken aback. Physicist Heinz Pagels recalled, "Here was a form of reasoning completely foreign to the usual way that theoretical physicists went about their business."[2]

The anthropic argument is a more sophisticated version of the old argument that God located the earth at just the right distance from the sun. If God had placed the earth too close, then it would be too hot to support life. If God had placed the earth too far, then it would be too cold. The fallacy of this argument is that millions of planets in the galaxy probably are sitting at the incorrect distance from their sun, and therefore life on them is impossible. However, some planets will, by pure accident, be at the right distance from their sun. Our planet is one of them, and hence we are here to discuss the question.

Eventually, most scientists become disillusioned with the anthropic principle because it has no predictive power, nor can it be tested. Pagels reluctantly concluded that "unlike the principles of physics, it affords no way to determine whether it is right or wrong; there is no way to test it. Unlike conventional physical principles, the anthropic principle is not subject to experimental falsification—the sure sign that it is not a scientific principle."[3] Physicist Alan Guth says bluntly, "Emotionally, the anthropic principle kind of rubs me the wrong way. . . . The anthropic principle is something that people do if they can't think of anything better to do."[4]

To Richard Feynman, the goal of a theoretical physicist is to "prove yourself wrong as fast as possible."[5] However, the anthropic principle is sterile and cannot be disproved. Or, as Weinberg said, "although science is clearly impossible without scientists, it is not clear that the universe is impossible without science."[6]

The debate over the anthropic principle (and hence, about God) was dormant for many years, until it was recently revived by Hawking's wave function of the universe. If Hawking is correct, then indeed there are an infinite number of parallel universes, many with different physical constants. In some of them, perhaps protons decay too rapidly, or stars cannot manufacture the heavy elements beyond iron, or the Big Crunch takes place too rapidly before life can begin, and so on. In fact, an infinite number of these parallel universes are dead, without the physical laws that can make life as we know it possible.

On one such parallel universe (ours), the laws of physics were compatible with life as we know it. The proof is that we are here today to discuss the matter. If this is true, then perhaps God does not have to be evoked to explain why life, precious as it is, is possible in our universe. However, this reopens the possibility of the weak anthropic principle— that is, that we coexist with many dead universes, and that ours is the only one compatible with life.

The second controversy stimulated by Hawking's wave function of the universe is much deeper and in fact is still unresolved. It is called the Schrödinger's cat problem.

Schrödinger's Cat Revisited

Because Hawking's theory of baby universes and wormholes uses the power of quantum theory, it inevitably reopens the still unresolved debates concerning its foundations. Hawking's wave function of the universe does not completely solve these paradoxes of quantum theory; it only expresses them in a startling new light.

Quantum theory, we recall, states that for every object there is a wave function that measures the probability of finding that object at a certain point in space and time. Quantum theory also states that you never really know the state of a particle until you have made an observation. Before a measurement is made, the particle can be in one of a variety of states, described by the Schrödinger wave function. Thus before an observation or measurement can be made, you can't really know the state of the particle. In fact, the particle exists in a nether state, a sum of *all possible states*, until a measurement is made.

When this idea was first proposed by Niels Bohr and Werner Heisenberg, Einstein revolted against this concept. "Does the moon exist just because a mouse looks at it?" he was fond of asking. According to the strict interpretation of quantum theory, the moon, before it is observed, doesn't really exist as we know it. The moon can be, in fact, in any one of an infinite number of states, including the state of being in the sky, of being blown up, or of not being there at all. It is the measurement process of looking at it that decides that the moon is actually circling the earth.

Einstein had many heated discussions with Niels Bohr challenging this unorthodox world view. (In one exchange, Bohr said to Einstein in exasperation, "You are not thinking. You are merely being logical!"[7]) Even Erwin Schrödinger (who initiated the whole discussion with his celebrated wave equation) protested this reinterpretation of his equation. He once lamented, "I don't like it, and I'm sorry I ever had anything to do with it."[8]

To challenge this revisionist interpretation, the critics asked, "Is a cat dead or alive before you look at it?"

To show how absurd this question is, Schroödinger placed an imaginary cat in a sealed box. The cat faces a gun, which is connected to a Geiger counter, which in turn is connected to a piece of uranium. The uranium atom is unstable and will undergo radioactive decay. If a uranium nucleus disintegrates, it will be picked up by the Geiger counter, which will then trigger the gun, whose bullet will kill the cat.

To decide whether the cat is dead or alive, we must open the box and observe the cat. However, what is the state of the cat before we open the box? According to quantum theory, we can only state that the cat is described by a wave function that describes the sum of a dead cat and a live cat.

To Schrödinger, the idea of thinking about cats that are neither dead nor alive was the height of absurdity, yet nevertheless the experimental confirmation of quantum mechanics forces us to this conclusion. At present, every experiment has verified quantum theory.

The paradox of Schrödinger's cat is so bizarre that one is often reminded of how Alice reacted to the vanishing of the Cheshire cat in Lewis Carroll's fable: " 'You'll see me there,' said the Cat, and vanished. Alice was not much surprised at this, she was getting so well used to queer things happening." Over the years, physicists, too, have gotten used to "queer" things happening in quantum mechanics.

There are at least three major ways that physicists deal with this complexity. First, we can assume that God exists. Because all "observations" imply an observer, then there must be some "consciousness" in the universe. Some physicists, like Nobel laureate Eugene Wigner, have insisted that quantum theory proves the existence of some sort of universal cosmic consciousness in the universe.

The second way of dealing with the paradox is favored by the vast majority of working physicists—to ignore the problem. Most physicists, pointing out that a camera without any consciousness can also make measurements, simply wish that this sticky, but unavoidable, problem would go away.

The physicist Richard Feynman once said, "I think it is safe to say that no one understands quantum mechanics. Do not keep saying to yourself, if you can possibly avoid it, 'But how can it be like that?' because you will go 'down the drain' into a blind alley from which nobody has yet escaped. Nobody knows how it can be like that."[9] In fact, it is often stated that of all the theories proposed in this century, the silliest is quantum theory. Some say that the only thing that quantum theory has going for it, in fact, is that it is unquestionably correct.

However, there is a third way of dealing with this paradox, called the *many-worlds theory*. This theory (like the anthropic principle) fell out of favor in the past decades, but is being revived again by Hawking's wave function of the universe.

Many Worlds

In 1957, physicist Hugh Everett raised the possibility that during the evolution of the universe, it continually "split" in half, like a fork in a road. In one universe, the uranium atom did not disintegrate and the cat was not shot. In the other, the uranium atom did disintegrate and the cat was shot. If Everett is correct, there are an infinite number of universes. Each universe is linked to every other through the network of forks in the road. Or, as the Argentinian writer Jorge Luis Borges wrote in *The Garden of Forking Paths,* "time forks perpetually toward innumerable futures."

Physicist Bryce DeWitt, one of the proponents of the many-worlds theory, describes the lasting impact it made on him: "Every quantum transition taking place on every star, in every galaxy, in every remote corner of the universe is splitting our local world on earth into myriads of copies of itself. I still recall vividly the shock I experienced on first encountering this multiworld concept."[10] The many-worlds theory postulates that *all* possible quantum worlds exist. In some worlds, humans exist as the dominant life form on earth. In other worlds, subatomic events took place that prevented humans from ever evolving on this planet.

As physicist Frank Wilczek noted,

It is said that the history of the world would be entirely different if Helen of Troy had had a wart at the tip of her nose. Well, warts can arise from mutations in single cells, often triggered by exposure to the ultraviolet rays of the sun. Conclusion: there are many, many worlds in which Helen of Troy *did* have a wart at the tip of her nose.[11]

Actually, the idea that there may be multiple universes is an old one. The philosopher St. Albertus Magnus once wrote, "Do there exist many worlds, or is there but a single world? This is one of the most noble and exalted questions in the study of Nature." However, the new twist on this ancient idea is that these many worlds resolve the Schrödinger cat paradox. In one universe, the cat may be dead; in another, the cat is alive.

As strange as Everett's many-worlds theory seems, one can show that it is mathematically equivalent to the usual interpretations of quantum theory. But traditionally, Everett's many-worlds theory has not been popular among physicists. Although it cannot be ruled out, the idea of an *infinite* number of equally valid universes, each fissioning in half at every instant in time, poses a philosophical nightmare for physicists, who love simplicity. There is a principle of physics called Occam's razor, which states that we should always take the simplest possible path and ignore more clumsy alternatives, especially if the alternatives can never be measured. (Thus Occam's razor dismisses the old "aether" theory, which stated that a mysterious gas once pervaded the entire universe. The aether theory provided a convenient answer to an embarrassing question: If light is a wave, and light can travel in a vacuum, then what is waving? The answer was that aether, like a fluid, was vibrating even in a vacuum. Einstein showed that the aether was unnecessary. However, he never said that the aether didn't exist. He merely said it was irrelevant. Thus by Occam's razor, physicists don't refer to the aether anymore.)

One can show that communication between Everett's many worlds is not possible. Therefore, each universe is unaware of the existence of the others. If experiments cannot test for the existence of these worlds, we should, by Occam's razor, eliminate them.

Somewhat in the same vein, physicists do not say categorically that angels and miracles cannot exist. Perhaps they do. But miracles, almost by definition, are not repeatable and therefore not measurable by experiment. Therefore, by Occam's razor, we must dismiss them (unless, of course, we can find a reproducible, measurable miracle or angel). One of the developers of the many-worlds theory, Everett's mentor John Wheeler, reluctantly rejected it because "it required too much metaphysical baggage to carry around."[12]

The unpopularity of the many-worlds theory, however, may subside as Hawking's wave function of the universe gains popularity. Everett's theory was based on single particles, with no possibility of communication between different universes as they fissioned. However, Hawking's theory, although related, goes much further: It is based on an infinite number of self-contained universes (and not just particles) and postulates the possibility of tunneling (via wormholes) between them.

Hawking has even undertaken the daunting task of calculating the solution to the wave function of the universe. He is confident that his approach is correct partly because the theory is well defined (if, as we mentioned, the theory is ultimately defined in ten dimensions). His goal is to show that the wave function of the universe assumes a large value near a universe that looks like ours. Thus our universe is the most likely universe, but certainly not the only one.

By now, there have been a number of international conferences on the wave function of the universe. However, as before, the mathematics involved in the wave function of the universe is beyond the calculational ability of any human on this planet, and we may have to wait years before any enterprising individual can find a rigorous solution to Hawking's equations.

Parallel Worlds

A major difference between Everett's many-worlds theory and Hawking's wave function of the universe is that Hawking's theory places wormholes that connect these parallel universes at the center of his theory. However, there is no need to wonder whether you will someday walk home from work, open the door, enter a parallel universe, and discover that your family never heard of you. Instead of rushing to meet you after a hard day's work, your family is thrown into a panic, scream about an intruder, and have you thrown in jail for illegal entry. This kind of scenario happens only on television or in the movies. In Hawking's approach, the wormholes do, in fact, constantly connect our universe with billions upon billions of parallel universes, but the size of these wormholes, on the average, is extremely small, about the size of the Planck length (about a 100 billion billion times smaller than a proton, too small for human travel). Furthermore, since large quantum transitions between these universes are infrequent, we may have to wait a long time, longer than the lifetime of the universe, before such an event takes place.

Thus it is perfectly consistent with the laws of physics (although *highly* unlikely) that someone may enter a twin universe that is precisely like our universe except for one small crucial difference, created at some point in time when the two universes split apart.

This type of parallel world was explored by John Wyndham in the story "Random Quest." Colin Trafford, a British nuclear physicist, is almost killed in 1954 when a nuclear experiment blows up. Instead of winding up in the hospital, he wakes up, alone and unhurt, in a remote part of London. He is relieved that everything appears normal, but soon discovers that something is very wrong. The newspaper headlines are all impossible. World War II never took place. The atomic bomb was never discovered.

World history has been twisted. Furthermore, he glances at a store shelf and notices his own name, with his picture, as the author of a best-selling book. He is shocked. An exact counterpart of himself exists in this parallel world as an author instead of a nuclear physicist!

Is he dreaming all this? Years ago, he once thought of becoming a writer, but instead he chose to become a nuclear physicist. Apparently in this parallel universe, different choices were made in the past.

Trafford scans the London telephone book and finds his name listed, but the address is wrong. Shaking, he decides to visit "his" home.

Entering "his" apartment, he is shocked to meet "his" wife—someone he has never seen before—a beautiful woman who is bitter and angry over "his" numerous affairs with other women. She berates "him" for his extramarital indiscretions, but she notices that her husband seems confused. His counterpart, Trafford finds out, is a cad and a womanizer. However, he finds it difficult to argue with a beautiful stranger he has never seen before, even if she happens to be "his" wife. Apparently, he and his counterpart have switched universes.

He gradually finds himself falling in love with "his" own wife. He cannot understand how his counterpart could ever have treated his lovely wife in such a despicable manner. The next few weeks spent together are the best of their lives. He decides to undo all the harm his counterpart inflicted on his wife over the years. Then, just as the two are rediscovering each other, he is suddenly wrenched back into his own universe, leaving "his" love behind. Thrown back into his own universe against his will, he begins a frantic quest to find "his" wife. He has discovered that most, but not all, people in his universe have a counterpart in the other. Surely, he reasons, "his" wife must have a counterpart in his own world.

He becomes obsessed, tracking down all the clues that he remembers from the twin universe. Using all his knowledge of history and physics, he concludes that two worlds diverged from each other because of some pivotal event in 1926 or 1927. A single event, he reasons, must have split the two universes apart.

He then meticulously traces the birth and death records of several families. He spends his remaining savings interviewing scores of people until he locates "his" wife's family tree. Eventually, he succeeds in tracking down "his" wife in his own universe. In the end, he marries her.

Attack of the Giant Wormholes

One Harvard physicist who has jumped into the fray concerning wormholes is Sidney Coleman. Resembling a cross between Woody Allen and Albert Einstein, he shuffles through the corridors of Jefferson Hall, trying to convince the skeptics of his latest theory of wormholes. With his Chaplinesque moustache, his hair swept back like Einstein's, and his oversize sweatshirt, Coleman stands out in any crowd. Now he claims to have solved the celebrated cosmological constant problem, which has puzzled physicists for the past 80 years.

His work even made the cover of *Discover Magazine,* with an article entitled "Parallel Universes: The New Reality—From Harvard's Wildest Physicist." He is also wild about science fiction; a serious science-fiction fan, he even co-founded Advent Publishers, which published books on science-fiction criticism.

At present, Coleman vigorously engages the critics who say that scientists won't be able to verify wormhole theories within our lifetime. If we believe in Thorne's wormholes, then we have to wait until someone discovers exotic matter or masters the Casimir effect. Until then, our time machines have no "engine" capable of shooting us into the past. Similarly, if we believe in Hawking's wormholes, then we have to travel in "imaginary time" in order to travel between wormholes. Either way, it a very sad state of affairs for the average theoretical physicist, who feels frustrated by the inadequate, feeble technology of the twentieth century and who can only dream of harnessing the Planck energy.

This is where Coleman's work comes in. He recently made the claim that the wormholes might yield a very tangible, very measurable result in the present, and not in some distant, unforeseeable future. As we pointed out earlier, Einstein's equations state that the matter–energy content of an object determines the curvature of space–time surrounding it. Einstein wondered whether the pure vacuum of empty space could contain energy. Is pure emptiness devoid of energy? This vacuum energy is measured by something called the *cosmological constant;* in principle, there is nothing to prevent a cosmological constant from appearing in the equations. Einstein thought this term was aesthetically ugly, but he could not rule it out on physical or mathematical grounds.

In the 1920s, when Einstein tried to solve his equations for the universe, he found, much to his chagrin, that the universe was expanding. Back then, the prevailing wisdom was that the universe was static and unchanging. In order to "fudge" his equations to prevent the expansion of the universe, Einstein inserted a tiny cosmological constant into this solution, chosen so it would just balance out the expansion, yielding a static universe by fiat. In 1929, when Hubble conclusively proved that the universe is indeed expanding, Einstein banished the cosmological constant and said it was the "greatest blunder of my life."

Today, we know that the cosmological constant is very close to zero. If there were a small negative cosmological constant, then gravity would be powerfully attractive and the entire universe might be, say, just a few feet across. (By reaching out with your hand, you should be able to grab the person in front of you, who happens to be yourself.) If there were a small positive cosmological constant, then gravity would be repulsive and everything would be flying away from you so fast that their light would never reach you. Since neither nightmarish scenario occurs, we are confident that the cosmological constant is extremely tiny or even zero.

But this problem resurfaced in the 1970s, when symmetry breaking was being intensively studied in the Standard Model and GUT theory. Whenever a symmetry is broken, a large amount of energy is dumped into the vacuum. In fact, the amount of energy flooding the vacuum is 10^{100} times larger than the experimentally observed amount. In all of physics, this discrepancy of 10^{100} is unquestionably the largest. Nowhere in physics do we see such a large divergence between theory (which predicts a large vacuum energy whenever a symmetry is broken) and experiment (which measures zero cosmological constant in the universe). This is where Coleman's wormholes comes in; they're needed to cancel the unwanted contributions to the cosmological constant.

According to Hawking, there may be an infinite number of alternative universes coexisting with ours, all of which are connected by an infinite web of interlocking wormholes. Coleman tried to add up the contribution from this infinite series. After the sum was performed, he found a surprising result: The wave function of the universe prefers to have zero cosmological constant, as desired. If the cosmological constant was zero, the wave function became exceptionally large, meaning that there was a high probability of finding a universe with zero cosmological constant. Moreover, the wave function of the universe quickly vanished if the cosmological constant became nonzero, meaning that there was zero probability for that unwanted universe. This was exactly what was needed to cancel the cosmological constant. In other words, the cosmological constant was zero because that was the most probable outcome. The only effect of having billions upon billions of parallel universes was to keep the cosmological constant zero in our universe.

Because this was such an important result, physicists immediately began to leap into the field. "When Sidney came out with this work, everyone jumped," recalls Stanford physicist Leonard Susskind.[13] In his typical puckish way, Coleman published this potentially important result with a bit of humor. "It is always possible that unknown to myself I am up to my neck in quicksand and sinking fast," he wrote.[14]

Coleman likes to impress audiences vividly with the importance of this problem, that the chances of canceling out a cosmological constant to one part in 10^{100} is fantastically small. "Imagine that over a ten-year period you spend millions of dollars without looking at your salary, and when you finally compare what you earn with what you spent, they balance out to the penny," he notes.[15] Thus his calculation, which shows that you can cancel the cosmological constant to one part in 10^{100}, is a highly nontrivial result. To add frosting to the cake, Coleman emphasizes that these wormholes also solve another problem: They help to determine the values of the fundamental constants of the universe. Coleman adds, "It was a completely different mechanism from any that had been considered. It was Batman swinging in on his rope."[16]

But criticisms also began to surface; the most persistent criticism was that he assumed that the wormholes were small, on the order of the Planck length, and that he forgot to sum over large wormholes. According to the critics, large wormholes should also be included in his sum. But since we don't see large, visible wormholes anywhere, it seems that his calculation has a fatal flaw.

Unfazed by this criticism, Coleman shot back in his usual way: choosing outrageous titles for his papers. To prove that large wormholes can be neglected in his calculation, he wrote a rebuttal to his critics with the title "Escape from the Menace of the Giant Wormholes." When asked about his titles, he replied, "If Nobel Prizes were given for titles, I'd have already collected mine."[17]

If Coleman's purely mathematical arguments are correct, they would give hard experimental evidence that wormholes are an essential feature of all physical processes, and not just some pipe dream. It would mean that wormholes connecting our universe with an infinite number of dead universes are essential to prevent our universe from wrapping itself up into a tight, tiny ball, or from exploding outward at fantastic rates. It would mean that wormholes are the essential feature making our universe relatively stable.

But as with most developments that occur at the Planck length, the final solution to these wormhole equations will have to wait until we have a better grasp of quantum gravity. Many of Coleman's equations require a means of eliminating the infinities common to all quantum theories of gravity, and this means using superstring theory. In particular, we may have to wait until we can confidently calculate finite quantum corrections to his theory. Many of these strange predictions will have to wait until we can sharpen our calculational tools.

As we have emphasized, the problem is mainly theoretical. We simply do not have the mathematical brainpower to break open these well-defined problems. The equations stare at us from the blackboard, but we are helpless to find rigorous, finite solutions to them at present. Once physicists have a better grasp of the physics at the Planck energy, then a whole new universe of possibilities opens up. Anyone, or any civilization, that truly masters the energy found at the Planck length will become the master of all fundamental forces. That is the next topic to which we will turn. When can we expect to become masters of hyperspace?

PART IV

Masters of Hyperspace

13

Beyond the Future

What does it mean for a civilization to be a million years old?
We have had radio telescopes and spaceships for a few decades; our technical civilization is a few hundred years old . . .
an advanced civilization millions of years old is as much
beyond us as we are beyond a bush baby or a macaque.

Carl Sagan

PHYSICIST Paul Davies once commented on what to expect once
we have solved the mysteries of the unification of all forces into a
single superforce. He wrote that

we could change the structure of space and time, tie our own knots in
nothingness, and build matter to order. Controlling the superforce would
enable us to construct and transmute particles at will, thus generating
exotic forms of matter. We might even be able to manipulate the dimensionality of space itself, creating bizarre artificial worlds with unimaginable
properties. Truly we should be lords of the universe.[1]

When can we expect to harness the power of hyperspace? Experimental verification of the hyperspace theory, at least indirectly, may
come in the twenty-first century. However, the energy scale necessary to
manipulate (and not just verify) ten-dimensional space–time, to become
"lords of the universe," is many centuries beyond today's technology.
As we have seen, enormous amounts of matter–energy are necessary to

perform near-miraculous feats, such as creating wormholes and altering the direction of time.

To be masters of the tenth dimension, either we encounter intelligent life within the galaxy that has already harnessed these astronomical energy levels, or we struggle for several thousand years before we attain this ability ourselves. For example, our current atom smashers or particle accelerators can boost the energy of a particle to over 1 trillion electron volts (the energy created if an electron were accelerated by 1 trillion volts). The largest accelerator is currently located in Geneva, Switzerland, and operated by a consortium of 14 European nations. But this energy pales before the energy necessary to probe hyperspace: 10^{19} billion electron volts, or a quadrillion times larger than the energy that might have been produced by the SSC.

A quadrillion (1 with 15 zeros after it) may seem like an impossibly large number. The technology necessary to probe this incredible energy may require atom smashers billions of miles long, or an entirely new technology altogether. Even if we were to liquidate the entire gross national product of the world and build a super-powerful atom smasher, we would not be able to come close to this energy. At first, it seems an impossible task to harness this level of energy.

However, this number does not seem so ridiculously large if we understand that technology expands *exponentially*, which is difficult for our minds to comprehend. To understand how fast exponential growth is, imagine a bacterium that splits in half every 30 minutes. If its growth is unimpeded, then within a few weeks this single bacterium will produce a colony that will weigh as much as the entire planet earth.

Although humans have existed on this planet for perhaps 2 million years, the rapid climb to modern civilization within the last 200 years was possible due to the fact that the growth of scientific knowledge is exponential; that is, its rate of expansion is proportional to how much is already known. The more we know, the faster we can know more. For example, we have amassed more knowledge since World War II than all the knowledge amassed in our 2-million-year evolution on this planet. In fact, the amount of knowledge that our scientists gain doubles approximately every 10 to 20 years.

Thus it becomes important to analyze our own development historically. To appreciate how technology can grow exponentially, let us analyze our own evolution, focusing strictly on the energy available to the average human. This will help put the energy necessary to exploit the ten-dimensional theory into proper historical perspective.

The Exponential Rise of Civilization

Today, we may think nothing about taking a Sunday drive in the country in a car with a 200-horsepower engine. But the energy available to the average human during most of our evolution on this planet was considerably less.

During this period, the basic energy source was the power of our own hands, about one-eighth of a horsepower. Humans roamed the earth in small bands, hunting and foraging for food in packs much like animals, using only the energy of their own muscles. From an energy point of view, this changed only within the last 100,000 years. With the invention of hand tools, humans could extend the power of their limbs. Spears extended the power of their arms, clubs the power of their fists, and knives the power of their jaws. In this period, their energy output doubled, to about one-quarter of a horsepower.

Within the past 10,000 or so years, the energy output of a human doubled once again. The main reason for this change was probably the end of the Ice Age, which had retarded human development for thousands of years.

Human society, which consisted of small bands of hunters and gatherers for hundreds of thousands of years, changed with the discovery of agriculture soon after the ice melted. Roving bands of humans, not having to follow game across the plains and forests, settled in stable villages where crops could be harvested around the year. Also, with the melting of the ice sheet came the domestication of animals such as horses and oxen; the energy available to a human rose to approximately 1 horsepower.

With the beginning of a stratified, agrarian life came the division of labor, until society underwent an important change: the transition to a slave society. This meant that one person, the slave owner, could command the energy of hundreds of slaves. This sudden increase in energy made possible inhuman brutality; it also made possible the first true cities, where kings could command their slaves to use large cranes, levers, and pulleys to erect fortresses and monuments to themselves. Because of this increase in energy, out of the deserts and forests rose temples, towers, pyramids, and cities.

From an energy point of view, for about 99.99% of the existence of humanity on this planet, the technological level of our species was only one step above that of animals. It has only been within the past few hundred years that humans have had more than 1 horsepower available to them.

A decisive change came with the Industrial Revolution. Newton's discovery of the universal law of gravity and motion made it possible to reduce mechanics to a set of well-defined equations. Thus Newton's classical theory of the gravitational force, in some sense, paved the way for the modern theory of machines. This helped to make possible the widespread use of steam-powered engines in the nineteenth century; with steam, the average human could command tens to hundreds of horsepowers. For example, the railroads opened up entire continents to development, and steamships opened up modern international trade. Both were energized by the power of steam, heated by coal.

It took over 10,000 years for humanity to create modern civilization over the face of Europe. With steam-driven and later oil-fired machines, the United States was industrialized within a century. Thus the mastery of just a single fundamental force of nature vastly increased the energy available to a human being and irrevocably changed society.

By the late nineteenth century, Maxwell's mastery of the electromagnetic force once again set off a revolution in energy. The electromagnetic force made possible the electrification of our cities and our homes, exponentially increasing the versatility and power of our machines. Steam engines were now being replaced by powerful dynamos.

Within the past 50 years, the discovery of the nuclear force has increased the power available to a single human by a factor of a million. Because the energy of chemical reactions is measured in electron volts, while the energy of fission and fusion is measured in millions of electron volts, we have a millionfold increase in the power available to us.

The lesson from analyzing the historical energy needs of humanity shows graphically how for only 0.01% of our existence we have manipulated energy levels beyond that of animals. Yet within just a few centuries, we have unleashed vast amounts of energy via the electromagnetic and nuclear forces. Let us now leave the past and begin a discussion of the future, using the same methodology, to understand the point at which we may harness the superforce.

Type I, II, and III Civilizations

Futurology, or the prediction of the future from reasonable scientific judgments, is a risky science. Some would not even call it a science at all, but something that more resembles hocus pocus or witchcraft. Futurology has deservedly earned this unsavory reputation because every "scientific" poll conducted by futurologists about the next decade has proved to be wildly off the mark. What makes futurology such a primitive science is that our brains think linearly, while knowledge progresses exponentially. For example, polls of futurologists have shown that they take known technology and simply double or triple it to predict the future. Polls taken in the 1920s showed that futurologists predicted that we would have, within a few decades, huge fleets of blimps taking passengers across the Atlantic.

But science also develops in unexpected ways. In the short run, when extrapolating within a few years, it is a safe bet that science will progress through steady, quantitative improvements on existing technology. However, when extrapolating over a few decades, we find that qualitative breakthroughs in new areas become the dominant factor, where new industries open up in unexpected places.

Perhaps the most famous example of futurology gone wrong is the predictions made by John von Neumann, the father of the modern electronic computer and one of the great mathematicians of the century. After the war, he made two predictions: first, that in the future computers would become so monstrous and costly that only large governments would be able to afford them, and second, that computers would be able to predict the weather accurately.

In reality, the growth of computers went in precisely the opposite direction: We are flooded with inexpensive, miniature computers that can fit in the palm of our hands. Computer chips have become so cheap and plentiful that they are an integral part of some modern appliances. Already, we have the "smart" typewriter (the word processor), and eventually we will have the "smart" vacuum cleaner, the "smart" kitchen, the "smart" television, and the like. Also, computers, no matter how powerful, have failed to predict the weather. Although the classical motion of individual molecules can, in principle, be predicted, the weather is so complex that even someone sneezing can create distortions that will ripple and be magnified across thousands of miles, eventually, perhaps, unleashing a hurricane.

With all these important caveats, let us determine when a civilization (either our own or one in outer space) may attain the ability to master the tenth dimension. Astronomer Nikolai Kardashev of the former Soviet Union once categorized future civilizations in the following way.

A Type I civilization is one that controls the energy resources of an entire planet. This civilization can control the weather, prevent earthquakes, mine deep in the earth's crust, and harvest the oceans. This civilization has already completed the exploration of its solar system.

A Type II civilization is one that controls the power of the sun itself.

This does not mean passively harnessing solar energy; this civilization mines the sun. The energy needs of this civilization are so large that it directly consumes the power of the sun to drive its machines. This civilization will begin the colonization of local star systems.

A Type III civilization is one that controls the power of an entire galaxy. For a power source, it harnesses the power of billions of star systems. It has probably mastered Einstein's equations and can manipulate space–time at will.

The basis of this classification is rather simple: Each level is categorized on the basis of the power source that energizes the civilization. Type I civilizations use the power of an entire planet. Type II civilizations use the power of an entire star. Type III civilizations use the power of an entire galaxy. This classification ignores any predictions concerning the detailed nature of future civilizations (which are bound to be wrong) and instead focuses on aspects that can be reasonably understood by the laws of physics, such as energy supply.

Our civilization, by contrast, can be categorized as a Type 0 civilization, one that is just beginning to tap planetary resources, but does not have the technology and resources to control them. A Type 0 civilization like ours derives its energy from fossil fuels like oil and coal and, in much of the Third World, from raw human labor. Our largest computers cannot even predict the weather, let alone control it. Viewed from this larger perspective, we as a civilization are like a newborn infant.

Although one might guess that the slow march from a Type 0 civilization to a Type III civilization might take millions of years, the extraordinary fact about this classification scheme is that this climb is an exponential one and hence proceeds much faster than anything we can readily conceive.

With all these qualifications, we can still make educated guesses about when our civilization will reach these milestones. Given the rate at which our civilization is growing, we might expect to reach Type I status within a few centuries.

For example, the largest energy source available to our Type 0 civilization is the hydrogen bomb. Our technology is so primitive that we can unleash the power of hydrogen fusion only by detonating a bomb, rather than controlling it in a power generator. However, a simple hurricane generates the power of hundreds of hydrogen bombs. Thus weather control, which is one feature of Type I civilizations, is at least a century away from today's technology.

Similarly, a Type I civilization has already colonized most of its solar system. By contrast, milestones in today's development of space travel are painfully measured on the scale of decades, and therefore qualitative leaps such as space colonization must be measured in centuries. For example, the earliest date for NASA's manned landing on the planet Mars is 2020. Therefore, the colonization of Mars may take place 40 to 50 years after that, and the colonization of the solar system within a century.

By contrast, the transition from a Type I to a Type II civilization may take only 1,000 years. Given the exponential growth of civilization, we may expect that within 1,000 years the energy needs of a civilization will become so large that it must begin to mine the sun to energize its machines.

A typical example of a Type II civilization is the Federation of Planets portrayed in the "Star Trek" series. This civilization has just begun to master the gravitational force—that is, the art of warping space–time via wormholes—and hence, for the first time, has the capability of reaching nearby stars. It has evaded the limit placed by the speed of light by mastering Einstein's theory of general relativity. Small colonies have been established on some of these systems, which the starship *Enterprise* is sworn to protect. The civilization's starships are powered by the collision of matter and antimatter. The ability to create large concentrations of antimatter suitable for space travel places that civilization many centuries to a millennium away from ours.

Advancing to a Type III civilization may take several thousand years or more. This is, in fact, the time scale predicted by Isaac Asimov in his classic Foundation Series, which describes the rise, fall, and re-emergence of a galactic civilization. The time scale involved in each of these transitions involves thousands of years. This civilization has harnessed the energy source contained within the galaxy itself. To it, warp drive, instead of being an exotic form of travel to the nearby stars, is the standard means of trade and commerce between sectors of the galaxy. Thus although it took 2 million years for our species to leave the safety of the forests and build a modern civilization, it may take only thousands of years to leave the safety of our solar system and build a galactic civilization.

One option open to a Type III civilization is harnessing the power of supernovae or black holes. Its starships may even be able to probe the galactic nucleus, which is perhaps the most mysterious of all energy sources. Astrophysicists have theorized that because of the enormous size of the galactic nucleus, the center of our galaxy may contain millions of black holes. If true, this would provide virtually unlimited amounts of energy.

At this point, manipulating energies a million billion times larger than present-day energies should be possible. Thus for a Type III civilization, with the energy output of uncountable star systems and perhaps the galactic nucleus at its disposal, the mastery of the tenth dimension becomes a real possibility.

Astrochicken

I once had lunch with physicist Freeman Dyson of the Institute for Advanced Study. Dyson is a senior figure in the world of physics who has tackled some of the most intellectually challenging and intriguing questions facing humanity, such as new directions in space exploration, the nature of extraterrestrial life, and the future of civilization.

Unlike other physicists, who dwell excessively in narrow, well-defined areas of specialization, Dyson's fertile imagination has roamed across the galaxy. "I cannot, as Bohr and Feynman did, sit for years with my whole mind concentrated upon one deep question. I am interested in too many different directions," he confessed.[2] Thin, remarkably spry, with the owlish expression of an Oxford don, and speaking with a trace of his British accent, he engaged in a long, wide-ranging lunch conversation with me, touching on many of the ideas that have fascinated him over the years.

Viewing the transition of our civilization to Type I status, Dyson finds that our primitive space program is headed in the wrong direction. The current trend is toward heavier payloads and greater lag time between space shots, which is severely retarding the exploration of space. In his writings, he has proposed a radical departure from this trend, based on what he calls the *Astrochicken*.

Small, lightweight, and intelligent, Astrochicken is a versatile space probe that has a clear advantage over the bulky, exorbitantly expensive space missions of the past, which have been a bottleneck to space exploration. "Astrochicken will weight a kilogram instead of Voyager's ton," he claims. "Astrochicken will not be built, it will be grown," he adds. "Astrochicken could be as agile as a hummingbird with a brain weighing no more than a gram."[3]

It will be part machine and part animal, using the most advanced developments in bioengineering. It will be small but powerful enough to explore the outer planets, such as Uranus and Neptune. It will not need huge quantities of rocket fuel; it will be bred and programmed to "eat" ice and hydrocarbons found in the rings surrounding the outer planet. Its genetically engineered stomach will then digest these materials into chemical fuel. Once its appetite has been satisfied, it will then rocket to the next moon or planet.

Astrochicken depends on technological breakthroughs in genetic engineering, artificial intelligence, and solar-electric propulsion. Given the remarkable progress in these ares, Dyson expects that the various technologies for Astrochicken may be available by the year 2016.

Taking the larger view of the development of civilization, Dyson also believes that, at the current rate of development, we may attain Type I status within a few centuries. He does not believe that making the transition between the various types of civilizations will be very difficult. He estimates that the difference in size and power separating the various types of civilizations is roughly a factor of 10 billion. Although this may seem like a large number, a civilization growing at the sluggish rate of 1 percent per year can expect to make the transition between the various civilizations within 2,500 years. Thus it is almost guaranteed that a civilization can steadily progress toward Type III status.

Dyson has written, "A society which happens to possess a strong expansionist drive will expand its habitat from a single planet (Type I) to a biosphere exploiting an entire star (Type II) within a few thousand years, and from a single star to an entire galaxy (Type III) within a few million years. A species which has once passed beyond Type II status is invulnerable to extinction by even the worst imaginable natural or artificial catastrophe."[4]

However, there is one problem. Dyson has concluded that the transition from a Type II to a Type III civilization may pose formidable physical difficulties, due mainly to the limitation imposed by the speed of light. The expansion of a Type II civilization will necessarily proceed at less than the speed of light, which he feel places a severe restriction on its development.

Will a Type II civilization break the light barrier and the bonds of special relativity by exploring the power of hyperspace? Dyson is not sure. Nothing can be ruled out, but the Planck length, he reminded me, is a fantastically small distance, and the energies required to probe down to that distance are unimaginable. Perhaps, he mused, the Planck length is a natural barrier facing all civilizations.

Type III Civilizations in Outer Space

If the long journey to reach Type III status seems remote for our own civilization, perhaps one day we will meet an extraterrestrial civilization that has already harnessed hyperspace for its needs and is willing to share its technology with us. The puzzle facing us, however, is that we do not see signs of any advanced civilization in the heavens, at least not in our solar system or even in our small sector of the galaxy. Our space probes, especially the *Viking* landing on Mars in the 1970s and the *Voyager* missions to Jupiter, Saturn, Uranus, and Neptune in the 1980s, have sent back discouraging information concerning the bleak, lifeless nature of our solar system.

The two most promising planets, Venus and Mars, have turned up no signs of life, let alone advanced civilizations. Venus, named after the goddess of love, was once envisioned by astronomers as well as romantics to be a lush, tropical planet. Instead, our space probes have found a harsh, barren planet, with a suffocating atmosphere of carbon dioxide, blistering temperatures exceeding 800°F, and toxic rains of sulfuric acid.

Mars, the focus of speculation even before Orson Welles caused panic in the country in 1938 during the Depression with his fictional broadcast about an invasion from that planet, has been equally disappointing. We know it to be a desolate, desert planet without traces of surface water. Ancient riverbeds and long-vanished oceans have left their distinctive mark on the surface of Mars, but we see no ruins or any indications of civilization.

Going beyond our solar system, scientists have analyzed the radio emissions from nearby stars with equally fruitless results. Dyson has stressed that any advanced civilization, by the Second Law of Thermodynamics, must necessarily generate large quantities of waste heat. Its energy consumption should be enormous, and a small fraction of that waste heat should be easily detected by our instruments. Thus, Dyson claims, by scanning the nearby stars, our instruments should be able find the telltale fingerprint of waste heat being generated by an advanced civilization. But no matter where we scan the heavens, we see no traces of waste heat or radio communications from Type I, II, or III civilizations. On our own earth, for example, we have mastered the art of radio and television within the past half-century. Thus an expanding sphere of radio waves, about 50 light-years in radius, surrounds our planet. Any star within 50 light-years of earth, if it contains intelligent life, should be able to detect our presence. Likewise, any Type II or III civilization should be broadcasting copious quantities of electromagnetic radiation continuously for the past several thousand years, so that any intelligent life within several thousand light-years of the civilization's planet should be able to detect its presence.

In 1978, astronomer Paul Horowitz scanned all sunlike star systems

(185 in all) within 80 light-years of our solar system, and found no traces of radio emissions from intelligent life. Astronomers Donald Goldsmith and Tobius Owen reported in 1979 a search of more than 600 star systems, also with negative results. This search, called SETI (search for extraterrestrial intelligence), has met with consistent failure. (Encouragingly, in a rare display of scientific generosity, in 1992 Congress appropriated $100 million to be spent over a 10-year period for the High Resolution Microwave Survey, which will scan the nearby stars for intelligent life. These funds will make it possible for the gigantic 305-meter fixed radio dish at Arecibo, Puerto Rico, to scan select stars systematically within 100 light-years of the earth. This will be complemented by the 34-meter movable radio antenna at Goldstone, California, which will sweep broad portions of the night sky. After years of negative results, astronomer Frank Drake of the University of California at Santa Cruz is cautiously optimistic that they will find some positive signs of intelligent life. He remarks, "Many human societies developed science independently through a combination of curiosity and trying to create a better life, and I think those same motivations would exist in other creatures.")

The puzzle deepens when we realize that the probability of intelligent life emerging within our galaxy is surprisingly large. Drake even derived a simple equation to calculate the number of planets with intelligent life forms in the galaxy.

Our galaxy, for example, contains about 200 billion stars. To get a ballpark figure for the number of stars with intelligent life forms, we can make the following very crude estimate. We can be conservative and say that 10% of these stars are yellow stars much like the sun, that 10% of those have planets orbiting them, that 10% of those have earthlike planets, that 10% of those have earthlike planets with atmospheres compatible with life, that 10% have earthlike atmospheres with life forms growing in them, and that 10% of those have some form of intelligent life. This means that one-millionth of the 200 billion stars in the galaxy will probably have some intelligent life form. This implies that a staggering 200,000 stars will have planets harboring some form of intelligent life. A slightly more optimistic set of values for Drake's equation shows that intelligent life might be, on the average, as close as 15 light-years from our sun.

With recent advanced computer techniques, scientists have been able to refine Drake's original back-of-the-envelope calculation. George W. Wetherill of the Carnegie Institution of Washington, for example, has run computer simulations of the early evolution of our solar system, beginning with a large, swirling disk of gas and dust around the sun. He lets the computer evolve the disk until small, rocky masses begin to coalesce out of the dust. Much to his pleasant surprise, he found that planets of approximately the size of the earth were easy to evolve out of these rocky cores. Most of the time, in fact, earth-size planets spontaneously coalesced with masses between 80% and 130% of the earth's distance from the sun. (Curiously, he also found that the formation of Jupiter-size planets far from the sun was important for the evolution of the earth-size planets. The Jupiter-size planets were essential to sweep out swarms of comets and debris that would eventually strike the earthlike planet, extinguishing any primitive life forms on it. Wetherill's computer simulations show that without a Jupiter-like planet to clean out these comets with its gigantic gravitational pull, these comets would hit the earthlike planet about 1,000 times more frequently than they do in reality, making a life-destroying impact every 100,000 years or so.)

Thus it is a compelling (but certainly not rigorous) conclusion that the laws of probability favor the presence of other intelligence within the galaxy. The fact that our galaxy is perhaps 10 billion years old means that there has been ample time for scores of intelligent life forms to have flourished within it. Type II and III civilizations, broadcasting for several hundred to several thousand years, should be sending out an easily detectable sphere of electromagnetic radiation measuring several hundred to several thousand light-years in diameter. Yet we see no signs of intelligent life forms in the heavens.

Why?

Several speculative theories have been advanced to explain why we have been unable to detect signs of intelligent life out to 100 light-years of our planet. None of them is particularly satisfying, and the final truth may be a combination of all of them.

One theory holds that Drake's equation may give us rough probabilities of how many planets contain intelligent life, but tells us nothing about when these planets attain this level of development. Given the astronomical time scales involved, perhaps Drake's equation predicts intelligent life forms that existed millions of years before us, or will exist millions of years after us.

For example, our solar system is approximately 4.5 billion years old. Life started on the earth about 3 to 4 billion years ago, but only within the past million years has intelligent life developed on the planet (and only within the past few decades has this civilization built radio stations capable of sending signals into outer space). However, 1 million years, on the time scale of billions of years, is but an instant of time. It is reasonable to assume that thousands of advanced civilizations existed before our distant ancestors even left the forest and have since perished, or that thousands more civilizations will develop long after ours has died. Either way, we would not be able to detect them via our instruments.

The second theory holds that the galaxy is, in fact, teeming with advanced forms of civilizations, but they are advanced enough to conceal their existence from our prying instruments. We would mean nothing to them because they are so many millions of years ahead of us. For example, if we stumble on an ant colony while walking in a field, our first impulse is certainly not to make contact with the ants, ask to see their leader, wave trinkets before their eyes, and offer them unparalleled prosperity and the fruits of our advanced technology. More likely, our first temptation is to ignore them (or perhaps even step on a few of them).

Puzzled by these long-standing questions, I asked Dyson if he thought we would soon be making contact with extraterrestrial life forms. His answer rather surprised me. He said, "I hope not." I thought it was strange that someone who had spent decades speculating about intelligent civilizations in outer space should have reservations about actually meeting them. Knowing British history, however, he must have had good reasons for not rushing in to embrace other civilizations. British civilization was probably only several hundred years more advanced than many of the civilizations, such as the Indian and the African, conquered by the British army and navy.

Although most science-fiction writers bewail the limitations on space exploration placed by the speed of light, Dyson takes the unorthodox view that perhaps this is a good thing. Viewing the often bloody history of colonialism throughout our own world history, perhaps it is a blessing in disguise, he muses, that various Type II civilizations will be separated by large distances and that the Planck energy is inaccessible. Looking at the bright side, he quipped, "At least, one can evade the tax collector."

Unfortunately, the meeting of two unequal civilizations has often had catastrophic implications for the weaker one. For example, the Aztec civilization had risen over thousands of years to great prominence in central Mexico. In some areas, its mastery of science, art, and technology rivaled the achievements of Europe. However, in the area of gunpowder and warships, the Aztecs were perhaps several centuries behind the Spanish. The sudden clash between a small, ragged band of 400 conquistadors and the advanced civilizations of the Aztecs ended in tragedy in 1521. Within a brief period of time, the Aztec people, with a population numbering in the millions, were systematically crushed and enslaved to work in the mines. Their treasuries were looted, their history was erased, and even the faintest memory of the great Aztec civilization was obliterated by waves of missionaries.

When we think of how we might react to visitors from outer space, it is sobering to read how the Aztecs reacted to the visitors from Spain: "They seized upon the gold as if they were monkeys, their faces gleaming. For clearly their thirst for gold was insatiable; they starved for it; they lusted for it; they wanted to stuff themselves with it as if they were pigs. So they went about fingering, taking up the streamers of gold, moving them back and forth, grabbing them to themselves, babbling, talking gibberish among themselves."*[5]

On a cosmic scale, the sudden interactions between civilizations could be even more dramatic. Because we are talking about astronomical time scales, it is likely that a civilization that is a million years ahead of us will find us totally uninteresting. Furthermore, there is probably little that our planet can offer these aliens in terms of natural resources that isn't simultaneously available in numerous other star systems.

In the "Star Trek" series, however, the Federation of Planets encounters other hostile civilizations, the Klingons and Romulans, which are *precisely* at the same stage of technological development as the Federation. This may increase the drama and tension of the series, but the odds of this happening are truly astronomical. More likely, as we venture off into the galaxy in starships, we will encounter civilizations at vastly different levels of technological development, some perhaps millions of years ahead of us.

The Rise and Fall of Civilizations

In addition to the possibilities that we may have missed other civilizations by millions of years and that other civilizations may not consider us worthy of notice, a third theory, which is more interesting, holds that thousands of intelligent life forms did arise from the swamp, but they were unable to negotiate a series of catastrophes, both natural and self-

---

*So perhaps we shouldn't be so enthusiastic about making contact with intelligent extraterrestrials. Scientists point out that on the earth, there are two types of animals: predators like cats, dogs, and tigers (which have eyes to the front of their face, so they can stereoscopically zero in on their target) and prey like rabbits and deer (which have eyes to the side of their face in order to look around 360 degrees for the predators). Typically, predators are more intelligent then prey. Tests show that cats are more intelligent than mice, and foxes are more intelligent than rabbits. Humans, with eyes to the front, are also predators. In our search for intelligent life in the heavens, we should keep in mind that the aliens we meet will probably also have evolved from predators.

inflicted. If this theory is correct, then perhaps someday our starships will find the ruins of ancient civilizations on far-off planets, or, more likely, our own civilization will be faced with these catastrophes. Instead of becoming "lords of the universe," we may follow the road to self-destruction. Thus the question we ask is: What is the fate of advanced civilizations? Will we (they) survive long enough to master the physics of the tenth dimension?

The rise of civilizations is not marked by a steady and sure growth in technology and knowledge. History shows us that civilizations rise, mature, and then disappear, sometimes without a trace. In the future, perhaps humanity will unleash a Pandora's box of technological horrors that threaten our very existence, from atomic bombs to carbon dioxide. Far from trumpeting the coming of the Age of Aquarius, some futurologists predict that we may be facing technological and ecological collapse. For the future, they conjure up the frightening image of humanity reduced to a pathetic, terrified Scrooge in Charles Dickens's fable, groveling on the ground of his own grave and pleading for a second chance.

Unfortunately, the bulk of humanity is largely uncaring, or unaware, of the potential disasters facing us. Some scientists have argued that perhaps humanity, considered as a single entity, can be compared to a teenager careening out of control. For example, psychologists tell us that teenagers act as if they are invulnerable. Their driving, drinking, and drug habits are graphic proof, they say, of the devil-may-care recklessness that pervades their life-style and outlook. The main cause of death among teenagers in this country is no longer disease, but accidents, probably caused by the fact that they think they will live forever.

If that is true, then we are abusing technology and the environment as if we will live forever, unaware of the catastrophes that lie in the future. Society as a whole may have a "Peter Pan complex," never wanting to grow up and face the consequences of its own irresponsibility.

To concretize our discussion, using the knowledge at our disposal, we can identify several important hurdles that must be crossed over during the next several aeons before we can become masters of the tenth dimension: the uranium barrier, ecological collapse, a new ice age, astronomical close encounters, Nemesis and extinction, and the death of the sun and the Milky Way galaxy.

The Uranium Barrier

Jonathan Schell, in his watershed book *The Fate of the Earth,* points out how perilously close we have come to mutual annihilation. Although the recent collapse of the Soviet Union has made possible sweeping arms cuts, there are still close to 50,000 nuclear weapons, both tactical and strategic, in the world today, and with deadly accurate rockets to deliver them. Humanity has finally mastered the possibility of total annihilation.

If the missiles do not destroy everyone in the opening shots of a nuclear war, we can still look forward to the agonizing death caused by nuclear winter, during which the soot and ash from burning cities slowly chokes off all the life-giving sunlight. Computer studies have shown that as few as 100 megatons of explosives may generate enough fire storms in the cities to cloud the atmosphere significantly. As temperatures plummet, crops fail, and cities freeze over, the last vestiges of civilization will be snuffed out like a candle.

Finally, there is the increasing danger of nuclear proliferation. United States intelligence estimates that India, which detonated its first bomb in 1974, now has a stockpile of about 20 atomic bombs. Arch-enemy Pakistan, these sources claim, has built four atomic bombs, one of which weighs no more than 400 pounds, at its secret Kahuta nuclear facility. An atomic worker at Israel's Dimona nuclear installation in the Negev desert claimed that he saw enough material to build 200 atomic bombs there. And South Africa admitted that it had made seven atomic bombs and apparently tested two atomic bombs in the late 1970s off its coast. The U.S. spy satellite *Vela* picked up the "fingerprint" of the atomic bomb, a characteristic, unmistakable double-flash, on two occasions off the coast of South Africa in the presence of Israeli warships. Nations like North Korea, South Korea, and Taiwan are poised at the brink of going nuclear. It's highly probable, given recent U.S. intelligence disclosures, that 20 nations will possess the bomb by the year 2000. The bomb will have proliferated into the hottest spots around the world, including the Middle East.

This situation is highly unstable, and will continue to become more so as nations compete for diminishing resources and spheres of influence. Not just our society, but every intelligent civilization in the galaxy building an industrial society, will discover element 92 (uranium) and with it the ability for mass destruction. Element 92 has the curious property of sustaining a chain reaction and releasing the vast amount of energy stored within its nucleus. With the ability to master element 92 comes the ability either to liberate our species from want, ignorance, and hunger, or to consume the planet in nuclear fire. The power of element 92, however, can be unleashed only when an intelligent species reaches a certain point of development as a Type 0 civilization. It depends on the size of its cohesive social unit and its state of industrial development.

Fire, for example, can be harnessed by isolated groups of intelligent individuals (such as a tribe). Smelting and primitive metallurgy, necessary for the manufacture of weapons, requires a larger social unit, perhaps numbering in the thousands (such as a small village). The development of the internal-combustion engine (for example, a car engine) requires the development of a complex chemical and industrial base, which can be accomplished by only a cohesive social unit numbering in the millions (for example, a nation-state).

The discovery of element 92 upsets this balance between the slow, steady rise of the cohesive social unit and its technological development. The releasing of nuclear energy dwarfs chemical explosives by a factor of a million, but the same nation-state that can harness the internal-combustion engine can also refine element 92. Thus a severe mismatch occurs, especially when the social development of this hypothetical civilization is still locked in the form of hostile nation-states. The technology for mayhem and destruction abruptly outpaces the slow development of social relations with the discovery of element 92.

It is natural to conclude, therefore, that Type 0 civilizations arose on numerous occasions within the past 5- to 10-billion-year history of our galaxy, but that they all eventually discovered element 92. If a civilization's technological capability outraced its social development, then, with the rise of hostile nation-states, there was a large chance that the civilization destroyed itself long ago in an atomic war.[6] Regrettably, if we live long enough to reach nearby stars in our sector of the galaxy, we may see the ashes of numerous, dead civilizations that settled national passions, personal jealousies, and racial hatreds with nuclear bombs.

As Heinz Pagels has said,

> The challenge to our civilization which has come from our knowledge of the cosmic energies that fuel the stars, the movement of light and electrons through matter, the intricate molecular order which is the biological basis of life, must be met by the creation of a moral and political order which will accommodate these forces or we shall be destroyed. It will try our deepest resources of reason and compassion.[7]

It seems likely, therefore, that advanced civilizations sprang up on numerous occasions within our galaxy, but that few of them negotiated the uranium barrier, especially if their technology outpaced their social development.

If we plot, for example, the rise of radio technology on a graph, we see that our planet evolved for 5 billion years before an intelligent species discovered how to manipulate the electromagnetic and nuclear forces. However, if we annihilate ourselves in a nuclear war, then this curve will become a spike and return to zero. Thus in order to communicate with an advanced civilization, we must scan at precisely the right era, to an accuracy of a few decades, before the civilization blows itself up. There is a vanishingly small "window" through which we may make contact with another living civilization, before it destroys itself. In Figure 13.1, we see the rise of alien civilizations throughout the galaxy represented as a series of peaks, each representing the rapid rise of a civilization and the even more rapid fall due to nuclear war. Scanning the heavens for intelligent life, therefore, may be a difficult task. Perhaps there have been many thousands of peaks within the past few billion years, with thousands of planets briefly mastering radio technology before blowing themselves up. Each brief peak, unfortunately, takes place at different cosmic times.

Ecological Collapse

Assuming that a Type 0 civilization can master uranium without destroying itself in a nuclear war, the next barrier is the possibility of ecological collapse.

We recall the earlier example of a single bacterium, which divides so frequently that it eventually outweighs the planet earth. However, in reality we do not see gigantic masses of bacteria on the earth—in fact, bacterial colonies usually do not even grow to the size of a penny. Laboratory bacteria placed in a dish filled with nutrients will indeed grow exponentially, but eventually die because they produce too much waste and exhaust the food supply. These bacterial colonies essentially suffocate in their own waste products.

Like bacterial colonies, we may also be exhausting our resources while drowning in the waste products that we relentlessly produce. Our oceans and the atmosphere are not limitless, but ultrathin films on the surface of the earth. The population of a Type 0 civilization, before it reaches Type I status, may soar to the billions, creating a strain on resources and exacerbating the problems of pollution. One of the most immediate dangers is the poisoning of the atmosphere, in the form of carbon dioxide, which traps sunlight and raises the average world temperature, possibly initiating a runaway greenhouse effect.

Since 1958, carbon dioxide concentrations in the air have increased 25%, mostly from oil and coal burning (45% of carbon dioxide comes from the United States and the former Soviet Union). This, in turn, may have accelerated the mean temperature rise of the earth. It took almost

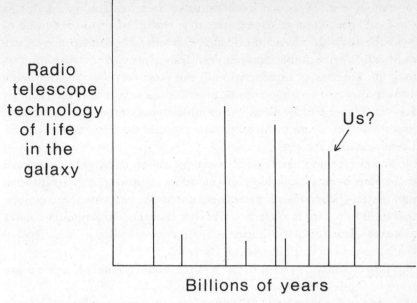

Figure 13.1. Why don't we see other intelligent life in the galaxy? Perhaps intelligent life forms that could build radio telescopes flourished millions of years in the past, but perished in a nuclear war. Our galaxy could have been teeming with intelligent life, but perhaps most are dead now. Will our civilization be any different?

a century, from 1880, to raise the mean world temperature 1°F. However, the mean temperature is now rising at almost 0.6°F per decade. By the year 2050, this translates into a rise of coastal waters by 1 to 4 feet, which could swamp nations like Bangladesh and flood areas like Los Angeles and Manhattan. Even more serious would be a devastation of the nation's food basket in the Midwest, the acceleration of the spread of deserts, and destruction of tropical rain forests, which in turn accelerates the greenhouse effect. Famine and economic ruin could spread on a global scale.

The fault lies in an uncoordinated planetary policy. Pollution takes place in millions of individual factories all over the planet, but the power to curb this unbridled pollution resides with a planetary policy, which is difficult, if not impossible, to enforce if the dominant cohesive social unit is the nation-state, numbering only in the hundreds of millions. In the short term, this may mean emergency policies and the sharp curtailment of the internal-combustion engine and coal and oil burning. The standard of living could also drop. It means additional hardships in developing nations, which need access to cheap sources of energy. In the long term, however, our society may be forced to resort to one of three possible solutions that do not give off carbon dioxide and are essentially inexhaustible: solar energy, fusion plants, and breeder reactors. Of these, solar and fusion hold the most promise. Fusion power (which fuses the hydrogen atoms found in sea water) and solar energy are still several decades away, but should provide ample energy supplies into the next few centuries, until society makes the transition to a Type I civilization.

The fault once again lies in the fact that the technology has outpaced social development. As long as pollution is produced by individual nation-states, while the measures necessary to correct this are planetary, there will be a fatal mismatch that invites disaster. The uranium barrier and ecological collapse will exist as life-threatening disasters for Type 0 civilizations until this mismatch is bridged.

Once a civilization passes Type 0 status, however, there is much more room for optimism. To reach Type I status requires a remarkable degree of social cooperation on a planetary scale. Aggregates on the order of tens to hundreds of millions of individuals are necessary to exploit the resources of uranium, internal combustion, and chemicals. However, aggregates on the order of billions are probably necessary truly to harness planetary resources. Thus the social organization of a Type I civilization must be very complex and very advanced, or else the technology cannot be developed.

By definition, a Type I civilization requires a cohesive social unit that is the entire planet's population. A Type I civilization by its very nature must be a planetary civilization. It cannot function on a smaller scale.

This can, in some sense, be compared to childbirth. The most dangerous period for a child is the first few months of life, when the transition to an external, potentially hostile environment places enormous biological strains on the baby. After the first year of life, the death rate plunges dramatically. Similarly, the most dangerous period for a civilization is the first few centuries after it has reached nuclear capability. It may turn out that once a civilization has achieved a planetary political system, the worst is over.

A New Ice Age

No one knows what causes an ice age, which has a duration measured in tens to hundreds of thousands of years. One theory is that it is caused by minute variations in the earth's rotation, which are too small to be noticed even over a period of centuries. These tiny effects, over hundreds of thousands of years, apparently accumulate to cause slight changes in the jet stream over the poles. Eventually, the jet streams are diverted, sending freezing polar air masses farther and farther south, causing temperatures to plummet around the globe, until an ice age begins. The ice ages did considerable damage to the ecology of the earth, wiping out scores of mammalian life forms and perhaps even isolating bands of humans on different continents, perhaps even giving rise to the various races, which is a relatively recent phenomenon.

Unfortunately, our computers are too primitive even to predict tomorrow's weather, let alone when the next ice age will strike. For example, computers are now entering their fifth generation. We sometimes forget that no matter how large or complex a fourth-generation computer is, it can only add two numbers at a time. This is an enormous bottleneck that is just beginning to be solved with fifth-generation computers, which have parallel processors that can perform several operations simultaneously.

It is highly likely that our civilization (if it successfully negotiates the uranium barrier and ecological collapse) will attain Type I status, and with it the ability to control the weather, within a few hundred years. If humanity reaches Type I status or higher before the next ice age occurs, then there is ample reason to believe that an ice age will not destroy humanity. Humans either will change the weather and prevent the ice age or will leave the earth.

Astronomical Close Encounters

On a time scale of several thousand to several million years, Types 0 and I civilizations have to worry about asteroid collisions and nearby supernovas.

Only within this century, with refined astronomical measurements, has it become apparent that the earth's orbit cuts across the orbits of many asteroids, making the possibility of near misses uncomfortably large. (One way for a Type 0 or I civilization to prevent a direct collision is to send rockets with hydrogen bombs to intercept and deflect the asteroid while it is still tens of millions of miles away from the earth. This method has, in fact, been proposed by international bodies of scientists.)

These near misses are more frequent than most people realize. The last one took place on January 3, 1993, and was actually photographed using radar by NASA astronomers. Photos of the asteroid Toutatis show that it consists of two rocky cores, each 2 miles in diameter. It came within 2.2 million miles of the planet earth. On March 23, 1989, an asteroid about half a mile across drifted even closer to the earth, about 0.7 million miles (roughly three times the distance from the earth to the moon).

In fact, it was also announced in late 1992 that a gigantic comet would hit the earth on exactly August 14, 2126, perhaps ending all life on the planet. Astronomer Brian Marsden of the Harvard–Smithsonian Center for Astrophysics estimated the chances of a direct hit as 1 in 10,000. The Swift–Tuttle comet (named after the two American astronomers who first spotted it during the Civil War) was soon dubbed the Doomsday Rock by the media. Soon-to-be-unemployed nuclear weapons physicists argued, perhaps in a self-serving way, that they should be allowed to build massive hydrogen bombs to blow it to smithereens when the time comes.

Bits and pieces of the Swift–Tuttle comet have already impacted on the earth. Making a complete revolution around the sun every 130 years, it sheds a considerable amount of debris, creating a river of meteors and particles in outer space. When the earth crosses this river, we have the annual Perseid meteor shower, which rarely fails to light up the sky with celestial fireworks. (We should also point out that predicting near misses of comets is a risky business. Because the heat of the sun's radiation causes the comet's icy surface to vaporize irregularly and sputter like thousands of small firecrackers, there are slight but important distortions in its trajectory. Not surprisingly, Marsden retracted his prediction a few weeks later as being incorrect. "We're safe for the next millennium," admitted Marsden.)

A NASA panel in January 1991 estimated that there are about 1,000 to 4,000 asteroids that cross the earth's orbit and are bigger than a half-mile across, sufficient to pose a threat to human civilization. However, only about 150 of these large asteroids have been adequately tracked by radar. Furthermore, there are estimated to be about 300,000 asteroids that cross the earth's orbit that are at least 300 feet across. Unfortunately, scientists hardly know the orbits of any of these smaller asteroids.

My own personal close encounter with an extraterrestrial object came when I was a senior at Harvard in the winter of 1967. A close friend of mine in my dormitory, who had a part-time job at the university observatory, told me a closely held secret: The astronomers there had detected a gigantic asteroid, several miles across, heading directly for the planet earth. Furthermore, although it was too early to tell, he informed me that their computers calculated it might strike the earth in June 1968, the time of our graduation. An object that size would crack the earth's crust, spew open billions of tons of molten magma, and send huge earthquakes and tidal waves around the world. As the months went by, I would get periodic updates on the course of the Doomsday asteroid. The astronomers at the observatory were obviously being careful not to cause any undue panic with this information.

Twenty years later, I had forgotten all about the asteroid, until I was browsing through an article on asteroid near misses. Sure enough, the article made reference to the asteroid of 1968. Apparently, the asteroid came within about 1 million miles of a direct impact with the earth.

More rare, but more spectacular than asteroid collisions are supernova bursts in the vicinity of the earth. A supernova releases enormous quantities of energy, greater than the output of hundreds of billions of stars, until eventually it outshines the entire galaxy itself. It creates a burst of x-rays, which would be sufficient to cause severe disturbances in any nearby star system. At the very minimum, a nearby supernova would create a gigantic EMP (electromagnetic pulse), similar to the one that would be unleashed by a hydrogen bomb detonated in outer space. The x-ray burst would eventually hit our atmosphere, smashing electrons out of atoms; the electrons would then spiral through the earth's magnetic field, creating enormous electric fields. These fields are sufficient to black out all electrical and communication devices for hundreds of miles, creating confusion and panic. In a large-scale nuclear war, the EMP would be sufficient to wipe out or damage any form of electronics over a wide area of the earth's population. At worst, in fact, a supernova burst in the vicinity of a star system might be sufficient to destroy all life.

Astronomer Carl Sagan speculates that such an event may have wiped out the dinosaurs:

> If there were by chance a supernova within ten or twenty light-years of the solar system some sixty-five million years ago, it would have sprayed an intense flux of cosmic rays into space, and some of these, entering the Earth's envelope of air, would have burned the atmospheric nitrogen. The oxides of nitrogen thus generated would have removed the protective layer of ozone from the atmosphere, increasing the flux of solar ultraviolet radiation at the surface and frying and mutating the many organisms imperfectly protected against intense ultraviolet light.

Unfortunately, the supernova would give little warning of its explosion. A supernova eruption takes place quite rapidly, and its radiation travels at the speed of light, so a Type I civilization would have to make a speedy escape into outer space. The only precaution that a civilization can take is to monitor carefully those nearby stars that are on the verge of going supernova.

The Nemesis Extinction Factor

In 1980, the late Luis Alvarez, his son Walter, and Frank Asaro and Helen Michel of the University of California at Berkeley proposed that a comet or an asteroid hit the earth 65 million years ago, thereby initiating vast atmospheric disturbances that led to the sudden extinction of the dinosaurs. By examining the rocky strata laid down by river beds 65 million years ago, they were able to determine the presence of unusually high amounts of iridium, which is rarely found on earth but commonly found in extraterrestrial objects, like meteors. The theory is quite plausible, since a comet 5 miles in diameter hitting the earth at about 20 miles per second (ten times faster than a speeding bullet) would have the force of 100 million megatons of TNT (or 10,000 times the world's total nuclear arsenal). It would create a crater 60 miles across and 20 miles deep, sending up enough debris to cut off all sunlight for an extended period of time. As temperatures fall dramatically, the vast majority of the species on this planet would be either killed off or seriously depleted.

In fact, it was announced in 1992 that a strong candidate for the dinosaur-killing comet or asteroid had been identified. It was already known that there is a large impact crater, measuring 110 miles across, in Mexico, in the Yucatán, near the village of Chicxulub Puerto. In 1981, geophysicists with the Mexican national petroleum company, Pemex, told geologists that they had picked up gravitational and magnetic anomalies that were circular in shape at the site. However, only after Alvarez's theory became popular did geologists actively analyze the remnants of that cataclysmic impact. Radioactive-dating methods using argon-39 have shown that the Yucatán crater is 64.98 ± 0.05 million years old. More impressively, it was shown that Mexico, Haiti, and even Florida are littered with small, glassy debris called *tektites,* which were probably silicates that were glassified by the impact of this large asteroid or comet. These glassy tektites can be found in sediment that was laid down between the Tertiary and Cretaceous periods. Analyses of five different tektite samples show an average age of 65.07 ± 0.10 million years. Given the accuracy of these independent measurements, geologists now have the "smoking gun" for the dinosaur-killing asteroid or comet.

But one of the astonishing features of life on earth is that the extinction of the dinosaurs is but one of several well-documented mass extinctions. Other mass extinctions were much worse than the one that ended the Cretaceous period 65 million years ago. The mass extinction that ended the Permian period, for example, destroyed fully 96% of all plant and animal species 250 million years ago. The trilobites, which ruled the oceans as one of earth's dominant life forms, mysteriously and abruptly perished during this great mass extinction. In fact, there have been five mass extinctions of animal and plant life. If one includes mass extinctions that are less well documented, a pattern becomes evident: Every 26 million years or so, there is a mass extinction. Paleontologists David Raup and John Sepkoski have shown that if we plot the number of known species on the earth at any given time, then the chart shows a sharp drop in the number of life forms on the earth every 26 million years, like clockwork. This can be shown to extend over ten cycles going back 260 million years (excluding two cycles).

In one extinction cycle, at the end of the Cretaceous period, 65 million years ago, most of the dinosaurs were killed off. In another extinction cycle, at the end of the Eocene period, 35 million years ago, many species of land mammals were extinguished. But the central puzzle to this is: What in heaven's name has a cycle time of 26 million years? A search through biological, geological, or even astronomical data suggests that nothing has a cycle time of 26 million years.

Richard Muller of Berkeley has theorized that our sun is actually part of a double-star system, and that our sister star (called Nemesis or the Death Star) is responsible for periodic extinctions of life on the earth. The conjecture is that our sun has a massive unseen partner that circles it every 26 million years. As it passes through the Oort cloud (a cloud of comets that supposedly exists beyond the orbit of Pluto), it brings with it an unwelcome avalanche of comets, some of which strike the earth, causing enough debris that the sunlight is blocked from reaching the earth's surface.

Experimental evidence for this unusual theory comes from the fact that the geological layers from the past, corresponding to the end of each extinction cycle, contain unusually large quantities of the element iridium. Since iridium is naturally found in extraterrestrial meteors, it is possible that these traces of iridium are remnants of the comets sent down by Nemesis. At present, we are half-way between extinction cycles, meaning that Nemesis, if it exists, is at its farthest point in its orbit (probably several light-years away). This would give us over 10 million years or so until its next arrival.*

Fortunately, by the time comets from the Oort cloud streak through the solar system again, we will have reached Type III status, meaning that we will have conquered not just the nearby stars, but travel through space–time.

The Death of the Sun

Scientists sometimes wonder what will eventually happen to the atoms of our bodies long after we are dead. The most likely possibility is that our molecules will eventually return to the sun.

Our sun is a middle-aged star. It is approximately 5 billion years old, and will probably remain a yellow star for another 5 billion years. When our sun exhausts its supply of hydrogen fuel, however, it will burn helium and become vastly inflated—a red giant. Its atmosphere will expand rapidly, eventually extending out to the orbit of Mars, and the earth's orbit will be entirely within the sun's atmosphere, so that the earth will be fried by the sun's enormous temperatures. The molecules making up our bodies, and in fact the earth itself, will be consumed by the solar atmosphere.

Sagan paints the following picture:

> Billions of years from now, there will be a last perfect day on Earth. . . . The Arctic and Antarctic icecaps will melt, flooding the coasts of the world. The high oceanic temperatures will release more water vapor into the air, increasing cloudiness, shielding the Earth from sunlight and delaying the end a little. But solar evolution is inexorable. Eventually the oceans will boil, the atmosphere will evaporate away to space and a catastrophe of the most immense proportions imaginable will overtake our planet.[8]

Thus, for those who wish to know whether the earth will be consumed in ice or fire, physics actually gives a definite answer. It will be consumed in fire. However, it is highly likely that humans, if we have survived that

---

*Another theory that might explain periodic extinctions on this vast time scale is the orbit of our solar system around the Milky Way galaxy. The solar system actually dips below and above the galactic plane in its orbit around the galaxy, much like carousel horses move up and down as a merry-go-round turns. As it dips periodically through the galactic plane, the solar system may encounter large quantities of dust that disturb the Oort cloud, bringing down a hail of comets.

long, will have long departed from the solar system. Unlike a supernova, there is ample warning of the demise of our sun.

The Death of the Galaxy

On a time scale of several billions of years, we must confront the fact that the Milky Way galaxy in which we live, will die. More precisely, we live on the Orion spiral arm of the Milky Way. When we gaze at the night sky and feel dwarfed by the immensity of the celestial lights dotting the heavens, we are actually looking at a tiny portion of the stars located on the Orion arm. The millions of stars that have inspired both lovers and poets for generations occupy only a tiny part of the Orion arm. The rest of the 200 billion stars within the Milky Way are so distant that they can barely be seen as a hazy ribbon that cuts across the night sky.

About 2 million light-years from the Milky Way is our nearest galactic neighbor, the great Andromeda galaxy, which is two to three times larger than our own galaxy. The two galaxies are hurtling toward each other at 125 kilometers per second, and should collide within 5 to 10 billion years. As astronomer Lars Hernquist at the University of California at Santa Cruz has said, this collision will be "analogous to a hostile takeover. Our galaxy will be consumed and destroyed."[9]

As seen from outer space, the Andromeda galaxy will appear to collide with and then slowly absorb the Milky Way galaxy. Computer simulations of colliding galaxies show that the gravitational pull of the larger galaxy will slowly overwhelm the gravity of the smaller galaxy, and after several rotations the smaller galaxy will be eaten up. But because the stars within the Milky Way galaxy are so widely separated by the vacuum of space, the number of collisions between stars will be quite low, on the order of several collisions per century. So our sun may avoid a direct collision for an extended period of time.

Ultimately, on this time scale of billions of years, we have a much more deadly fate, the death of the universe itself. Clever forms of intelligent life may find ways to build space arks to avoid most natural catastrophes, but how can we avoid the death of the universe, when space itself is our worst enemy?

The Aztecs believed that the end of the world would come when the sun one day falls from the sky. They foretold that this would come "when the Earth has become tired . . . , when the seed of Earth has ended." The stars would be shaken from the heavens.

Perhaps they were close to the truth.

One can hope that by the time our sun begins to flicker out, humanity will have long since left the solar system and reached for the stars. (In fact, in Asimov's Foundation series, the location of our original star system has been lost for thousands of years.) However, inevitably, *all* the stars in the heavens will flicker out as their nuclear fuel is exhausted. On a scale of tens to hundreds of billions of years, we are facing the death of the universe itself. Either the universe is open, in which case it will expand forever until temperatures gradually reach near absolute zero, or the universe is closed, in which case the expansion will be reversed and the universe will die in a fiery Big Crunch. Even for a Type III civilization, this is a daunting threat to its existence. Can mastery of hyperspace save civilization from its ultimate catastrophe, the death of the universe?

14

The Fate of the Universe

Some say the world will end in fire.
Some say in ice.
From what I've tasted of desire
I hold with those who favor fire.
                                        Robert Frost

It ain't over 'til it's over.
                                        Yogi Berra

WHETHER a civilization, either on earth or in outer space, can
reach a point in its technological development to harness the
power of hyperspace depends partly, as we have seen, on negotiating a
series of disasters typical of Type 0 civilizations. The danger period is
the first several hundred years after the dawn of the nuclear age, when
a civilization's technological development has far outpaced its social and
political maturity in handling regional conflicts.

By the time a civilization has attained Type III status, it will have
achieved a planetary social structure advanced enough to avoid self-anni-
hilation and a technology powerful enough to avoid an ecological or a
natural disaster, such as an ice age or solar collapse. However, even a
Type III civilization will have difficulty avoiding the ultimate catastrophe:
the death of the universe itself. Even the mightiest and most sophisticated of the Type III civilization's starships will be unable to escape the
final destiny of the universe.

That the universe itself must die was known to nineteenth-century

301

scientists. Charles Darwin, in his *Autobiography,* wrote of his anguish when he realized this profound but depressing fact: "Believing as I do that man in the distant future will be a far more perfect creature than he now is, it is an intolerable thought that he and all other sentient beings are doomed to complete annihilation after such long-continued slow progress."[1]

The mathematician and philosopher Bertrand Russell wrote that the ultimate extinction of humanity is a cause of "unyielding despair." In what must be one of the most depressing passages ever written by a scientist, Russell noted:

> That man is the product of causes which had no prevision of the end they were achieving; that his origin, his growth, his hopes and fears, his loves and his beliefs, are but the outcome of accidental collocations of atoms; that no fire, no heroism, no intensity of thought or feeling, can preserve a life beyond the grave; that all the labors of the ages, all the devotion, all the inspiration, all the noonday brightness of human genius, are destined to extinction in the vast death of the solar system; and the whole temple of Man's achievement must inevitably be buried beneath the debris of a universe in ruins—all these things, if not quite beyond dispute, are yet so nearly certain, that no philosophy which rejects them can hope to stand. Only within the scaffolding of these truths, only on the firm foundation of unyielding despair, can the soul's habitation be safely built.[2]

Russell wrote this passage in 1923, decades before the advent of space travel. The death of the solar system loomed large in his mind, a rigorous conclusion of the laws of physics. Within the confines of the limited technology of his time, this depressing conclusion seemed inescapable. Since that time, we have learned enough about stellar evolution to know that our sun will eventually become a red giant and consume the earth in nuclear fire. However, we also understand the basics of space travel. In Russell's time, the very thought of large ships capable of placing humans on the moon or the planets was universally considered to be the thinking of a madman. However, with the exponential growth of technology, the prospect of the death of the solar system is not such a fearsome event for humanity, as we have seen. By the time our sun turns into a red giant, humanity either will have long perished into nuclear dust or, hopefully, will have found its rightful place among the stars.

Still, it is a simple matter to generalize Russell's "unyielding despair" from the death of our solar system to the death of the entire universe. In that event, it appears that no space ark can transport humanity out of harm's way. The conclusion seems irrefutable; physics predicts that all intelligent life forms, no matter how advanced, will eventually perish when the universe itself dies.

According to Einstein's general theory of relativity, the universe either will continue to expand forever in a Cosmic Whimper, in which case the universe reaches near absolute zero temperatures, or will contract into a fiery collapse, the Big Crunch. The universe will die either in "ice," with an open universe, or in "fire," with a closed universe. Either way, a Type III civilization is doomed because temperatures will approach either absolute zero or infinity.

To tell which fate awaits us, cosmologists use Einstein's equations to calculate the total amount of matter–energy in the universe. Because the matter in Einstein's equation determines the amount of space–time curvature, we must know the average matter density of the universe in order to determine if there is enough matter and energy for gravitation to reverse the cosmic expansion of the original Big Bang.

A critical value for the average matter density determines the ultimate fate of the universe and all intelligent life within it. If the average density of the universe is less than 10^{-29} gram per cubic centimeter, which amounts to 10 milligrams of matter spread over the volume of the earth, then the universe will continue to expand forever, until it becomes a uniformly cold, lifeless space. However, if the average density is larger than this value, then there is enough matter for the gravitational force of the universe to reverse the Big Bang, and suffer the fiery temperatures of the Big Crunch.

At present, the experimental situation is confused. Astronomers have several ways of measuring the mass of a galaxy, and hence the mass of the universe. The first is to count the number of stars in a galaxy, and multiply that number by the average weight of each star. Calculations performed in this tedious fashion show that the average density is less than the critical amount, and that the universe will continue to expand forever. The problem with this calculation is that it omits matter that is not luminous (for example, dust clouds, black holes, cold dwarf stars).

There is also a second way to perform this calculation, which is to use Newton's laws. By calculating the time it takes for stars to move around a galaxy, astronomers can use Newton's laws to estimate the total mass of the galaxy, in the same way that Newton used the time it took for the moon to orbit the earth to estimate the mass of the moon and earth.

The problem is the mismatch between these two calculations. In fact, astronomers know that up to 90% of the mass of a galaxy is in the form of hidden, undetectable "missing mass" or "dark matter," which is not luminous but has weight. Even if we include an approximate value for the mass of nonluminous interstellar gas, Newton's laws predict that the galaxy is far heavier than the value calculated by counting stars.

Until astronomers resolve the question of this missing mass or dark matter, we cannot resolve the question of whether the universe will contract and collapse into a fiery ball or will expand forever.

Entropy Death

Assume, for the moment, that the average density of the universe is less than the critical value. Since the matter–energy content determines the curvature of space–time, we find that there is not enough matter–energy to make the universe recollapse. It will then expand limitlessly until its temperature reaches almost absolute zero. This increases *entropy* (which measures the total amount of chaos or randomness in the universe). Eventually, the universe dies in an entropy death.

The English physicist and astronomer Sir James Jeans wrote about the ultimate death of the universe, which he called the "heat death," as early as the turn of the century: "The second law of thermodynamics predicts that there can be but one end to the universe—a 'heat death' in which [the] temperature is so low as to make life impossible."[3]

To understand how entropy death occurs, it is important to understand the three laws of thermodynamics, which govern all chemical and nuclear processes on the earth and in the stars. The British scientist and author C. P. Snow had an elegant way of remembering the three laws:

1. *You cannot win* (that is, you cannot get something for nothing, because matter and energy are conserved).
2. *You cannot break even* (you cannot return to the same energy state, because there is always an increase in disorder; entropy always increases).
3. *You cannot get out of the game* (because absolute zero is unattainable).

For the death of the universe, the most important is the Second Law, which states that any process creates a net increase in the amount of disorder (entropy) in the universe. The Second Law is actually an integral part of our everyday lives. For example, consider pouring cream into a cup of coffee. Order (separate cups of cream and coffee) has naturally changed into disorder (a random mixture of cream and coffee). However, reversing entropy, extracting order from disorder, is exceedingly difficult. "Unmixing" the liquid back into separate cups of cream and coffee is impossible without an elaborate chemistry laboratory. Also, a lighted cigarette can fill an empty room with wisps of smoke, increasing entropy in that room. Order (tobacco and paper) has again turned into disorder (smoke and charcoal). Reversing entropy—that is, forcing the smoke back into the cigarette and turning the charcoal back into unburned tobacco—is impossible even with the finest chemistry laboratory on the planet.

Similarly, everyone knows that it's easier to destroy than to build. It may take a year to construct a house, but only an hour or so to destroy it in a fire. It took almost 5,000 years to transform roving bands of hunters into the great Aztec civilization, which flourished over Mexico and Central America and built towering monuments to its gods. However, it only took a few months for Cortez and the conquistadors to demolish that civilization.

Entropy is relentlessly increasing in the stars as well as on our planet. Eventually, this means that the stars will exhaust their nuclear fuel and die, turning into dead masses of nuclear matter. The universe will darken as the stars, one by one, cease to twinkle.

Given our understanding of stellar evolution, we can paint a rather dismal picture of how the universe will die. All stars will become black holes, neutron stars, or cold dwarf stars (depending on their mass) within 10^{24} years as their nuclear furnaces shut down. Entropy increases as stars slide down the curve of binding energy, until no more energy can be extracted by fusing their nuclear fuel. Within 10^{32} years, all protons and neutrons in the universe will probably decay. According to the GUTs, the protons and neutrons are unstable over that vast time scale. This means that eventually all matter as we know it, including the earth and the solar system, will dissolve into smaller particles, such as electrons and neutrinos. Thus intelligent beings will have to face the unpleasant possibility that the protons and neutrons in their bodies will disintegrate. The bodies of intelligent organisms will no longer be made of the familiar 100 chemical elements, which are unstable over that immense period of time. Intelligent life will have to find ways of creating new bodies made of energy, electrons, and neutrinos.

After a fantastic 10^{100} (a googol) years, the universe's temperature will reach near absolute zero. Intelligent life in this dismal future will face the prospect of extinction. Unable to huddle next to stars, they will freeze to death. But even in a desolate, cold universe at temperatures near absolute zero, there is one last remaining flickering source of energy: black holes. According to cosmologist Stephen Hawking, black holes are not completely black, but slowly leak energy into outer space over an extended period of time.

In this distant future, black holes may become "life preservers" because they slowly evaporate energy. Intelligent life would necessarily congregate next to these black holes and extract energy from them to keep their machines functioning. Intelligent civilizations, like shivering homeless people huddled next to a fading fire, would be reduced to pathetic outposts of misery clinging to a black hole.[4]

But what, we may ask, happens after 10^{100} years, when the evaporating black holes will have exhausted most of their own energy? Astronomers John D. Barrow of the University of Sussex and Joseph Silk of the University of California at Berkeley caution that this question may ultimately have no answer with present-day knowledge. On that time scale, quantum theory, for example, leaves open the possibility that our universe may "tunnel" into another universe.

The probabilities for these kinds of events are exceedingly small; one would have to wait a time interval larger than the lifetime of our present universe, so we need not worry that reality will suddenly collapse in our lifetime, bringing with it a new set of physical laws. However, on the scale of 10^{100} years, these kinds of rare cosmic quantum events can no longer be ruled out.

Barrow and Silk add, "Where there is quantum theory there is hope. We can never be completely sure this cosmic heat death will occur because we can never predict the future of a quantum mechanical universe with complete certainty; for in an infinite quantum future anything that can happen, eventually will."[5]

Escape Through a Higher Dimension

The Cosmic Whimper is indeed a dismal fate awaiting us if the average density of the universe is too low. Now assume that the average density is larger than the critical value. This means that the expansion process will contract within tens of billions of years, and the universe will end in fire, not ice.

In this scenario, there is enough matter and hence a strong enough gravitational pull in the universe to halt the expansion, and then the universe will begin to slowly recollapse, bringing the distant galaxies together again. Starlight will become "blue shifted," instead of red shifted, indicating that the stars are rapidly approaching one another. The temperatures once again will rise to astronomical limits. Eventually, the heat will become sufficiently great to vaporize all matter into a gas.

Intelligent beings will find that their planets' oceans have boiled away and that their atmospheres have turned into a searing furnace. As their planets begin to disintegrate, they will be forced to flee into outer space in giant rockets.

Even the sanctuary of outer space may prove to be inhospitable, however. Temperatures will eventually rise past the point where atoms are stable, and electrons will be ripped off their nuclei, creating a plasma (like that found in our sun). At this point, intelligent life may have to build gigantic shields around their ships and use their entire energy output to keep their shields from disintegrating from the intense heat.

As temperatures continue to rise, the protons and neutrons in the nucleus will be ripped apart. Eventually, the protons and neutrons themselves will be torn apart into quarks. As in a black hole, the Big Crunch devours everything. Nothing survives it. Thus it seems impossible that ordinary matter, let alone intelligent life, can survive the violent disruption.

However, there is one possible escape. If all of space–time is collapsing into a fiery cataclysm, then the only way to escape the Big Crunch is to leave space and time—escape via hyperspace. This may not be as far-fetched as it sounds. Computer calculations performed with Kaluza–Klein and superstring theories have shown that moments after Creation, the four-dimensional universe expanded at the expense of the six-dimensional universe. Thus the ultimate fate of the four- and the six-dimensional universes are linked.

Assuming that this basic picture is correct, our six-dimensional twin universe may gradually expand, as our own four-dimensional universe collapses. Moments before our universe shrinks to nothing, intelligent life may realize that the six-dimensional universe is opening up, and find a means to exploit that fact.

Interdimensional travel is impossible today because our sister universe has shrunk down to the Planck scale. However, in the final stages of a collapse, the sister universe may open up, making dimensional travel possible once again. If the sister universe expands enough, then matter and energy may escape into it, making an escape hatch possible for any intelligent beings smart enough to calculate the dynamics of space–time.

The late Columbia University physicist Gerald Feinberg speculated on this long shot of escaping the ultimate compression of the universe through extra dimensions:

At present, this is no more than a science fiction plot. However, if there are more dimensions than those we know, or four-dimensional space–times in addition to the one we inhabit, then I think it very likely that there are physical phenomena that provide connections between them. It seems plausible that if intelligence persists in the universe, it will, in much less time than the many billions of years before the Big Crunch, find out whether there is anything to this speculation, and if so how to take advantage of it.[6]

Colonizing the Universe

Almost all scientists who have investigated the death of the universe, from Bertrand Russell to current cosmologists, have assumed that intelligent life will be almost helpless in the face of the inevitable, final death throes of the universe. Even the theory that intelligent beings can tunnel through hyperspace and avoid the Big Crunch assumes that these beings are passive victims until the final moments of the collapse.

However, physicists John D. Barrow of the University of Sussex and Frank J. Tipler of Tulane University, in their book *The Anthropic Cosmological Principle*, have departed from conventional wisdom and concluded just the opposite: that intelligent life, over billions of years of evolution, will play an active role in the final moments of our universe. They take the rather unorthodox view that technology will continue to rise exponentially over billions of years, constantly accelerating in proportion to existing technology. The more star systems that intelligent beings have colonized, the more star systems they can colonize. Barrow and Tipler argue that over several billion years, intelligent beings will have completely colonized vast portions of the visible universe. But they are conservative; they do not assume that intelligent life will have mastered the art of hyperspace travel. They assume only that their rockets will travel at near-light velocities.

This scenario should be taken seriously for several reasons. First, rockets traveling at near-light velocities (propelled, say, by photon engines using the power of large laser beams) may take hundreds of years to reach distant star systems. But Barrow and Tipler believe that intelligent beings will thrive for billions of years, which is sufficient time to colonize their own and neighboring galaxies even with sub-light-speed rockets.

Without assuming hyperspace travel, Barrow and Tipler argue that intelligent beings will send millions of small "von Neumann probes" into the galaxy at near-light speeds to find suitable star systems for colonization. John von Neumann, the mathematical genius who developed the first electronic computer at Princeton University during World War II, proved rigorously that robots or automatons could be built with the ability to program themselves, repair themselves, and even create carbon copies of themselves. Thus Barrow and Tipler suggest that the von Neumann probes will function largely independently of their creators. These small probes will be vastly different from the current generation of *Viking* and *Pioneer* probes, which are little more than passive, preprogrammed machines obeying orders from their human masters. The von Neumann probes will be similar to Dyson's Astrochicken, except vastly more powerful and intelligent. They will enter new star systems, land on planets, and mine the rock for suitable chemicals and metals. They will then create a small industrial complex capable of manufacturing numerous robotic copies of themselves. From these bases, more von Neumann probes will be launched to explore even more star systems.

Being self-programming automatons, these probes will not need instructions from their mother planet; they will explore millions of star systems entirely on their own, pausing only to periodically radio back their findings. With millions of these von Neumann probes scattered throughout the galaxy, creating millions of copies of themselves as they "eat" and "digest" the chemicals on each planet, an intelligent civilization will be able to cut down the time wasted exploring uninteresting star systems. (Barrow and Tipler even consider the possibility that von Neumann probes from distant civilizations have already entered our own solar system. Perhaps the monolith featured so mysteriously in *2001: A Space Odyssey* was a von Neumann probe.)

In the "Star Trek" series, for example, the exploration of other star systems by the Federation is rather primitive. The exploration process depends totally on the skills of humans aboard a small number of starships. Although this scenario may make for intriguing human-interest dramas, it is a highly inefficient method of stellar exploration, given the large number of planetary systems that are probably unsuitable for life. Von Neumann probes, although they may not have the interesting adventures of Captain Kirk or Captain Picard and their crews, would be more suitable for galactic exploration.

Barrow and Tipler make a second assumption that is crucial to their argument: The expansion of the universe will eventually slow down and reverse itself over tens of billions of years. During the contraction phase of the universe, the distance between galaxies will decrease, making it vastly easier for intelligent beings to continue the colonization of the

galaxies. As the contraction of the universe accelerates, the rate of colonization of neighboring galaxies will also accelerate, until the entire
universe is eventually colonized.

Even though Barrow and Tipler assume that intelligent life will populate the entire universe, they are still at a loss to explain how any life
form will be able to withstand the unbelievably large temperatures and
pressures created by the final collapse of the universe. They concede
that the heat created by the contraction phase will be great enough to
vaporize any living being, but perhaps the robots that they have created
will be sufficiently heat resistant to withstand the final moments of the
collapse.

Re-Creating the Big Bang

Along these lines, Isaac Asimov has conjectured how intelligent beings
might react to the final death of the universe. In "The Last Question,"
Asimov asks the ancient question of whether the universe must inevitably
die, and what will happen to all intelligent life when we reach Doomsday.
Asimov, however, assumes that the universe will die in ice, rather than
in fire, as the stars cease to burn hydrogen and temperatures plummet
to absolute zero.

The story begins in the year 2061, when a colossal computer has
solved the earth's energy problems by designing a massive solar satellite
in space that can beam the sun's energy back to earth. The AC (analog
computer) is so large and advanced that its technicians have only the
vaguest idea of how it operates. On a $5 bet, two drunken technicians
ask the computer whether the sun's eventual death can be avoided or,
for that matter, whether the universe must inevitably die. After quietly
mulling over this question, the AC responds: INSUFFICIENT DATA FOR A
MEANINGFUL ANSWER.

Centuries into the future, the AC has solved the problem of hyperspace travel, and humans begin colonizing thousands of star systems.
The AC is so large that it occupies several hundred square miles on each
planet and so complex that it maintains and services itself. A young
family is rocketing through hyperspace, unerringly guided by the AC, in
search of a new star system to colonize. When the father casually men-
tions that the stars must eventually die, the children become hysterical.
"Don't let the stars die," plead the children. To calm the children, he
asks the AC if entropy can be reversed. "See," reassures the father, reading the AC's response, the AC can solve everything. He comforts them

by saying, "It will take care of everything when the time comes, so don't worry." He never tells the children that the AC actually prints out: INSUF-FICIENT DATA FOR A MEANINGFUL ANSWER.

Thousands of years into the future, the Galaxy itself has been colonized. The AC has solved the problem of immortality and harnesses the energy of the Galaxy, but must find new galaxies for colonization. The AC is so complex that it is long past the point where anyone understands how it works. It continually redesigns and improves its own circuits. Two members of the Galactic Council, each hundreds of years old, debate the urgent question of finding new galactic energy sources, and wonder if the universe itself is running down. Can entropy be reversed? they ask. The AC responds: INSUFFICIENT DATA FOR A MEANINGFUL ANSWER.

Millions of years into the future, humanity has spread across the uncountable galaxies of the universe. The AC has solved the problem of releasing the mind from the body, and human minds are free to explore the vastness of millions of galaxies, with their bodies safely stored on some long forgotten planet. Two minds accidentally meet each other in outer space, and casually wonder where among the uncountable galaxies humans originated. The AC, which is now so large that most of it has to be housed in hyperspace, responds by instantly transporting them to an obscure galaxy. They are disappointed. The galaxy is so ordinary, like millions of other galaxies, and the original star has long since died. The two minds become anxious because billions of stars in the heavens are slowly meeting the same fate. The two minds ask, can the death of the universe itself be avoided? From hyperspace, the AC responds: INSUF-FICIENT DATA FOR A MEANINGFUL ANSWER.

Billions of years into the future, humanity consists of a trillion, trillion, trillion immortal bodies, each cared for by automatons. Humanity's collective mind, which is free to roam anywhere in the universe at will, eventually fuses into a single mind, which in turn fuses with the AC itself. It no longer makes sense to ask what the AC is made of, or where in hyperspace it really is. "The universe is dying," thinks Man, collectively. One by one, as the stars and galaxies cease to generate energy, temperatures throughout the universe approach absolute zero. Man desperately asks if the cold and darkness slowly engulfing the galaxies mean its eventual death. From hyperspace, the AC answers: INSUFFICIENT DATA FOR A MEANINGFUL ANSWER.

When Man asks the AC to collect the necessary data, it responds: I WILL DO SO. I HAVE BEEN DOING SO FOR A HUNDRED BILLION YEARS. MY PRED-ECESSORS HAVE BEEN ASKED THIS QUESTION MANY TIMES. ALL THE DATA I HAVE REMAINS INSUFFICIENT.

A timeless interval passes, and the universe has finally reached its ultimate death. From hyperspace, the AC spends an eternity collecting data and contemplating the final question. At last, the AC discovers the solution, even though there is no longer anyone to give the answer. The AC carefully formulates a program, and then begins the process of reversing Chaos. It collects cold, interstellar gas, brings together the dead stars, until a gigantic ball is created.

Then, when its labors are done, from hyperspace the AC thunders: LET THERE BE LIGHT!

And there was light—

And on the seventh day, He rested.

15

Conclusion

The known is finite, the unknown infinite; intellectually we
stand on an islet in the midst of an illimitable ocean of inexpl-
icability. Our business in every generation is to reclaim a little
more land.

Thomas H. Huxley

PERHAPS the most profound discovery of the past century in physics
has been the realization that nature, at its most fundamental level,
is simpler than anyone thought. Although the mathematical complexity
of the ten-dimensional theory has soared to dizzying heights, opening
up new areas of mathematics in the process, the basic concepts driving
unification forward, such as higher-dimensional space and strings, are
basically simple and geometric.

Although it is too early to tell, future historians of science, when
looking back at the tumultuous twentieth century, may view one of the
great conceptual revolutions to be the introduction of higher-dimensional space–time theories, such as superstring and Kaluza–Klein-type
theories. As Copernicus simplified the solar system with his series of
concentric circles and dethroned the central role of the earth in the
heavens, the ten-dimensional theory promises to vastly simplify the laws
of nature and dethrone the familiar world of three dimensions. As we
have seen, the crucial realization is that a three-dimensional description
of the world, such as the Standard Model, is "too small" to unite all the
fundamental forces of nature into one comprehensive theory. Jamming

the four fundamental forces into a three-dimensional theory creates an ugly, contrived, and ultimately incorrect description of nature.

Thus the main current dominating theoretical physics in the past decade has been the realization that the fundamental laws of physics appear simpler in higher dimensions, and that all physical laws appear to be unified in ten dimensions. These theories allow us to reduce an enormous amount of information into a concise, elegant fashion that unites the two greatest theories of the twentieth century: quantum theory and general relativity. Perhaps it is time to explore some of the many implications that the ten-dimensional theory has for the future of physics and science, the debate between reductionism and holism in nature, and the aesthetic relation among physics, mathematics, religion, and philosophy.

Ten Dimensions and Experiment

When caught up in the excitement and turmoil accompanying the birth of any great theory, there is a tendency to forget that ultimately all theories must be tested against the bedrock of experiment. No matter how elegant or beautiful a theory may appear, it is doomed if it disagrees with reality.

Goethe once wrote, "Gray is the dogma, but green is the tree of life." History has repeatedly borne out the correctness of his pungent observation. There are many examples of old, incorrect theories that stubbornly persisted for years, sustained only by the prestige of foolish but well-connected scientists. At times, it even became politically risky to oppose the power of ossified, senior scientists. Many of these theories have been killed off only when some decisive experiment exposed their incorrectness.

For example, because of Hermann von Helmholtz's fame and considerable influence in nineteenth-century Germany, his theory of electromagnetism was much more popular among scientists than Maxwell's relatively obscure theory. But no matter how well known Helmholtz was, ultimately experiment confirmed the theory of Maxwell and relegated Helmholtz's theory to obscurity. Similarly, when Einstein proposed his theory of relativity, many politically powerful scientists in Nazi Germany, like Nobel laureate Philip Lenard, hounded him until he was driven out of Berlin in 1933. Thus the yeoman's work in any science, and especially physics, is done by the experimentalist, who must keep the theoreticians honest.

Victor Weisskopf, a theoretical physicist at MIT, once summarized the relationship between theoretical and experimental science when he observed that there are three kinds of physicists: the machine builders (who build the atom smashers that make the experiment possible), the experimentalists (who plan and execute the experiment), and the theoreticians (who devise the theory to explain the experiment). He then compared these three classes to Columbus's voyage to America. He observed that

> the machine builders correspond to the captains and ship builders who really developed the techniques at that time. The experimentalists were those fellows on the ships that sailed to the other side of the world and then jumped upon the new islands and just wrote down what they saw. The theoretical physicists are those fellows who stayed back in Madrid and told Columbus that he was going to land in India.[1]

If, however, the laws of physics become united in ten dimensions only at energies far beyond anything available with our present technology, then the future of experimental physics is in jeopardy. In the past, every new generation of atom smashers has brought forth a new generation of theories. This period may be coming to a close.

Although everyone expected new surprises if the SSC became operational by about the year 2000, some were betting that it would simply reconfirm the correctness of our present-day Standard Model. Most likely, the decisive experiments that will prove or disprove the correctness of the ten-dimensional theory cannot be performed anytime in the near future. We may be entering a long dry spell where research in ten-dimensional theories will become an exercise in pure mathematics. All theories derive their power and strength from experiment, which is like fertile soil that can nourish and sustain a field of flowering plants once they take root. If the soil becomes barren and dry, then the plants will wither along with it.

David Gross, one of the originators of the heterotic string theory, has compared the development of physics to the relationship between two mountain climbers:

> It used to be that as we were climbing the mountain of nature, the experimentalists would lead the way. We lazy theorists would lag behind. Every once in a while they would kick down an experimental stone which would bounce off our heads. Eventually we would get the idea and we would follow the path that was broken by the experimentalists. . . . But now we theorists might have to take the lead. This is a much more lonely enter-

prise. In the past we always knew where the experimentalists were and thus what we should aim for. Now we have no idea how large the mountain is, nor where the summit is.

Although experimentalists have traditionally taken the lead in breaking open new territory, the next era in physics may be an exceptionally difficult one, forcing theoreticians to assume the lead, as Gross notes.

The SSC probably would have found new particles. The Higgs particles may have been discovered, or "super" partners of the quarks may have shown up, or maybe a sublayer beneath the quarks may have been revealed. However, the basic forces binding these particles will, if the theory holds up, be the same. We may have seen more complex Yang–Mills fields and gluons coming forth from the SSC, but these fields may represent only larger and larger symmetry groups, representing fragments of the even larger $E(8) \times E(8)$ symmetry coming from string theory.

In some sense, the origin of this uneasy relation between theory and experiment is due to the fact that this theory represents, as Witten has noted, "21st century physics that fell accidentally into the 20th century."[2] Because the natural dialectic between theory and experiment was disrupted by the fortuitous accidental discovery of the theory in 1968, perhaps we must wait until the twenty-first century, when we expect the arrival of new technologies that will hopefully open up a new generation of atom smashers, cosmic-ray counters, and deep space probes. Perhaps this is the price we must pay for having a forbidden "sneak preview" into the physics of the next century. Perhaps by then, through indirect means, we may experimentally see the glimmer of the tenth dimension in our laboratories.

Ten Dimensions and Philosophy: Reductionism versus Holism

Any great theory has equally great repercussions on technology and the foundations of philosophy. The birth of general relativity opened up new areas of research in astronomy and practically created the science of cosmology. The philosophical implications of the Big Bang have sent reverberations throughout the philosophical and theological communities. A few years ago, this even led to leading cosmologists having a special audience with the pope at the Vatican to discuss the implications of the Big Bang theory on the Bible and Genesis.

Similarly, quantum theory gave birth to the science of subatomic particles and helped fuel the current revolution in electronics. The transistor—the linchpin of modern technological society—is a purely quantum-mechanical device. Equally profound was the impact that the Heisenberg Uncertainty Principle has had on the debate over free will and determinism, affecting religious dogma on the role of sin and redemption for the church. Both the Catholic Church and the Presbyterian Church, with a large ideological stake in the outcome of this controversy over predestination, have been affected by this debate over quantum mechanics. Although the implications of the ten-dimensional theory are still unclear, we ultimately expect that the revolution now germinating in the world of physics will have a similar far-reaching impact once the theory becomes accessible to the average person.

In general, however, most physicists feel uncomfortable talking about philosophy. They are supreme pragmatists. They stumble across physical laws not by design or ideology, but largely through trial and error and shrewd guesses. The younger physicists, who do the lion's share of research, are too busy discovering new theories to waste time philosophizing. Younger physicists, in fact, look askance at older physicists if they spend too much time sitting on distinguished policy committees or pontificating on the philosophy of science.

Most physicists feel that, outside of vague notions of "truth" and "beauty," philosophy has no business intruding on their private domain. In general, they argue, reality has always proved to be much more sophisticated and subtle than any preconceived philosophy. They remind us of some well-known figures in science who, in their waning years, took up embarrassingly eccentric philosophical ideas that led down blind alleys.

When confronted with sticky philosophical questions, such as the role of "consciousness" in performing a quantum measurement, most physicists shrug their shoulders. As long as they can calculate the outcome of an experiment, they really don't care about its philosophical implications. In fact, Richard Feynman almost made a career trying to expose the pompous pretenses of certain philosophers. The greater their puffed-up rhetoric and erudite vocabulary, he thought, the weaker the scientific foundation of their arguments. (When debating the relative merits of physics and philosophy, I am sometimes reminded of the note written by an anonymous university president who analyzed the differences between them. He wrote, "Why is it that you physicists always require so much expensive equipment? Now the Department of Mathematics requires nothing but money for paper, pencils, and waste paper baskets and the Department of Philosophy is still better. It doesn't even ask for waste paper baskets."[3])

Nevertheless, although the average physicist is not bothered by philosophical questions, the greatest of them were. Einstein, Heisenberg, and Bohr spent long hours in heated discussions, wrestling late into the night with the meaning of measurement, the problems of consciousness, and the meaning of probability in their work. Thus it is legitimate to ask how higher-dimensional theories reflect on this philosophical conflict, especially regarding the debate between "reductionism" and "holism."

Heinz Pagels once said, "We are passionate about our experience of reality, and most of us project our hopes and fears onto the universe."[4] Thus it is inevitable that philosophical, even personal questions will intrude into the discussion on higher-dimensional theories. Inevitably, the revival of higher dimensions in physics will rekindle the debate between "reductionism" and "holism" that has flared, on and off, for the past decade.

Webster's Collegiate Dictionary defines *reductionism* as a "procedure or theory that reduces complex data or phenomena to simple terms." This has been one of the guiding philosophies of subatomic physics—to reduce atoms and nuclei to their basic components. The phenomenal experimental success, for example, of the Standard Model in explaining the properties of hundreds of subatomic particles shows that there is merit in looking for the basic building blocks of matter.

Webster's Collegiate Dictionary defines *holism* as the "theory that the determining factors esp. in living nature are irreducible wholes." This philosophy maintains that the Western philosophy of breaking things down into their components is overly simplistic, that one misses the larger picture, which may contain vitally important information. For example, think of an ant colony containing thousands of ants that obeys complex, dynamic rules of social behavior. The question is: What is the best way to understand the behavior of an ant colony? The reductionist would break the ants into their constituents: organic molecules. However, one may spend hundreds of years dissecting ants and analyzing their molecular makeup without finding the simplest clues as to how an ant colony behaves. The obvious way is to analyze the behavior of an ant colony as an integral whole, without breaking it down.

Similarly, this debate has sparked considerable controversy within the area of brain research and artificial intelligence. The reductionist approach is to reduce the brain to its ultimate units, the brain cells, and try to reassemble the brain from them. A whole school of research in artificial intelligence held that by creating elemental digital circuits we could build up increasingly complex circuits, until we created artificial intelligence. Although this school of thought had initial success in the 1950s by modeling "intelligence" along the lines of modern digital computers, it proved disappointing because it could not mimic even the simplest of brain functions, such as recognizing patterns in a photograph.

The second school of thought has tried to take a more holistic approach to the brain. It attempts to define the functions of the brain and create models that treat the brain as a whole. Although this has proved more difficult to initiate, it holds great promise because certain brain functions that we take for granted (for example, tolerance of error, weighing of uncertainty, and making creative associations between different objects) are built into the system from the start. Neural network theory, for example, uses aspects of this organic approach.

Each side of this reductionist–holistic debate takes a dim view of the other. In their strenuous attempts to debunk each other, they sometimes only diminish themselves. They often talk past each other, not addressing each other's main points.

The latest twist in the debate is that the reductionists have, for the past few years, declared victory over holism. Recently, there has been a flurry of claims in the popular press by the reductionists that the successes of the Standard Model and the GUT theory are vindications of reducing nature to smaller and more basic constituents. By probing down to the elemental quarks, leptons, and Yang–Mills fields, physicists have finally isolated the basic constituents of all matter. For example, physicist James S. Trefil of the University of Virginia takes a swipe at holism when he writes about the "Triumph of Reductionism":

> During the 1960s and 1970s, when the complexity of the particle world was being made manifest in one experiment after another, some physicists broke faith with the reductionist philosophy and began to look outside of the Western tradition for guidance. In his book *The Tao of Physics*, for example, Fritjhof Capra argued that the philosophy of reductionism had failed and that it was time to take a more holistic, mystical view of nature. . . . [T]he 1970s [however] can be thought of as the period in which the great traditions of Western scientific thought, seemingly imperiled by the advances of twentieth-century science, have been thoroughly vindicated. Presumably, it will take a while for this realization to percolate away from a small group of theoretical physicists and become incorporated into our general world view.[5]

The disciples of holism, however, turn this debate around. They claim that the idea of unification, perhaps the greatest theme in all of physics, is holistic, not reductionist. They point to how reductionists would sometimes snicker behind Einstein's back in the last years of his life, saying that he was getting senile trying to unite all the forces of the world. The discovery of unifying patterns in physics was an idea pioneered by Einstein, not the reductionists. Furthermore, the inability of the reductionists to offer a convincing resolution of the Schrödinger's cat paradox shows that they have simply chosen to ignore the deeper, philosophical questions. The reductionists may have had great success with quantum field theory and the Standard Model, but ultimately that success is based on sand, because quantum theory, in the final analysis, is an incomplete theory.

Both sides, of course, have merit. Each side is merely addressing different aspects of a difficult problem. However, taken to extremes, this debate sometimes degenerates into a battle between what I call belligerent science versus know-nothing science.

Belligerent science clubs the opposition with a heavy, rigid view of science that alienates rather than persuades. Belligerent science seeks to win points in a debate, rather than win over the audience. Instead of appealing to the finer instincts of the lay audience by presenting itself as the defender of enlightened reason and sound experiment, it comes off as a new Spanish Inquisition. Belligerent science is science with a chip on its shoulder. Its scientists accuse the holists of being soft-headed, of getting their physics confused, of throwing pseudoscientific gibberish to cover their ignorance. Thus belligerent science may be winning the individual battles, but is ultimately losing the war. In every one-on-one skirmish, belligerent science may trounce the opposition by parading out mountains of data and learned Ph.D.s. However, in the long run, arrogance and conceit may eventually backfire by alienating the very audience that it is trying to persuade.

Know-nothing science goes to the opposite extreme, rejecting experiment and embracing whatever faddish philosophy happens to come along. Know-nothing science sees unpleasant facts as mere details, and the overall philosophy as everything. If the facts do not seem to fit the philosophy, then obviously something is wrong with the facts. Know-nothing science comes in with a preformed agenda, based on personal fulfillment rather than objective observation, and tries to fit in the science as an afterthought.

This split between these two factions first appeared during the Vietnam War, when the flower generation was appalled by the massive, excessive use of deadly technology against a peasant nation. But perhaps the area in which this legitimate debate has flared up most recently is personal health. For example, well-paid lobbyists for the powerful agri-business and food industry in the 1950s and 1960s exerted considerable influence on Congress and the medical establishment, preventing a thorough examination of the harmful effects of cholesterol, tobacco, animal fats, pesticides, and certain food additives on heart disease and cancer, which have now been thoroughly documented.

A recent example is the scandal that surrounded the uproar over the pesticide Alar in apples. When the environmentalists at the National Resources Defense Council announced that current levels of pesticides in apples could kill upward of 5,000 children, they sparked concern among consumers and indignation within the food industry, which denounced them as alarmists. Then it was revealed that the report used figures and data from the federal government to arrive at these conclusions. This, in turn, implied that the Food and Drug Administration was sacrificing 5,000 children in the interests of "acceptable risk."

In addition, the revelations about the widespread possible contamination of our drinking water by lead, which can cause serious neurological problems in children, only served to lower the prestige of science in the minds of most Americans. The medical profession, the food industry, and the chemical industry have begun to earn the distrust of wide portions of society. These and other scandals have also contributed to the national flareup of faddish health diets, most of which are well intentioned, but some of which are not scientifically sound.

Higher Synthesis in Higher Dimensions

These two philosophical viewpoints, apparently irreconcilable, must be viewed from the larger perspective. They are antagonistic only when viewed in their extreme form.

Perhaps a higher synthesis of both viewpoints lies in higher dimensions. Geometry, almost by definition, cannot fit the usual reductionist mode. By studying a tiny strand of fiber, we cannot possibly understand an entire tapestry. Similarly, by isolating a microscopic region of a surface, we cannot determine the overall structure of the surface. Higher dimensions, by definition, imply that we must take the larger, global viewpoint.

Similarly, geometry is not purely holistic, either. Simply observing that a higher-dimensional surface is spherical does not provide the information necessary to calculate the properties of the quarks contained within it. The precise way in which a dimension curls up into a ball determines the nature of the symmetries of the quarks and gluons living on that surface. Thus holism by itself does not give us the data necessary to turn the ten-dimensional theory into a physically relevant theory.

The geometry of higher dimensions, in some sense, forces us to realize the unity between the holistic and reductionist approaches. They are simply two ways of approaching the same thing: geometry. They are two sides of the same coin. From the vantage point of geometry, it makes no difference whether we approach it from the reductionist point of view (assembling quarks and gluons in a Kaluza–Klein space) or the holistic approach (taking a Kaluza–Klein surface and discovering the symmetries of the quarks and gluons).

We may prefer one approach over the other, but this is only for historical or pedagogical purposes. For historical reasons, we may stress the reductionist roots of subatomic physics, emphasizing how particle physicists over a period of 40 years pieced together three of the fundamental forces by smashing atoms, or we may take a more holistic approach and claim that the final unification of quantum forces with gravity implies a deep understanding of geometry. This leads us to approach particle physics through Kaluza–Klein and string theories and to view the Standard Model as a consequence of curling up higher-dimensional space.

The two approaches are equally valid. In our book *Beyond Einstein: The Cosmic Quest for the Theory of the Universe,* Jennifer Trainer and I took a more reductionist approach and described how the discoveries of phenomena in the visible universe eventually led to a geometric description of matter. In this book, we took the opposite approach, beginning with the invisible universe and taking the concept of how the laws of nature simplify in higher dimensions as our basic theme. However, both approaches yield the same result.

By analogy, we can discuss the controversy over the "left" brain and "right" brain. The neurologists who originally made the experimental discovery that the left and right hemispheres of our brain perform distinctly different functions became distressed that their data were grossly misrepresented in the popular press. Experimentally, they found that when someone is shown a picture, the left eye (or right brain) pays more attention to particular details, while the right eye (or left brain) more easily grasps the entire photo. However, they became disturbed when popularizers began to say that the left brain was the "holistic brain" and the right brain was the "reductionist brain." This took the distinction between the two brains out of context, resulting in many bizarre interpretations of how one should organize one's thoughts in daily life.

A more correct approach to brain function, they found, was that the brain necessarily uses both halves in synchrony, that the dialectic between both halves of the brain is more important than the specific function of each half individually. The truly interesting dynamics take place when both halves of the brain interact in harmony.

Similarly, anyone who sees the victory of one philosophy over the other in recent advances in physics is perhaps reading too much into the experimental data. Perhaps the safest conclusion that we can reach is that science benefits most from the intense interaction between these two philosophies.

Let us see concretely how this takes place, analyzing how the theory of higher dimensions gives us a resolution between diametrically opposed philosophies, using two examples, Schrödinger's cat and the *S* matrix theory.

Schrödinger's Cat

The disciples of holism sometimes attack reductionism by hitting quantum theory where it is weakest, on the question of Schrödinger's cat. The reductionists cannot give a reasonable explanation of the paradoxes of quantum mechanics.

The most embarrassing feature of quantum theory, we recall, is that an observer is necessary to make a measurement. Thus before the observation is made, cats can be either dead or alive and the moon may or may not be in the sky. Usually, this would be considered crazy, but quantum mechanics has been verified repeatedly in the laboratory. Since the process of making an observation requires an observer, and since an observer requires consciousness, then the disciples of holism claim that a cosmic consciousness must exist in order to explain the existence of any object.

Higher-dimensional theories do not resolve this difficult question completely, but they certainly put it in a new light. The problem lies in the distinction between the observer and the observed. However, in quantum gravity we write down the wave function of the entire universe. There is no more distinction between the observer and the observed; quantum gravity allows for the existence of only the wave function of everything.

In the past, such statements were meaningless because quantum gravity did not really exist as a theory. Divergences would crop up every time someone wanted to do a physically relevant calculation. So the concept of a wave function for the entire universe, although appealing, was meaningless. However, with the coming of the ten-dimensional theory, the meaning of the wave function of the entire universe becomes a relevant concept once again. Calculations with the wave function of the universe can appeal to the fact that the theory is ultimately a ten-dimensional theory, and is hence renormalizable.

This partial solution to the question of observation once again takes the best of both philosophies. On the one hand, this picture is reductionist because it adheres closely to the standard quantum-mechanical explanation of reality, without recourse to consciousness. On the other hand, it is also holistic because it begins with the wave function of the entire universe, which is the ultimate holistic expression! This picture does not make the distinction between the observer and the observed. In this picture, everything, including all objects and their observers, is included in the wave function.

This is still only a partial solution because the cosmic wave function itself, which describes the entire universe, does not live in any definite state, but is actually a composite of all possible universes. Thus the problem of indeterminacy, first discovered by Heisenberg, is now extended to the entire universe.

The smallest unit that one can manipulate in these theories is the universe itself, and the smallest unit that one can quantize is the space of all possible universes, which includes both dead cats and live cats. Thus in one universe, the cat is indeed dead; but in another, the cat is alive. However, both universes reside in the same home: the wave function of the universe.

A Child of S-Matrix Theory

Ironically, in the 1960s, the reductionist approach looked like a failure; the quantum theory of fields was hopelessly riddled with divergences found in the perturbation expansion. With quantum physics in disarray, a branch of physics called S-matrix (scattering matrix) theory broke off from the mainstream and began to germinate. Originally founded by Heisenberg, it was further developed by Geoffrey Chew at the University of California at Berkeley. S-matrix theory, unlike reductionism, tried to look at the scattering of particles as an inseparable, irreducible whole.

In principle, if we know the S matrix, we know everything about particle interactions and how they scatter. In this approach, how particles bump into one another is everything; the individual particle is nothing. S-matrix theory said that the self-consistency of the scattering matrix, and *self-consistency alone,* was sufficient to determine the S matrix. Thus fundamental particles and fields were banished forever from the Eden of S-matrix theory. In the final analysis, only the S matrix had any physical meaning.

As an analogy, let us say that we are given a complex, strange-looking machine and are asked to explain what it does. The reductionist will immediately get a screw driver and take the machine apart. By breaking down the machine to thousands of tiny pieces, the reductionist hopes to find out how the machine functions. However, if the machine is too complicated, taking it apart only makes matters worse.

The holists, however, do not want to take the machine apart for several reasons. First, analyzing thousands of gears and screws may not give us the slightest hint of what the overall machine does. Second, trying to explain how each tiny gear works may send us on a wild-goose chase. The correct way, they feel, is to look at the machine as a whole. They turn the machine on and ask how the parts move and interact with one another. In modern language, this machine is the S matrix, and this philosophy became the S-matrix theory.

In 1971, however, the tide shifted dramatically in favor of reductionism with Gerard 't Hooft's discovery that the Yang–Mills field can provide a self-consistent theory of subatomic forces. Suddenly, each of the particle interactions came tumbling down like huge trees in a forest. The Yang–Mills field gave uncanny agreement with the experimental data from atom smashers, leading to the establishment of the Standard Model, while S-matrix theory became entangled in more and more obscure mathematics. By the late 1970s, it seemed like a total, irreversible victory of reductionism over holism and the S-matrix theory. The reductionists began to declare victory over the prostrate body of the holists and the S matrix.

The tide, however, shifted once again in the 1980s. With the failure of the GUTs to yield any insight into gravitation or yield any experimentally verifiable results, physicists began to look for new avenues of research. This departure from GUTs began with a new theory, which owed its existence to the S-matrix theory.

In 1968, when S-matrix theory was in its heyday, Veneziano and Suzuki were deeply influenced by the philosophy of determining the S matrix in its entirety. They hit on the Euler beta function because they were searching for a mathematical representation of the entire S matrix. If they had looked for reductionist Feynman diagrams, they never would have stumbled on one of the great discoveries of the past several decades.

Twenty years later, we see the flowering of the seed planted by the

S-matrix theory. The Veneziano–Suzuki theory gave birth to string theory, which in turn has been reinterpreted via Kaluza–Klein as a ten-dimensional theory of the universe.

Thus we see that the ten-dimensional theory straddles both traditions. It was born as a child of a holistic S-matrix theory, but contains the reductionist Yang–Mills and quark theories. In essence, it has matured enough to absorb both philosophies.

Ten Dimensions and Mathematics

One of the intriguing features of superstring theory is the level to which the mathematics has soared. No other theory known to science uses such powerful mathematics at such a fundamental level. In hindsight, this is necessarily so, because any unified field theory first must absorb the Riemannian geometry of Einstein's theory and the Lie groups coming from quantum field theory, and then must incorporate an even higher mathematics to make them compatible. This new mathematics, which is responsible for the merger of these two theories, is *topology*, and it is responsible for accomplishing the seemingly impossible task of abolishing the infinities of a quantum theory of gravity.

The abrupt introduction of advanced mathematics into physics via string theory has caught many physicists off guard. More than one physicists has secretly gone to the library to check out huge volumes of mathematical literature to understand the ten-dimensional theory. CERN physicist John Ellis admits, "I find myself touring through the bookshops trying to find encyclopedias of mathematics so that I can mug up on all these mathematical concepts like homology and homotopy and all this sort of stuff which I never bothered to learn before!"[6] To those who have worried about the ever-widening split between mathematics and physics in this century, this is a gratifying, historic event in itself.

Traditionally, mathematics and physics have been inseparable since the time of the Greeks. Newton and his contemporaries never made a sharp distinction between mathematics and physics; they called themselves natural philosophers, and felt at home in the disparate worlds of mathematics, physics, and philosophy.

Gauss, Riemann, and Poincaré all considered physics to be of the utmost importance as a source of new mathematics. Throughout the eighteenth and nineteenth centuries, there was extensive cross-pollination between mathematics and physics. But after Einstein and Poincaré, the development of mathematics and physics took a sharp turn. For the past 70 years, there has been little, if any, real communication between mathematicians and physicists. Mathematicians explored the topology of *N*-dimensional space, developing new disciplines such as algebraic topology. Furthering the work of Gauss, Riemann, and Poincaré, mathematicians in the past century developed an arsenal of abstract theorems and corollaries that have no connection to the weak or strong forces. Physics, however, began to probe the realm of the nuclear force, using three-dimensional mathematics known in the nineteenth century.

All this changed with the introduction of the tenth dimension. Rather abruptly, the arsenal of the past century of mathematics is being incorporated into the world of physics. Enormously powerful theorems in mathematics, long cherished only by mathematicians, now take on physical significance. At last, it seems as though the diverging gap between mathematics and physics will be closed. In fact, even the mathematicians have been startled at the flood of new mathematics that the theory has introduced. Some distinguished mathematicians, such as Isadore A. Singer of MIT, have stated that perhaps superstring theory should be treated as a branch of mathematics, independent of whether it is physically relevant.

No one has the slightest inkling why mathematics and physics are so intertwined. The physicist Paul A. M. Dirac, one of the founders of quantum theory, stated that "mathematics can lead us in a direction we would not take if we only followed up physical ideas by themselves."[7]

Alfred North Whitehead, one of the greatest mathematicians of the past century, once said that mathematics, at the deepest level, is inseparable from physics at the deepest level. However, the precise reason for the miraculous convergence seems totally obscure. No one has even a reasonable theory to explain why the two disciplines should share concepts.

It is often said that "mathematics is the language of physics." For example, Galileo once said, "No one will be able to read the great book of the Universe if he does not understand its language, which is that of mathematics."[8] But this begs the question of why. Furthermore, mathematicians would be insulted to think that their entire discipline is being reduced to mere semantics.

Einstein, noting this relationship, remarked that pure mathematics might be one avenue to solve the mysteries of physics: "It is my conviction that pure mathematical construction enables us to discover the concepts and the laws connecting them, which gives us the key to the understanding of nature. . . . In a certain sense, therefore, I hold it true that pure thought can grasp reality, as the ancients dreamed."[9] Heisenberg echoed this belief: "If nature leads us to mathematical forms of great simplicity and beauty ... that no one has previously encountered, we cannot help thinking that they are 'true,' that they reveal a genuine feature of nature."

Nobel laureate Eugene Wigner once even penned an essay with the candid title "The Unreasonable Effectiveness of Mathematics in the Natural Sciences."

Physical Principles versus Logical Structures

Over the years, I have observed that mathematics and physics have obeyed a certain dialectical relationship. Physics is not just an aimless, random sequence of Feynman diagrams and symmetries, and mathematics is not just a set of messy equations, but rather physics and mathematics obey a definite symbiotic relationship.

Physics, I believe, is ultimately based on a small set of *physical principles*. These principles can usually be expressed in plain English without reference to mathematics. From the Copernican theory, to Newton's laws of motion, and even Einstein's relativity, the basic physical principles can be expressed in just a few sentences, largely independent of any mathematics. Remarkably, only a handful of fundamental physical principles are sufficient to summarize most of modern physics.

Mathematics, by contrast, is the set of all possible *self-consistent structures,* and there are vastly many more logical structures than physical principles. The hallmark of any mathematical system (for example, arithmetic, algebra, or geometry) is that its axioms and theorems are consistent with one another. Mathematicians are mainly concerned that these systems never result in a contradiction, and are less interested in discussing the relative merits of one system over another. Any self-consistent structure, of which there are many, is worthy of study. As a result, mathematicians are much more fragmented than physicists; mathematicians in one area usually work in isolation from mathematicians in other areas.

The relationship between physics (based on physical principles) and mathematics (based on self-consistent structures) is now evident: To solve a physical principle, physicists may require many self-consistent structures. Thus *physics automatically unites many diverse branches of mathematics.* Viewed in this light, we can understand how the great ideas in theoretical physics evolved. For example, both mathematicians and physicists claim Isaac Newton as one of the giants of their respective professions. However, Newton did not begin the study of gravitation starting with mathematics. By analyzing the motion of falling bodies, he was led to believe that the moon was continually falling toward the earth, but never collided with it because the earth curved beneath it; the curvature of the earth compensated for the falling of the moon. He was therefore led to postulate a physical principle: the universal law of gravitation.

However, because he was at a loss to solve the equations for gravity, Newton began a 30-year quest to construct from scratch a mathematics powerful enough to calculate them. In the process, he discovered many self-consistent structures, which are collectively called *calculus*. From this viewpoint, the physical principle came first (law of gravitation), and then came the construction of diverse self-consistent structures necessary to solve it (such as analytic geometry, differential equations, derivatives, and integrals). In the process, the physical principle united these diverse self-consistent structures into a coherent body of mathematics (the calculus).

The same relationship applies to Einstein's theory of relativity. Einstein began with physical principles (such as the constancy of the speed of light and the equivalence principle for gravitation) and then, by searching through the mathematical literature, found the self-consistent structures (Lie groups, Riemann's tensor calculus, differential geometry) that allowed him to solve these principles. In the process, Einstein discovered how to link these branches of mathematics into a coherent picture.

String theory also demonstrates this pattern, but in a startlingly different fashion. Because of its mathematical complexity, string theory has linked vastly different branches of mathematics (such as Riemann surfaces, Kac–Moody algebras, super Lie algebras, finite groups, modular functions, and algebraic topology) in a way that has surprised the mathematicians. As with other physical theories, it automatically reveals the relationship among many different self-consistent structures. However, the underlying physical principle behind string theory is unknown. Physicists hope that once this principle is revealed, new branches of mathematics will be discovered in the process. In other words, the reason why the string theory cannot be solved is that twenty-first-century mathematics has not yet been discovered.

One consequence of this formulation is that a physical principle that unites many smaller physical theories must automatically unite many seemingly unrelated branches of mathematics. This is precisely what string theory accomplishes. In fact, of all physical theories, string theory unites by far the largest number of branches of mathematics into a single coherent picture. Perhaps one of the by-products of the physicists' quest for unification will be the unification of mathematics as well.

Of course, the set of logically consistent mathematical structures is many times larger than the set of physical principles. Therefore, some mathematical structures, such as number theory (which some mathematicians claim to be the purest branch of mathematics), have never been incorporated into any physical theory. Some argue that this situation may always exist: Perhaps the human mind will always be able to conceive of logically consistent structures that cannot be expressed through any physical principle. However, there are indications that string theory may soon incorporate number theory into its structure as well.

Science and Religion

Because the hyperspace theory has opened up new, profound links between physics and abstract mathematics, some people have accused scientists of creating a new theology based on mathematics; that is, we have rejected the mythology of religion, only to embrace an even stranger religion based on curved space–time, particle symmetries, and cosmic expansions. While priests may chant incantations in Latin that hardly anyone understands, physicists chant arcane superstring equations that even fewer understand. The "faith" in an all-powerful God is now replaced by "faith" in quantum theory and general relativity. When scientists protest that our mathematical incantations can be checked in the laboratory, the response is that Creation cannot be measured in the laboratory, and hence these abstract theories like the superstring can never be tested.

This debate is not new. Historically, scientists have often been asked to debate the laws of nature with theologians. For example, the great British biologist Thomas Huxley was the foremost defender of Darwin's theory of natural selection against the church's criticisms in the late nineteenth century. Similarly, quantum physicists have appeared on radio debates with representatives of the Catholic Church concerning whether the Heisenberg Uncertainty Principle negates free will, a question that may determine whether our souls will enter heaven or hell.

But scientists usually are reluctant to engage in theological debates about God and Creation. One problem, I have found, is that "God" means many things to many people, and the use of loaded words full of unspoken, hidden symbolism only clouds the issue. To clarify this problem somewhat, I have found it useful to distinguish carefully between two types of meanings for the word *God*. It is sometimes helpful to differentiate between the God of Miracles and the God of Order.

When scientists use the word *God*, they usually mean the God of Order. For example, one of the most important revelations in Einstein's early childhood took place when he read his first books on science. He immediately realized that most of what he had been taught about religion could not possibly be true. Throughout his career, however, he clung to the belief that a mysterious, divine Order existed in the universe. His life's calling, he would say, was to ferret out his thoughts, to determine whether he had any choice in creating the universe. Einstein repeatedly referred to this God in his writings, fondly calling him "the Old Man." When stumped with an intractable mathematical problem, he would often say, "God is subtle, but not malicious." Most scientists, it is safe to say, believe that there is some form of cosmic Order in the universe. However, to the nonscientist, the word *God* almost universally refers to the God of Miracles, and this is the source of miscommunication between scientists and nonscientists. The God of Miracles intervenes in our affairs, performs miracles, destroys wicked cities, smites enemy armies, drowns the Pharaoh's troops, and avenges the pure and noble.

If scientists and nonscientists fail to communicate with each other over religious questions, it is because they are talking past each other, referring to entirely different Gods. This is because the foundation of science is based on observing reproducible events, but miracles, by definition, are not reproducible. They happen only once in a lifetime, if at all. Therefore, the God of Miracles is, in some sense, beyond what we know as science. This is not to say that miracles cannot happen, only that they are outside what is commonly called *science.*

Biologist Edward O. Wilson of Harvard University has puzzled over this question and asked whether there is any scientific reason why humans cling so fiercely to their religion. Even trained scientists, he found, who are usually perfectly rational about their scientific specialization, lapse into irrational arguments to defend their religion. Furthermore, he observes, religion has been used historically as a cover to wage hideous wars and perform unspeakable atrocities against infidels and heathens. The sheer ferocity of religious or holy wars, in fact, rivals the worst crime that any human has ever committed against any other.

Religion, notes Wilson, is universally found in every human culture ever studied on earth. Anthropologists have found that all primitive tribes have an "origin" myth that explains where they came from. Furthermore, this mythology sharply separates "us" from "them," provides a cohesive (and often irrational) force that preserves the tribe, and suppresses divisive criticism of the leader.

This is not an aberration, but the norm of human society. Religion,

Wilson theorizes, is so prevalent because it provided a definite evolutionary advantage for those early humans who adopted it. Wilson notes that animals that hunt in packs obey the leader because a pecking order based on strength and dominance has been established. But roughly 1 million years ago, when our apelike ancestors gradually became more intelligent, individuals could rationally begin to question the power of their leader. Intelligence, by its very nature, questions authority by reason, and hence could be a dangerous, dissipative force on the tribe. Unless there was a force to counteract this spreading chaos, intelligent individuals would leave the tribe, the tribe would fall apart, and all individuals would eventually die. Thus, according to Wilson, a selection pressure was placed on intelligent apes to suspend reason and blindly obey the leader and his myths, since doing otherwise would challenge the tribe's cohesion. Survival favored the intelligent ape who could reason rationally about tools and food gathering, but also favored the one who could suspend that reason when it threatened the tribe's integrity. A mythology was needed to define and preserve the tribe.

To Wilson, religion was a very powerful, life-preserving force for apes gradually becoming more intelligent, and formed a "glue" that held them together. If correct, this theory would explain why so many religions rely on "faith" over common sense, and why the flock is asked to suspend reason. It would also help to explain the inhuman ferocity of religious wars, and why the God of Miracles always seems to favor the victor in a bloody war. The God of Miracles has one powerful advantage over the God of Order. The God of Miracles explains the mythology of our purpose in the universe; on this question, the God of Order is silent.

Our Role in Nature

Although the God of Order cannot give humanity a shared destiny or purpose, what I find personally most astonishing about this discussion is that we humans, who are just beginning our ascent up the technological scale, should be capable of making such audacious claims concerning the origin and fate of the universe.

Technologically, we are just beginning to leave the earth's gravitational pull; we have only begun to send crude probes to the outer planets. Yet imprisoned on our small planet, with only our minds and a few instruments, we have been able to decipher the laws that govern matter billions of light-years away. With infinitesimally small resources, without even leaving the solar system, we have been able to determine what happens deep inside the nuclear furnaces of a star or inside the nucleus itself.

According to evolution, we are intelligent apes who have only recently left the trees, living on the third planet from a minor star, in a minor spiral arm of a minor galaxy, in a minor group of galaxies near the Virgo supercluster. If the inflation theory is correct, then our entire visible universe is but an infinitesimal bubble in a much larger cosmos. Even then, given the almost insignificant role that we play in the larger universe, it seems amazing that we should be capable of making the claim to have discovered the theory of everything.

Nobel laureate Isidor I. Rabi was once asked what event in his life first set him on the long journey to discover the secrets of nature. He replied that it was when he checked out some books on the planets from the library. What fascinated him was that the human mind is capable of knowing such cosmic truths. The planets and the stars are so much larger than the earth, so much more distant than anything ever visited by humans, yet the human mind is able to understand them.

Physicist Heinz Pagels recounted his pivotal experience when, as a child, he visited the Hayden Planetarium in New York. He recalled,

> The drama and power of the dynamic universe overwhelmed me. I learned that single galaxies contain more stars than all the human beings who have ever lived. . . . The reality of the immensity and duration of the universe caused a kind of 'existential shock' that shook the foundations of my being. Everything that I had experienced or known seemed insignificant placed in that vast ocean of existence.[10]

Instead of being overwhelmed by the universe, I think that perhaps one of the deepest experiences a scientist can have, almost approaching a religious awakening, is to realize that we are children of the stars, and that our minds are capable of understanding the universal laws that they obey. The atoms within our bodies were forged on the anvil of nucleosynthesis within an exploding star aeons before the birth of the solar system. Our atoms are older than the mountains. We are literally made of star dust. Now these atoms, in turn, have coalesced into intelligent beings capable of understanding the universal laws governing that event.

What I find fascinating is that the laws of physics that we have found on our tiny, insignificant planet are the same as the laws found everywhere else in the universe, yet these laws were discovered without our ever having left the earth. Without mighty starships or dimensional windows, we have been able to determine the chemical nature of the stars and decode the nuclear processes that take place deep in their cores.

Finally, if ten-dimensional superstring theory is correct, then a civilization thriving on the farthest star will discover precisely the same truth about our universe. It, too, will wonder about the relation between marble and wood, and come to the conclusion that the traditional three-dimensional world is "too small" to accommodate the known forces in its world.

Our curiosity is part of the natural order. Perhaps we as humans want to understand the universe in the same way that a bird wants to sing. As the great seventeenth-century astronomer Johannes Kepler once said, "We do not ask for what useful purpose the birds do sing, for song is their pleasure since they were created for singing. Similarly, we ought not to ask why the human mind troubles to fathom the secrets of the heavens." Or, as the biologist Thomas H. Huxley said in 1863, "The question of all questions for humanity, the problem which lies behind all others and is more interesting than any of them is that of the determination of man's place in Nature and his relation to the Cosmos."

Cosmologist Stephen Hawking, who has spoken of solving the problem of unification within this century, has written eloquently about the need to explain to the widest possible audience the essential physical picture underlying physics:

> [If] we do discover a complete theory, it should in time be understandable in broad principle by everyone, not just a few scientists. Then we shall all, philosophers, scientists, and just ordinary people, be able to take part in the discussion of the question of why it is that we and the universe exist. If we find the answer to that, it would be the ultimate triumph of human reason—for then we would know the mind of God.[11]

On a cosmic scale, we are still awakening to the larger world around us. Yet the power of even our limited intellect is such that we can abstract the deepest secrets of nature.

Does this give meaning or purpose to life?

Some people seek meaning in life through personal gain, through personal relationships, or through personal experiences. However, it seems to me that being blessed with the intellect to divine the ultimate secrets of nature gives meaning enough to life.

Notes

Preface

1. The subject is so new that there is yet no universally accepted term used by theoretical physicists when referring to higher-dimensional theories. Technically speaking, when physicists address the theory, they refer to a specific theory, such as Kaluza–Klein theory, supergravity, or superstring, although *hyperspace* is the term popularly used when referring to higher dimensions, and *hyper-* is the correct scientific prefix for higher-dimensional geometric objects. I have adhered to popular custom and used the word *hyperspace* to refer to higher dimensions.

Chapter I

1. Heinz Pagels, *Perfect Symmetry: The Search for the Beginning of Time* (New York: Bantam, 1985), 324.

2. Peter Freund, interview with author, 1990.

3. Quoted in Abraham Pais, *Subtle Is the Lord: The Science and the Life of Albert Einstein* (Oxford: Oxford University Press, 1982), 235.

4. This incredibly small distance will continually reappear throughout this book. It is the fundamental length scale that typifies any quantum theory of gravity. The reason for this is quite simple. In any theory of gravity, the strength of the gravitational force is measured by Newton's constant. However, physicists use a simplified set of units where the speed of light c is set equal to one. This means that 1 second is equivalent to 186,000 miles. Also, Planck's constant divided by 2π is also set equal to one, which sets a numerical relationship between seconds and ergs of energy. In these strange but convenient units, everything, including Newton's constant, can be reduced to centimeters. When we calculate the length associated with Newton's constant, it is precisely the Planck length, or 10^{-33} centimeter, or 10^{19} billion electron volts. Thus all quantum gravitational

effects are measured in terms of this tiny distance. In particular, the size of these unseen higher dimensions is the Planck length.

5. Linda Dalrymple Henderson, *The Fourth Dimension and Non-Euclidean Geometry in Modern Art* (Princeton, N.J.: Princeton University Press, 1983), xix.

Chapter 2

1. E. T. Bell, *Men of Mathematics* (New York: Simon and Schuster, 1937), 484.

2. Ibid., 487. This incident most likely sparked Riemann's early interest in number theory. Years later, he would make a famous speculation about a certain formula involving the zeta function in number theory. After 100 years of grappling with "Riemann's hypothesis," the world's greatest mathematicians have failed to offer any proof. Our most advanced computers have failed to give us a clue, and Riemann's hypothesis has now gone down in history as one of the most famous unproven theorems in number theory, perhaps in all of mathematics. Bell notes, "Whoever proves or disproves it will cover himself with glory" (ibid., 488).

3. John Wallis, *Der Barycentrische Calcul* (Leipzig, 1827), 184.

4. Although Riemann is credited as having been the driving creative force who finally shattered the confines of Euclidean geometry, by rights, the man who should have discovered the geometry of higher dimensions was Riemann's aging mentor, Gauss himself.

In 1817, almost a decade before Riemann's birth, Gauss privately expressed his deep frustration with Euclidean geometry. In a prophetic letter to his friend the astronomer Heinrich Olbers, he clearly stated that Euclidean geometry is mathematically incomplete.

In 1869, mathematician James J. Sylvester recorded that Gauss had seriously considered the possibility of higher-dimensional spaces. Gauss imagined the properties of beings, which he called "bookworms," that could live entirely on two-dimensional sheets of paper. He then generalized this concept to include "beings capable of realizing space of four or a greater number of dimensions" (quoted in Linda Dalrymple Henderson, *The Fourth Dimension and Non-Euclidean Geometry in Modern Art* [Princeton, N.J.: Princeton University Press, 1983], 19).

But if Gauss was 40 years ahead of anyone else in formulating the theory of higher dimensions, then why did he miss this historic opportunity to shatter the bonds of three-dimensional Euclidean geometry? Historians have noted Gauss's tendency to be conservative in his work, his politics, and his personal life. In fact, he never once left Germany, and spent almost his entire life in one city. This also affected his professional life.

In a revealing letter written in 1829, Gauss confessed to his friend Friedrich Bessel that he would never publish his work on non-Euclidean geometry for fear of the controversy it would raise among the "Boeotians." Mathematician Morris Kline wrote, "[Gauss] said in a letter to Bessel of January 27, 1829, that he probably would never publish his findings in this subject because he feared ridicule, or as he put it, he feared the clamor of the Boeotians, a figurative reference to a dull-witted Greek tribe" (*Mathematics and the Physical World* [New York: Crowell, 1959], 449). Gauss was so intimidated by the old guard, the narrow-minded "Boeotians" who believed in the sacred nature of three dimensions, that he kept secret some of his finest work.

In 1869, Sylvester, in an interview with Gauss's biographer Sartorious von Waltershausen, wrote that "this great man used to say that he had laid aside several questions which he had treated analytically, and hoped to apply to them geometrical methods in a future state of existence, when his conceptions of space should have become amplified and extended; for as we can conceive beings (like infinitely attenuated book-worms in an infinitely thin sheet of paper) which possess only the notion of space of two dimensions, so we may imagine beings capable of realizing space of four or a greater number of dimensions" (quoted in Henderson, *Fourth Dimension and Non-Euclidean Geometry in Modern Art*, 19).

Gauss wrote to Olbers, "I am becoming more and more convinced that the (physical) necessity of our (Euclidean) geometry cannot be proved, at least not by human reason nor for human reason. Perhaps in another life we will be able to obtain insight into the nature of space, which is now unattainable. Until then, we must place geometry not in the same class with arithmetic, which is purely a priori, but with mechanics" (quoted in Morris Kline, *Mathematical Thought from Ancient to Modern Times* [New York: Oxford University Press, 1972], 872).

In fact, Gauss was so suspicious of Euclidean geometry that he even conducted an ingenious experiment to test it. He and his assistants scaled three mountain peaks: Rocken, Hohehagen, and Inselsberg. From each mountain peak, the other two peaks were clearly visible. By drawing a triangle between the three peaks, Gauss was able to experimentally measure the interior angles. If Euclidean geometry is correct, then the angle should have summed to 180 degrees. To his disappointment, he found that the sum was exactly 180 degrees (plus or minus 15 minutes). The crudeness of his measuring equipment did not allow him to conclusively show that Euclid was wrong. (Today, we realize that this experiment would have to be performed between three different star systems to detect a sizable deviation from Euclid's result.)

We should also point out that the mathematicians Nikolaus I. Lobachevski and János Bolyai independently discovered the non-Euclidean mathematics defined on curved surfaces. However, their construction was limited to the usual lower dimensions.

5. Quoted in Bell, *Men of Mathematics*, 497.

6. The British mathematician William Clifford, who translated Riemann's famous speech for *Nature* in 1873, amplified many of Riemann's seminal ideas and was perhaps the first to expand on Riemann's idea that the bending of space is responsible for the force of electricity and magnetism, thus crystallizing Riemann's work. Clifford speculated that the two mysterious discoveries in mathematics (higher-dimensional spaces) and physics (electricity and magnetism) are

really the same thing, that the force of electricity and magnetism is caused by the bending of higher-dimensional space.

This is the first time that anyone had speculated that a "force" is nothing but the bending of space itself, preceding Einstein by 50 years. Clifford's idea that electromagnetism was caused by vibrations in the fourth dimension also preceded the work of Theodr Kaluza, who would also attempt to explain electromagnetism with a higher dimension. Clifford and Riemann thus anticipated the discoveries of the pioneers of the twentieth century, that the meaning of higher-dimensional space is in its ability to give a simple and elegant description of forces. For the first time, someone correctly isolated the true physical meaning of higher dimensions, that a theory about *space* actually gives us a unifying picture of *forces*.

These prophetic views were recorded by mathematician James Sylvester, who wrote in 1869, "Mr. W. K. Clifford has indulged in some remarkable speculations as to the possibility of our being able to infer, from certain unexplained phenomena of light and magnetism, the fact of our level space of three dimensions being in the act of undergoing in space of four dimensions . . . a distortion analogous to the rumpling of a page" (quoted in Henderson, *Fourth Dimension and Non-Euclidean Geometry in Modern Art,* 19).

In 1870, in a paper with the intriguing title "On the Space-Theory of Matter," he says explicitly that "this variation of the curvature of space is what really happens in that phenomenon which we call the *motion of matter,* whether ponderable or ethereal" (William Clifford, "On the Space-Theory of Matter," *Proceedings of the Cambridge Philosophical Society* 2 [1876]: 157–158).

7. More precisely, in N dimensions the Riemann metric tensor $g_{\mu\nu}$ is an $N \times N$ matrix, which determines the distance between two points, such that the infinitesimal distance between two points is given by $ds^2 = \Sigma dx^\mu\, g_{\mu\nu}\, dx^\nu$. In the limit of flat space, the Riemann metric tensor becomes diagonal, that is, $g_{\mu\nu} = \delta_{\mu\nu}$, and hence the formalism reduces back to the Pythagorean Theorem in N dimensions. The deviation of the metric tensor from $\delta_{\mu\nu}$, roughly speaking, measures the deviation of the space from flat space. From the metric tensor, we can construct the Riemann curvature tensor, represented by $R^\beta_{\mu\nu\alpha}$.

The curvature of space at any given point can be measured by drawing a circle at that point and measuring the area inside that circle. In flat two-dimensional space, the area inside the circle is πr^2. However, if the curvature is positive, as in a sphere, the area is less than πr^2. If the curvature is negative, as in a saddle or trumpet, the area is greater than πr^2.

Strictly speaking, by this convention, the curvature of a crumpled sheet of paper is zero. This is because the areas of circles drawn on this crumpled sheet of paper still equal πr^2. In Riemann's example of force created by the crumpling of a sheet of paper, we implicitly assume that the paper is distorted and stretched as well as folded, so that the curvature is nonzero.

8. Quoted in Bell, *Men of Mathematics,* 501.

9. Ibid., 14.

10. Ibid.

11. In 1917, physicist Paul Ehrenfest, a friend of Einstein, wrote a paper entitled "In What Way Does It Become Manifest in the Fundamental Laws of Physics that Space has Three Dimensions?" Ehrenfest asked himself whether the stars and planets are possible in higher dimensions. For example, the light of a candle gets dimmer as we move farther away from it. Similarly, the gravitational pull of a star gets weaker as we go farther away. According to Newton, gravity gets weaker by an inverse square law. If we double the distance away from a candle or star, the light or gravitational pull gets four times weaker. If we triple the distance, it gets nine times weaker.

If space were four dimensional, then candlelight or gravity would get weaker much more rapidly, as the inverse cube. Doubling the distance from a candle or star would weaken the candlelight or gravity by a factor of eight.

Can solar systems exist in such a four-dimensional world? In principle, yes, but the planets' orbits would not be stable. The slightest vibration would collapse the orbits of the planets. Over time, all the planets would wobble away from their usual orbits and plunge into the sun.

Similarly, the sun would not be able to exist in higher dimensions. The force of gravity tends to crush the sun. It balances out the force of fusion, which tends to blow the sun apart. Thus the sun is a delicate balancing act between nuclear forces that would cause it to explode and gravitational forces that would condense it down to a point. In a higher-dimensional universe, this delicate balance would be disrupted, and stars might spontaneously collapse.

12. Henderson, *Fourth Dimension and Non-Euclidean Geometry in Modern Art*, 22.

13. Zollner had been converted to spiritualism in 1875 when he visited the laboratory of Crookes, the discoverer of the element thalium, inventor of the cathode ray tube, and editor of the learned *Quarterly Journal of Science.* Crookes's cathode ray tube revolutionized science; anyone who watches television, uses a computer monitor, plays a video game, or has been x-rayed owes a debt to Crookes's famous invention.

Crookes, in turn, was no crank. In fact, he was a lion of British scientific society, with a wall full of professional honors. He was knighted in 1897 and received the Order of Merit in 1910. His deep interest in spiritualism was sparked by the tragic death of his brother Philip of yellow fever in 1867. He became a prominent member (and later president) of the Society for Psychical Research, which included an astonishing number of important scientists in the late nineteenth century.

14. Quoted in Rudy Rucker, *The Fourth Dimension* (Boston: Houghton Mifflin, 1984), 54.

15. To imagine how knots can be unraveled in dimensions beyond three, imagine two rings that are intertwined. Now take a two-dimensional cross section of this configuration, such that one ring lies on this plane while the other ring becomes a point (because it lies perpendicular to the plane). We now have a point inside a circle. In higher dimensions, we have the freedom of moving this

dot completely outside the circle without cutting any of the rings. This means that the two rings have now completely separated, as desired. This means that knots in dimensions higher than three can always be untied because there is "enough room." But also notice that we cannot remove the dot from the ring if we are in three-dimensional space, which is the reason why knots stay knotted only in the third dimension.

Chapter 3

1. A. T. Schofield wrote, "We conclude, therefore, that a higher world than ours is not only conceivably possible, but probable; secondly that such a world may be considered as a world of four dimensions; and thirdly, that the spiritual world agrees largely in its mysterious laws . . . with what by analogy would be the laws, language, and claims of a fourth dimension" (quoted in Rudy Rucker, *The Fourth Dimension* [Boston: Houghton Mifflin, 1984], 56).

2. Arthur Willink wrote, "When we have recognized the existence of Space of Four Dimensions there is no greater strain called for in the recognition of the existence of Space of Five Dimensions, and so on up to Space of an infinite number of Dimensions" (quoted in ibid., 200).

3. H. G. Wells, *The Time Machine: An Invention* (London: Heinemann, 1895), 3.

4. Linda Dalrymple Henderson, *The Fourth Dimension and Non-Euclidean Geometry in Modern Art* (Princeton, N.J.: Princeton University Press, 1983), xxi.

5. Ibid. According to Henderson, "[T]he fourth dimension attracted the notice of such literary figures as H. G. Wells, Oscar Wilde, Joseph Conrad, Ford Madox Ford, Marcel Proust, and Gertrude Stein. Among musicians, Alexander Scriabin, Edgar Varese, and George Antheil were actively concerned with the fourth dimension, and were encouraged to make bold innovations in the name of a higher reality" (ibid., xix–xx).

6. Lenin's *Materialism and Empiro-Criticism* is important today because it deeply affected modern Soviet and Eastern European science. For example, Lenin's celebrated phrase "the inexhaustibility of the electron" signified the dialectical notion that we find new sublayers and contradictions whenever we probe deeply into the heart of matter. For example, galaxies are composed of smaller star systems, which in turn contain planets, which are composed of molecules, which are made of atoms, which contain electrons, which, in turn, are "inexhaustible." This is a variation of the "worlds within worlds" theory.

7. Vladimir Lenin, *Materialism and Empiro-Criticism*, in Karl Marx, Friedrich Engels, and Vladimir Lenin, *On Dialectical Materialism* (Moscow: Progress, 1977), 305–306.

8. Ibid.

9. Quoted in Rucker, *Fourth Dimension*, 64.

10. Imagine a Flatlander building a sequence of six adjacent squares, in the

shape of a cross. To a Flatlander, the squares are rigid. They cannot be twisted or rotated along any of the sides connecting the squares. Now imagine, however, that we grab the squares and decide to fold up the series of squares, forming a cube. The joints connecting the squares, which were rigid in two dimensions, can be easily folded in three dimensions. In fact, the folding operation can be performed smoothly without a Flatlander even noticing that the folding is taking place.

Now, if a Flatlander were inside the cube, he would notice a surprising thing. Each square leads to another square. There is no "outside" to the cube. Each time a Flatlander moves from one square to the next, he smoothly (without his knowledge) bends 90 degrees in the third dimension and enters the next square. From the outside, the house is just an ordinary square. However, to someone entering the square, he would find a bizarre sequence of squares, each square leading impossibly to the next square. To him, it would seem impossible that the interior of a single square could house a series of six squares.

Chapter 4

1. Jacob Bronowski, *The Ascent of Man* (Boston: Little, Brown, 1974), 247

2. Quoted in Abraham Pais, *Subtle Is the Lord: The Science and the Life of Albert Einstein* (Oxford: Oxford University Press, 1982), 131.

3. Normally, it is absurd to think that two people can each be taller than the other. However, in this situation we have two people, each correctly thinking that the other has been compressed. This is not a true contradiction because it takes *time* in which to perform a measurement, and time as well as space has been distorted. In particular, events that appear simultaneous in one frame are not simultaneous when viewed in another frame.

For example, let's say that people on the platform take out a ruler and, as the train passes by, drop the measuring stick onto the platform. As the train goes by, they drop the two ends of the stick so that the ends hit the platform simultaneously. In this way, they can prove that the entire length of the compressed train, from front to back, is only 1 foot long.

Now consider the same measuring process from the point of view of the passengers on the train. They think they are at rest and see the compressed subway station coming toward them, with compressed people about to drop a compressed ruler onto the platform. At first it seems impossible that such a tiny ruler would be able to measure the entire length of the train. However, when the ruler is dropped, the ends of the ruler do *not* hit the floor simultaneously. One end of the ruler hits the floor just as the station goes by the front end of the train. Only when the station has moved completely by the length of the entire train does the second end of the ruler finally hit the floor. In this way, the same ruler has measured the entire length of the train in either frame.

The essence of this "paradox," and many others that appear in relativity theory, is that the measuring process takes time, and that both space and time become distorted in different ways in different frames.

4. Maxwell's equations look like this (we set $c = 1$):

$$\nabla \cdot \mathbf{E} = \rho$$

$$\nabla \times \mathbf{B} - \frac{\partial \mathbf{E}}{\partial t} = \mathbf{j}$$

$$\nabla \cdot \mathbf{B} = 0$$

$$\nabla \times \mathbf{E} + \frac{\partial \mathbf{B}}{\partial t} = 0$$

The second and last lines are actually vector equations representing three equations each. Therefore, there are eight equations in Maxwell's equations.

We can rewrite these equations relativistically. If we introduce the Maxwell tensor $F_{\mu\nu} = \partial_\mu A_\nu - \partial_\nu A_\mu$, then these equations reduce to one equation:

$$\partial_\mu F^{\mu\nu} = j^\nu$$

which is the relativistic version of Maxwell's equations.

5. Quoted in Pais, *Subtle Is the Lord*, 239.

6. Ibid., 179.

7. Einstein's equations look like this:

$$R_{\mu\nu} - \tfrac{1}{2} g_{\mu\nu} R = -\frac{8\pi}{c^2} G T_{\mu\nu}$$

where $T_{\mu\nu}$ is the energy–momentum tensor that measures the matter–energy content, while $R_{\mu\nu}$ is the contracted Riemann curvature tensor. This equation says that the energy–momentum tensor determines the amount of curvature present in hyperspace.

8. Quoted in Pais, *Subtle Is the Lord*, 212.

9. Quoted in K. C. Cole, *Sympathetic Vibrations: Reflections on Physics as a Way of Life* (New York: Bantam, 1985), 29.

10. A hypersphere can be defined in much the same way as a circle or sphere. A circle is defined as the set of points that satisfy the equation $x^2 + y^2 = r^2$ in the x–y plane. A sphere is defined as the set of points that satisfy $x^2 + y^2 + z^2 = r^2$ in x–y–z space. A four-dimensional hypersphere is defined as the set of points that satisfy $x^2 + y^2 + z^2 + u^2 = r^2$ in x–y–z–u space. This procedure can easily be extended to N-dimensional space.

11. Quoted in Abdus Salam, "Overview of Particle Physics," in *The New Physics*, ed. Paul Davies (Cambridge: Cambridge University Press, 1989), 487.

12. Theodr Kaluza, "Zum Unitätsproblem der Physik," *Sitzungsberichte Preussische Akademie der Wissenschaften* 96 (1921): 69.

13. In 1914, even before Einstein proposed his theory of general relativity,

physicist Gunnar Nordstrom tried to unify electromagnetism with gravity by introducing a five-dimensional Maxwell theory. If one examines his theory, one finds that it correctly contains Maxwell's theory of light in four dimensions, but it is a scalar theory of gravity, which is known to be incorrect. As a consequence, Nordstrom's ideas were largely forgotten. In some sense, he published too soon. His paper was written 1 year before Einstein's theory of gravity was published, and hence it was impossible for him to write down a five-dimensional Einstein-type theory of gravity.

Kaluza's theory, in contrast to Nordstrom's, began with a metric tensor $g_{\mu\nu}$ defined in five-dimensional space. Then Kaluza identified $g_{\mu 5}$ with the Maxwell tensor A_μ. The old four-dimensional Einstein metric was then identified by Kaluza's new metric only if μ and ν did not equal 5. In this simple but elegant way, both the Einstein field and the Maxwell field were placed inside Kaluza's five-dimensional metric tensor.

Also, apparently Heinrich Mandel and Gustav Mie proposed five-dimensional theories. Thus the fact that higher dimensions were such a dominant aspect of popular culture probably helped to cross-pollinate the world of physics. In this sense, the work of Riemann was coming full circle.

14. Peter Freund, interview with author, 1990.

15. Ibid.

Chapter 5

1. Quoted in K. C. Cole, *Sympathetic Vibrations: Reflections on Physics as a Way of Life* (New York: Bantam, 1985), 204.

2. Quoted in Nigel Calder, *The Key to the Universe* (New York: Penguin, 1977), 69.

3. Quoted in R. P. Crease and C. C. Mann, *The Second Creation* (New York: Macmillan, 1986), 326.

4. Ibid., 293.

5. William Blake, "Tyger! Tyger! burning bright," from "Songs of Experience," in *The Poems of William Blake,* ed. W. B. Yeats (London: Routledge, 1905).

6. Quoted in Heinz Pagels, *Perfect Symmetry: The Search for the Beginning of Time* (New York: Bantam, 1985), 177.

7. Quoted in Cole, *Sympathetic Vibrations,* 229.

8. Quoted in John Gribben, *In Search of Schrödinger's Cat* (New York: Bantam, 1984), 79.

Chapter 6

1. Quoted in R. P. Crease and C. C. Mann, *The Second Creation* (New York: Macmillan, 1986), 411.

2. Quoted in Nigel Calder, *The Key to the Universe* (New York: Penguin, 1977), 15.

3. Quoted in Crease and Mann, *Second Creation*, 418.

4. Heinz Pagels, *Perfect Symmetry: The Search for the Beginning of Time* (New York: Bantam, 1985), 327.

5. Quoted in Crease and Mann, *Second Creation*, 417.

6. Peter van Nieuwenhuizen, "Supergravity," in *Supersymmetry and Supergravity,* ed. M. Jacob (Amsterdam: North Holland, 1986), 794.

7. Quoted in Crease and Mann, *Second Creation*, 419.

Chapter 7

1. Quoted in K. C. Cole, "A Theory of Everything," *New York Times Magazine*, 18 October 1987, 20.

2. John Horgan, "The Pied Piper of Superstrings," *Scientific American*, November 1991, 42, 44.

3. Quoted in Cole, "Theory of Everything," 25.

4. Edward Witten, Interview, in *Superstrings: A Theory of Everything?* ed. Paul Davies and J. Brown (Cambridge: Cambridge University Press, 1988), 90–91.

5. David Gross, Interview, in *Superstrings*, ed. Davies and Brown, 150.

6. Witten, Interview, in *Superstrings*, ed. Davies and Brown, 95.

Witten stresses that Einstein was led to postulate the general theory of relativity starting from a physical principle, the equivalence principle (that the gravitational mass and inertial mass of an object are the same, so that all bodies, no matter how large, fall at the same rate on the earth). However, the counterpart of the equivalence principle for string theory has not yet been found.

As Witten points out, "It's been clear that string theory does, in fact, give a logically consistent framework, encompassing both gravity and quantum mechanics. At the same time, the conceptual framework in which this should be properly understood, analogous to the principle of equivalence that Einstein found in his theory of gravity, hasn't yet emerged" (ibid., 97).

This is why, at present, Witten is formulating what are called *topological field theories*—that is, theories that are totally independent of the way we measure distances. The hope is that these topological field theories may correspond to some "unbroken phase of string theory"—that is, string theory beyond the Planck length.

7. Gross, Interview, in *Superstrings,* ed. Davies and Brown, 150.

8. Horgan, "Pied Piper of Superstrings," 42.

9. Let us examine compactification in terms of the full heterotic string, which has two kinds of vibrations: one vibrating in the full 26-dimensional space–time, and the other in the usual ten-dimensional space time. Since $26 - 10 = 16$, we now assume that 16 of the 26 dimensions have curled up—that is, "com-

pactified" into some manifold—leaving us with a ten-dimensional theory. Anyone walking along any of these 16 directions will wind up precisely at the same spot.

It was Peter Freund who suggested that the symmetry group of this 16-dimensional compactified space was the group $E(8) \times E(8)$. A quick check shows that this symmetry is vastly larger and includes the symmetry group of the Standard Model, given by $SU(3) \times SU(2) \times U(1)$.

In summary, the key relation is $26 - 10 = 16$, which means that if we compactify 16 of the original 26 dimensions of the heterotic string, we are left with a 16-dimensional compact space with a leftover symmetry called $E(8) \times E(8)$. However, in Kaluza–Klein theory, when a particle is forced to live on a compactified space, it must necessarily inherit the symmetry of that space. This means that the vibrations of the string must rearrange themselves according to the symmetry group $E(8) \times E(8)$.

As a result, we can conclude that group theory reveals to us that this group is much larger than the symmetry group appearing in the Standard Model, and can thus include the Standard Model as a small subset of the ten-dimensional theory.

10. Although the supergravity theory is defined in 11 dimensions, the theory is still too small to accommodate all particle interactions. The largest symmetry group for supergravity is $O(8)$, which is too small to accommodate the Standard Model's symmetries.

At first, it appears that the 11-dimensional supergravity has more dimensions, and hence more symmetry, than the ten-dimensional superstring. This is an illusion because the heterotic string begins by compactifying 26-dimensional space down to ten-dimensional space, leaving us with 16 compactified dimensions, which yields the group $E(8) \times E(8)$. This is more than enough to accommodate the Standard Model.

11. Witten, Interview, in *Superstrings,* ed. Davies and Brown, 102.

12. Note that other alternative nonperturbative approaches to string theory have been proposed, but they are not as advanced as string field theory. The most ambitious is "universal moduli space," which tries to analyze the properties of string surfaces with an infinite number of holes in them. (Unfortunately, no one knows how to calculate with this kind of surface.) Another is the renormalization group method, which can so far reproduce only surfaces without any holes (tree-type diagrams). There is also the matrix models, which so far can be defined only in two dimensions or less.

13. To understand this mysterious factor of two, consider a light beam that has two physical modes of vibration. Polarized light can vibrate, say, either horizontally or vertically. However, a relativistic Maxwell field A_μ has four components, where $\mu = 1,2,3,4$. We are allowed to subtract two of these four components using the gauge symmetry of Maxwell's equations. Since $4 - 2 = 2$, the original four Maxwell fields have been reduced by two. Similarly, a relativistic string vibrates in 26 dimensions. However, two of these vibratory modes can be

removed when we break the symmetry of the string, leaving us with 24 vibratory modes, which are the ones that appear in the Ramanujan function.

14. Quoted in Godfrey H. Hardy, *Ramanujan* (Cambridge: Cambridge University Press, 1940), 3.

15. Quoted in James Newman, *The World of Mathematics* (Redmond, Wash.: Tempus Books, 1988), 1: 363.

16. Hardy, *Ramanujan*, 9.

17. Ibid., 10.

18. Ibid., 11.

19. Ibid., 12.

20. Jonathan Borwein and Peter Borwein, "Ramanujan and Pi," *Scientific American,* February 1988, 112.

Chapter 8

1. David Gross, Interview, in *Superstrings: A Theory of Everything?* ed. Paul Davies and J. Brown (Cambridge: Cambridge University Press, 1988), 147.

2. Sheldon Glashow, *Interactions* (New York: Warner, 1988), 335.

3. Ibid., 333.

4. Ibid., 330.

5. Steven Weinberg, *Dreams of a Final Theory* (New York: Pantheon, 1992), 218–219.

6. Quoted in John D. Barrow and Frank J. Tipler, *The Anthropic Cosmological Principle* (Oxford: Oxford University Press, 1986), 327.

7. Quoted in F. Wilczek and B. Devine, *Longing for the Harmonies* (New York: Norton, 1988), 65.

8. John Updike, "Cosmic Gall," in *Telephone Poles and Other Poems* (New York: Knopf, 1960).

9. Quoted in K. C. Cole, "A Theory of Everything," *New York Times Magazine,* 18 October 1987, 28.

10. Quoted in Heinz Pagels, *Perfect Symmetry: The Search for the Beginning of Time* (New York: Bantam, 1985), 11.

11. Quoted in K. C. Cole, *Sympathetic Vibrations: Reflections on Physics as a Way of Life* (New York: Bantam, 1985), 225.

Chapter 9

1. Quoted in E. Harrison, *Masks of the Universe* (New York: Macmillan, 1985), 211.

2. Quoted in Corey S. Powell, "The Golden Age of Cosmology," *Scientific American,* July 1992, 17.

3. The orbifold theory is actually the creation of several individuals, including L. Dixon, J. Harvey, and Edward Witten of Princeton.

4. Years ago, mathematicians asked themselves a simple question: Given a curved surface in N-dimensional space, how many kinds of vibrations can exist on it? For example, think of pouring sand on a drum. When the drum is vibrated at a certain frequency, the particles of sands dance on the drum surface and form beautiful symmetrical patterns. Different patterns of sand particles correspond to different frequencies allowed on the drum surface. Similarly, mathematicians have calculated the number and kind of resonating vibrations allowed on the surface of a curved N-dimensional surface. They even calculated the number and kind of vibrations that an electron could have on such a hypothetical surface. To the mathematicians, this was a cute intellectual exercise. No one thought it could possibly have any physical consequence. After all, electrons, they thought, don't vibrate on N-dimensional surfaces.

This large body of mathematical theorems can now be brought to bear on the problem of GUT families. Each GUT family, if string theory is correct, must be a reflection of some vibration on an orbifold. Since the various kinds of vibrations have been cataloged by mathematicians, all physicists have to do is look in a math book to tell them how many identical families there are! Thus the origin of the family problem is *topology*. If string theory is correct, the origin of these three duplicate families of GUT particles cannot be understood unless we expand our consciousness to ten dimensions.

Once we have curled up the unwanted dimensions into a tiny ball, we can then compare the theory with experimental data. For example, the lowest excitation of the string corresponds to a closed string with a very small radius. The particles that occur in the vibration of a small closed string are precisely those found in supergravity. Thus we retrieve all the good results of supergravity, without the bad results. The symmetry group of this new supergravity is $E(8) \times E(8)$, which is much larger than the symmetry of the Standard Model or even the GUT theory. Therefore, the superstring contains both the GUT and the supergravity theory (without many of the bad features of either theory). Instead of wiping out its rivals, the superstring simply eats them up.

The problem with these orbifolds, however, is that we can construct hundreds of thousands of them. We have an embarrassment of riches! Each one of them, in principle, describes a consistent universe. How do we tell which universe is the correct one? Among these thousands of solutions, we find many that predict exactly three generations or families of quarks and leptons. We can also predict thousands of solutions where there are many more than three generations. Thus while GUTs consider three generations to be too many, many solutions of string theory consider three generations to be too few!

5. David Gross, Interview, in *Superstrings: A Theory of Everything?* ed. Paul Davies and J. Brown (Cambridge: Cambridge University Press, 1988), 142–143.

6. Ibid.

Chapter 10

1. More precisely, the Pauli exclusion principle states that no two electrons can occupy the same quantum state with the same quantum numbers. This means that a white dwarf can be approximated as a Fermi sea, or a gas of electrons obeying the Pauli principle.

Since electrons cannot be in the same quantum state, a net repulsive force prevents them from being compressed down to a point. In a white dwarf star, it is this repulsive force that ultimately counteracts the gravitational force.

The same logic applies to neutrons in a neutron star, since neutrons also obey the Pauli exclusion principle, although the calculation is more complicated because of other nuclear and general relativistic effects.

2. John Michell, in *Philosophical Transactions of the Royal Society* 74 (1784): 35.

3. Quoted in Heinz Pagels, *Perfect Symmetry: The Search for the Beginning of Time* (New York: Bantam, 1985), 57.

Chapter 11

1. Quoted in Anthony Zee, *Fearful Symmetry* (New York: Macmillan, 1986), 68.

2. K. Gödel, "An Example of a New Type of Cosmological Solution of Einstein's Field Equations of Gravitation," *Reviews of Modern Physics* 21 (1949): 447.

3. F. Tipler, "Causality Violation in Asymptotically Flat Space-Times," *Physical Review Letters* 37 (1976): 979.

4. M. S. Morris, K. S. Thorne, and U. Yurtsever, "Wormholes, Time Machines, and the Weak Energy Condition," *Physical Review Letters* 61 (1988): 1446.

5. M. S. Morris and K. S. Thorne, "Wormholes in Spacetime and Their Use for Interstellar Travel: A Tool for Teaching General Relativity," *American Journal of Physics* 56 (1988): 411.

6. Fernando Echeverria, Gunnar Klinkhammer, and Kip S. Thorne, "Billiard Balls in Wormhole Spacetimes with Closed Timelike Curves: Classical Theory," *Physical Review D* 44 (1991): 1079.

7. Morris, Thorne, and Yurtsever, "Wormholes," 1447.

Chapter 12

1. Steven Weinberg, "The Cosmological Constant Problem," *Reviews of Modern Physics* 61 (1989): 6.

2. Heinz Pagels, *Perfect Symmetry: The Search for the Beginning of Time* (New York: Bantam, 1985), 377.

3. Ibid., 378.

4. Quoted in Alan Lightman and Roberta Brawer, *Origins: The Lives and*

Worlds of Modern Cosmologists (Cambridge, Mass.: Harvard University Press, 1990), 479.

5. Richard Feynman, Interview, in *Superstrings: A Theory of Everything?* ed. Paul Davies and J. Brown (Cambridge: Cambridge University Press, 1988), 196.

6. Weinberg, "Cosmological Constant Problem," 7.

7. Quoted in K. C. Cole, *Sympathetic Vibrations: Reflections on Physics as a Way of Life* (New York: Bantam, 1985), 204.

8. Quoted in John Gribben, *In Search of Schrödinger's Cat* (New York: Bantam, 1984), vi.

9. Quoted in Heinz Pagels, *The Cosmic Code* (New York: Bantam, 1982), 113.

10. Quoted in E. Harrison, *Masks of the Universe* (New York: Macmillan, 1985), 246.

11. F. Wilczek and B. Devine, *Longing for the Harmonies* (New York: Norton, 1988), 129.

12. Pagels, *Cosmic Code*, 155.

13. Quoted in David Freedman, "Parallel Universes: The New Reality—From Harvard's Wildest Physicist," *Discover Magazine*, July 1990, 52.

14. Ibid., 48.

15. Ibid., 49.

16. Ibid., 51.

17. Ibid., 48.

Chapter 13

1. Paul Davies, *Superforce: The Search for a Grand Unified Theory of Nature* (New York: Simon and Schuster, 1984), 168.

2. Freeman Dyson, *Disturbing the Universe* (New York: Harper & Row, 1979), 76.

3. Freeman Dyson, *Infinite in All Directions* (New York: Harper & Row, 1988), 196–197.

4. Dyson, *Disturbing the Universe*, 212.

5. Carl Sagan, *Cosmos* (New York: Random House, 1980), 306–307.

6. In fact, aeons ago it was even easier to self-destruct. In order to make an atomic bomb, the fundamental problem facing any species is to separate uranium-235 from its more abundant twin, uranium-238, which cannot sustain a chain reaction. Only the uranium-235 will sustain a chain reaction. But uranium-235 is only 0.3% of naturally occurring uranium. To sustain a runaway chain reaction, you need an enrichment level of at least 20%. In fact, weapons-grade uranium has a 90% or more enrichment rate. (This is the reason why uranium mines do not suffer from spontaneous nuclear detonations. Because naturally occurring uranium in a uranium mine is only 0.3% enriched, it contains far too low a concentration of U-235 to sustain a runaway nuclear chain reaction.)

Because uranium-235 is relatively short-lived compared with its more abundant twin, uranium-238, aeons ago, the naturally occurring enrichment rate in our universe was much larger than 0.3%.

In other words, it was far easier then for any civilization to fabricate an atomic bomb because the naturally occurring enrichment rate was much larger than it is today.

7. Heinz Pagels, *The Cosmic Code* (New York: Bantam, 1982), 309.

8. Sagan, *Cosmos*, 231.

9. Quoted in Melinda Beck and Daniel Glick, "And If the Comet Misses," *Newsweek*, 23 November 1992, 61.

Chapter 14

1. Quoted in John D. Barrow and Frank J. Tipler, *The Anthropic Cosmological Principle* (Oxford: Oxford University Press, 1986), 167.

2. Quoted in Heinz Pagels, *Perfect Symmetry: The Search for the Beginning of Time* (New York: Bantam, 1985), 382.

3. Ibid., 234.

4. Astronomers John D. Barrow of the University of Sussex in England and Joseph Silk of the University of California at Berkeley see some hope in this dismal scenario. They write, "If life, in any shape or form, is to survive this ultimate environmental crisis, then the universe must satisfy certain basic requirements. The basic prerequisite for intelligence to survive is a source of energy.

"The anisotropies in the cosmic expansion, the evaporating black holes, the remnant naked singularities are all life preservers of a sort. . . . An infinite amount of information is potentially available in an open universe, and its assimilation would be the principal goal of any surviving noncorporeal intelligence" (*The Left Hand of Creation* [New York: Basic Books, 1983], 226).

5. Ibid.

6. Gerald Feinberg, *Solid Clues* (New York: Simon and Schuster, 1985), 95.

Chapter 15

1. Quoted in Heinz Pagels, *The Cosmic Code* (New York: Bantam Books, 1982), 173–174.

2. Edward Witten, Interview, in *Superstrings: A Theory of Everything?* ed. Paul Davies and J. Brown (Cambridge: Cambridge University Press, 1988), 102.

3. Quoted in John D. Barrow and Frank J. Tipler, *The Anthropic Cosmological Principle* (Oxford: Oxford University Press, 1986), 185.

4. Pagels, *Cosmic Code*, 382.

5. James Trefil, *The Moment of Creation* (New York: Macmillan, 1983), 220.

6. John Ellis, Interview, in *Superstrings,* ed. Davies and Brown, 161.

7. Quoted in R. P. Crease and C. C. Mann, *The Second Creation* (New York: Macmillan, 1986), 77.

8. Quoted in Anthony Zee, *Fearful Symmetry* (New York: Macmillan, 1986), 122.

9. Ibid., 274.

10. Heinz Pagels, *Perfect Symmetry: The Search for the Beginning of Time* (New York: Bantam, 1985), xiii.

11. Stephen Hawking, *A Brief History of Time* (New York: Bantam, 1988), 175.

References and Suggested Reading

Abbot, E. A. *Flatland: A Romance of Many Dimensions*. New York: New American Library, 1984.

Barrow, J. D., and F. J. Tipler. *The Anthropic Cosmological Principle*. Oxford: Oxford University Press, 1986.

Bell, E. T. *Men of Mathematics*. New York: Simon and Schuster, 1937.

Calder, N. *The Key to the Universe*. New York: Penguin, 1977.

Chester, M. *Particles*. New York: Macmillan, 1978.

Crease, R., and C. Mann. *The Second Creation*. New York: Macmillan, 1986.

Davies, P. *The Forces of Nature*. Cambridge: Cambridge University Press, 1979.

Davies, P. *Superforce: The Search for a Grand Unified Theory of Nature*. New York: Simon and Schuster, 1984.

Davies, P., and J. Brown, eds. *Superstrings: A Theory of Everything?* Cambridge: Cambridge University Press, 1988.

Dyson, F. *Disturbing the Universe*. New York: Harper & Row, 1979.

Dyson F. *Infinite in All Directions*. New York: Harper & Row, 1988.

Feinberg, G. *Solid Clues*. New York: Simon and Schuster, 1985.

Feinberg, G. *What Is the World Made Of?* New York: Doubleday, 1977.

French, A. P. *Einstein: A Centenary Volume*. Cambridge, Mass.: Harvard University Press, 1979.

Gamow, G. *The Birth and Death of Our Sun*. New York: Viking, 1952.

Glashow, S. L. *Interactions*. New York: Warner, 1988.

Gribben, J. *In Search of Schrödinger's Cat*. New York: Bantam, 1984.

Hawking, S. W. *A Brief History of Time*. New York: Bantam, 1988.

Heisenberg, W. *Physics and Beyond*. New York: Harper Torchbooks, 1971.

Henderson, L. D. *The Fourth Dimension and Non-Eudidean Geometry in Modern Art*. Princeton, N.J.: Princeton University Press, 1983.

Kaku, M. *Introduction to Superstrings*. New York: Springer-Verlag, 1988.

Kaku, M., and J. Trainer. *Beyond Einstein: The Cosmic Quest for the Theory of the Universe*. New York: Bantam, 1987.

Kaufmann, W. J. *Black Holes and Warped Space–Time*. San Francisco: Freeman, 1979.

Lenin, V. *Materialism and Empiro-Criticism.* In K. Marx, F. Engels, and V. Lenin, *On Dialectical Materialism.* Moscow: Progress, 1977.

Pagels, H. *The Cosmic Code.* New York: Bantam, 1982.

Pagels, H. *Perfect Symmetry: The Search for the Beginning of Time.* New York: Bantam, 1985.

Pais, A. *Subtle Is the Lord: The Science and the Life of Albert Einstein.* Oxford: Oxford University Press, 1982.

Penrose, R. *The Emperor's New Mind.* Oxford: Oxford University Press, 1989.

Polkinghorne, J. C. *The Quantum World.* Princeton, N.J.: Princeton University Press, 1984.

Rucker, R. *Geometry, Relativity, and the Fourth Dimension.* New York: Dover, 1977.

Rucker, R. *The Fourth Dimension.* Boston: Houghton Mifflin, 1984.

Sagan, C. *Cosmos.* New York: Random House, 1980.

Silk, J. *The Big Bang: The Creation and Evolution of the Universe.* 2nd ed. San Francisco: Freeman, 1988.

Trefil, J. S. *From Atoms to Quarks.* New York: Scribner, 1980.

Trefil, J. S. *The Moment of Creation.* New York: Macmillan, 1983.

Weinberg, S. *The First Three Minutes: A Modern View of the Origin of the Universe.* New York: Basic Books, 1988.

Wilczek, F., and B. Devine. *Longing for the Harmonies.* New York: Norton, 1988.

Zee, A. *Fearful Symmetry.* New York: Macmillan, 1986.

Index

ABOUT THE AUTHOR

Michio Kaku is professor of theoretical physics at the City College of the City University of New York. He graduated from Harvard and received his Ph.D. from the University of California, Berkeley. He is author of *Beyond Einstein* (with Jennifer Trainer), *Quantum Field Theory: A Modern Introduction,* and *Introduction to Superstrings.* He has also hosted a weekly hour-long science program on radio for the past ten years.